13.99 · 80293

Advanced GNVQ Eng
Mathematics
Volume 1
Mandatory Unit

71520

Also by J. C. Yates

National Engineering Mathematics Volume 1
National Engineering Mathematics Volume 2
National Engineering Mathematics Volume 3

Advanced GNVQ Engineering Mathematics
Volume 1
Mandatory Unit

J.C. Yates
Senior Mathematics Lecturer
Wigan and Leigh College

MACMILLAN

First published 1995 by
MACMILLAN PRESS LTD
Houndmills, Basingstoke, Hampshire RG21 2XS
and London
Companies and representatives
throughout the world

ISBN 0–333–63650–3

A catalogue record for this book is available
from the British Library.

Copy-edited and typeset by Povey–Edmondson
Okehampton and Rochdale, England

Printed in Great Britain by
Biddles Ltd
Guildford and Kings Lynn

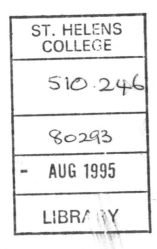

Contents

Acknowledgements

For the help I have received whilst writing this volume I am truly grateful. I offer my sincere thanks to Alan Smalley for his review of Chapter 7 on vectors. Also I wish to thank Alan Skinner for his review comments.

The author and publishers are grateful to the following organisations and individuals for permission to reproduce illustrative material: J. Allan Cash Ltd, p. 298, British Aerospace, p. 203, Farnell Electronic Components Limited, p. 67, J. H. Fenner & Co. Ltd, p. 91, D. W. Molyneux, p. 176 Northern Electric plc, p. 69, R.P.I. Conservatories, p. 235, Saab Great Britain Ltd, pp. 2, 33, 378, Texas Instruments, p. 2.

Author's Note

This volume covers all the mandatory Mathematics Unit for the GNVQ in Engineering. The coherent chapter order is a recommended teaching/ learning scheme, ideal both in formal settings and in self-study. To maintain the learner's interest algebra and graph work are interleaved where appropriate to produce a complete, understandable, mathematics package.

Each chapter starts with the major element(s), performance criteria and range statements. For variety of mathematics more than one element is often included within a chapter. Following each Introduction is an Assignment. This is a specimen example of how the mathematics within the chapter can be applied. The Assignment provides a common thread for the chapter, linking together the theory and techniques as they develop. At appropriate stages each Assignment is re-visited for further attention. Each chapter is completed with a selection of multiple choice questions leading you towards the style of a GNVQ test. It is a point for self-assessment prior to sitting the actual test.

The electronic calculator is discussed in Chapter 1 and used throughout the text.

The Answers section contains selected hints to some questions to provide extra help short of a complete solution.

The following table maps the elements in each chapter:

Element	1	2	3	4	5	6	7	8	9	10	11	12	13	14
Use algebra to solve engineering problems	✓	✓	✓					✓	✓	✓				
Use trigonometry to solve engineering problems				✓	✓	✓								
Use graphs to solve engineering problems		✓			✓		✓		✓	✓		✓	✓	✓	✓
Use calculus to solve engineering problems														✓	✓
Use phasors and vectors to solve engineering problems							✓	✓						

Greek letters used

Mathematics needs more letters than those provided by the alphabet. That is why the following Greek letters have been used in the text:

α	alpha
β	beta
γ	gamma
δ	delta
θ	theta
μ	mu
σ	sigma
π	pi
φ	phi
ω	omega
Σ	sigma – this is capital sigma
Ω	omega – this is capital omega

1 Using Your Calculator

Introduction

There are many different calculators currently available at relatively low cost. Each year, with the introduction of new function keys, they appear to be capable of performing more and more complicated calculations. Which one should you choose? Because there are so many features for use in different applications there is no right or wrong answer to this question. Perhaps you should include some of the following points in your decision to buy:

1) Is there a recommended calculator for your course that will be suitable throughout all the subject areas?
2) Do you want power from a battery or a solar cell, or a combination of these?
3) Is it a scientific calculator?
4) Price.

The majority of the calculators available today are produced by a small number of manufacturers. They have a history of reliability. A more expensive one need not be a better one for you. The **keys** (or buttons) on a calculator are either numbers or functions. Simply, if the key is not one of 0, 1, 2, 3, 4, 5, 6, 7, 8 or 9 then it is a function key. A **function** (or **operation**) **key** carries out some task on the numbers in a calculation. More function keys and a different display are the most likely reasons for the extra cost. In fact a calculator that is too complicated may take longer to learn to use and cause you frustration. Do not attempt to learn about all the functions at once. Throughout this book with the aid of the manual that comes with your calculator you should learn about the majority of them.

▬▬ ASSIGNMENT ▬▬

During this chapter we will consider a vehicle travelling along a road with a particular speed, then the brakes are applied. We will use the calculator to work out the distances travelled from the braking point at various times. In fact our first attempt at the method will be with **whole numbers** (**integers**). Once we have learned more about the importance of decimal places, we will repeat the calculation.

Calculator layout

There are many different layouts for the calculator keyboard. The photograph shows part of a typical one. We will include other keys later in the book. The positions of the ON| and OFF| keys change from one calculator to another. Some will switch off automatically after lying unused for a number of minutes to save the battery.

The main part of the picture involves the figure keys of

$\underline{0}|$ $\underline{1}|$ $\underline{2}|$ $\underline{3}|$ $\underline{4}|$ $\underline{5}|$ $\underline{6}|$ $\underline{7}|$ $\underline{8}|$ $\underline{9}|$

together with the arithmetic keys of

$\underline{\times}|$ $\underline{\div}|$ $\underline{+}|$ $\underline{-}|$ $\underline{=}|$ $\underline{.}|$

There are two other keys that, at first glance, look to be very similar. These are the $\underline{AC}|$ and $\underline{C}|$ keys. Pressing the $\underline{AC}|$ key will clear everything except the contents of the memory. The use of the $\underline{C}|$ key is less dramatic. It will only delete the effect of the last key pressed, so preserving the earlier part of the calculation.

███████ **Example 1.1** ████████████████████████████████████

i) 5×3 is evaluated by pressing the keys

$\underline{5}|$ $\underline{\times}|$ $\underline{3}|$ $\underline{=}|$

so that the final display is 15.

ii) $12 \div 3$ uses the keys

$\underline{12}|$ $\underline{\div}|$ $\underline{3}|$ $\underline{=}|$

to give a final display of 4.

iii) $7 + 9$ uses the keys

$\underline{7}|$ $\underline{+}|$ $\underline{9}|$ $\underline{=}|$

to give a final display of 16.

iv) $11 - 4$ uses the keys

$\underline{11}|$ $\underline{-}|$ $\underline{4}|$ $\underline{=}|$

to give a final display of 7.

v) Suppose that when attempting $11 - 4$ you realise at a late stage that the calculation should have been $11 - 5$. If you had pressed

$\underline{11}|$ $\underline{-}|$ $\underline{4}|$

and realised your mis-read now press $\underline{C}|$. This will delete the $\underline{4}|$ you entered and display 0. You may complete the calculation with $\underline{5}|$ $\underline{=}|$ to display the correct answer of 6. The full order is

$\underline{11}|$ $\underline{-}|$ $\underline{4}|$ $\underline{C}|$ $\underline{5}|$ $\underline{=}|$

___

$\underline{+/-}|$ is the key that changes the **sign before a value**.

███████ **Example 1.2** ████████████████████████████████████

To enter -5 you press $\underline{5}|$ $\underline{+/-}|$ which will change the initial display from 5 to the required -5.

___

Some calculators may not have function keys for brackets. You should work out that part of the calculation in the brackets before any other.

Examples 1.3

i) To work out $(13 + 9)\,4$ the order of key operations is

   $13 \rfloor \quad + \rfloor \quad 9 \rfloor \quad = \rfloor \quad \times \rfloor \quad 4 \rfloor \quad = \rfloor$

   After we have pressed the first $= \rfloor$ key the screen will display 22, finally displaying the correct answer of 88 when we have pressed the second $= \rfloor$ key.

ii) To work out $3\,(13 - 9)$ the order of key operations is

   $13 \rfloor \quad - \rfloor \quad 9 \rfloor \quad = \rfloor \quad \times \rfloor \quad 3 \rfloor \quad = \rfloor$

   Obeying the rule of working out the bracket first, the screen will display 4 after we have pressed the first $= \rfloor$ key and then 12 after we have pressed the second $= \rfloor$ key.

iii) In the evaluation of $\dfrac{96}{2 \times 6}$ though no brackets are obvious we understand that they exist around (2×6) in the denominator (bottom line of the fraction). It is this part of the calculation that we attempt first. The order of key operations is

   $2 \rfloor \quad \times \rfloor \quad 6 \rfloor \quad = \rfloor$

   The screen display of 12 is noted and the screen cleared using the function key $\underline{AC} \rfloor$. The calculation continues

   $96 \rfloor \quad \div \rfloor \quad 12 \rfloor \quad = \rfloor$

   Now the final answer of 8 is displayed on the screen.

iv) We can use the answer from part iii) above in the evaluation of $\left(\dfrac{96}{2 \times 6} - 5 \right) 7.$

   Already we have a solution of 8 from $\dfrac{96}{2 \times 6}$ so that the problem is reduced to $(8-5)7$. The order of key operations is

   $8 \rfloor \quad - \rfloor \quad 5 \rfloor \quad = \rfloor \quad \times \rfloor \quad 7 \rfloor \quad = \rfloor$

   3 will be displayed after the first $= \rfloor$ and the final answer of 21 after the second $= \rfloor$.

EXERCISE 1.1

Use your calculator to work out the following:

1 15×6

2 -15×6

3 $213 \div 3$

4 $213 \div (-3)$

5 $(-213) \div (-3)$

6 $(10 + 5)6$

7 $6(10 + 5)$

8 $(10 + 5 - 4)2$

9 $(10 - 5 + 4)2$

10 $\dfrac{213 \times 6}{3}$

11 $\dfrac{75}{3 \times 5}$

12 $4 + \left(\dfrac{75}{3 \times 5}\right)$

13 $(-10 + 5 - 3)/2$

14 $\dfrac{10 + 5}{3}$

15 $\dfrac{120}{10 - 5}$

16 $\dfrac{120}{10 - 5} + 7$

17 $\dfrac{120}{10 - 5} - 7$

18 $\left(\dfrac{120}{10 - 5} - 7\right)3$

19 $160 \div \left(\dfrac{75}{3 \times 5}\right)$

20 $\dfrac{160 \times 3 \times 5}{75}$

$\boxed{x^2}$ is used to square whatever number is displayed.

■■■■ Examples 1.4 ■■■■■■■■■■■■■■■■■■■■■■■■■■■■■

i) To work out 6^2 the order of key operations is

$\underline{\ 6\ |}\quad \underline{x^2\,|}$

displaying the answer 36.

ii) To work out $(-5)^2$ the order is

$\underline{\ 5\ |}\quad \underline{\,^{+/\!-}|}\quad \underline{x^2\,|}$

displaying the answer 25.

iii) This example places the minus sign in a slightly different position to the previous one by leaving out the brackets. -5^2 means that only the 5 is squared and then the minus sign is put in front of the squared value. The full order of keys, to get a correct display of -25, is

$\underline{\ 5\ |}\quad \underline{x^2\,|}\quad \underline{\,^{+/\!-}|}$

Whenever a positive number or a negative number is squared the answer will always be a positive value. This means that the opposite operation of taking the square root of a positive value should give two answers: a positive number and a negative number. However the calculator is only able to display one of these numbers, usually the positive one. So, for example, the square roots of 49 are 7 and -7 though the display will only show 7.

■■■■ **Example 1.5** ■■■■■■■■■■■■■■■

i) To work out $\left(\dfrac{21-5}{8}\right)^2$ the order of key operations is

$\underline{21\,|}$ $\underline{-\,|}$ $\underline{5\,|}$ $\underline{=\,|}$ $\underline{\div\,|}$ $\underline{8\,|}$ $\underline{=\,|}$ $\underline{x^2\,|}$

After the first $=|$ the display will be 16 and after the second $=|$ the display will change to 2. Once all the operations are complete the final answer is 4.

ii) To work out $\left(\dfrac{21-5}{8}\right)^2 + 19$ the order of key operations uses those from part (i) to obtain 4 and then simply adds 19 to give an answer of 23.

iii) To work out $\left(\dfrac{21-5}{8}\right)^2 + 19^2$ again the order of key operations uses the result from part (i) which should be noted. Press the $\underline{AC|}$ button to clear the calculator. Now we are in a position to work out 19^2 by pressing

$\underline{19\,|}$ $\underline{x^2\,|}$

The screen will display 361 and then we may add our earlier result of 4 to give a final answer of 365.

iv) To work out $19\left(\dfrac{21-5}{8}\right)^2$ we remember that only the bracket is squared. Then we multiply that value by 19. The order of key operations is

$\underline{21\,|}$ $\underline{-\,|}$ $\underline{5\,|}$ $\underline{=\,|}$ $\underline{\div\,|}$ $\underline{8\,|}$ $\underline{=\,|}$

$\underline{x^2\,|}$ $\underline{\times\,|}$ $\underline{19\,|}$ $\underline{=\,|}$

This means that the display from part (i) of 4 is multiplied by 19 to give a final answer of 76.

The key to calculate the square root is $\underline{\sqrt{}\,|}$

■■■■ **Example 1.6** ■■■■■■■■■■■■■■■

To work out the square root of 49 the order of key operations is

$\underline{49\,|}$ $\underline{\sqrt{}\,|}$

displaying 7 and implying that -7 is also another answer.

Because $\sqrt{}$ is the opposite operation to square you can check on your calculator that both $7^2 = 49$
         and $(-7)^2 = 49$.

The square root of a negative value does not exist in real terms, and any attempt at this operation will display -E- on the calculator screen.

What appears under the $\sqrt{}$ should be treated as though it is in a bracket, and so should be worked out first.

Examples 1.7

i) $\sqrt{67} - 3$ is worked out using the key operations

$$\boxed{67} \quad \boxed{-} \quad \boxed{3} \quad \boxed{=} \quad \boxed{\sqrt{}}$$

to give the answer 8.

ii) $\sqrt{\dfrac{67-3}{4}}$ is worked out using the key operations

$$\boxed{67} \quad \boxed{-} \quad \boxed{3} \quad \boxed{=} \quad \boxed{\div} \quad \boxed{4} \quad \boxed{=} \quad \boxed{\sqrt{}}$$

to give an answer of 4.

iii) $\dfrac{\sqrt{67}-3}{4}$ has the $\sqrt{}$ symbol in a slightly different position which needs a different order of key operations

$$\boxed{67} \quad \boxed{-} \quad \boxed{3} \quad \boxed{=} \quad \boxed{\sqrt{}} \quad \boxed{\div} \quad \boxed{4} \quad \boxed{=}$$

to give an answer of 2.

$\boxed{x^y}$ is the key that raises a number, x, to a power, y.

The order of pressing keys is important as the following examples show clearly.

Examples 1.8

i) 5^4 is worked out using the order of key operations

$$\boxed{5} \quad \boxed{x^y} \quad \boxed{4} \quad \boxed{=}$$

to display the answer 625.

ii) 4^5 uses the key operations

$$\boxed{4} \quad \boxed{x^y} \quad \boxed{5} \quad \boxed{=}$$

to display the answer 1024.

iii) $(-5)^4$ involves a combination of several key operations that we have mentioned so far in this chapter;

$$\boxed{5} \quad \boxed{+/-} \quad \boxed{x^y} \quad \boxed{4} \quad \boxed{=}$$

to display the answer 625.

iv) -5^4 requires a slightly different order of key operations because the minus sign applies only after 5 has been raised to the power 4. The order is

$$\boxed{5} \quad \boxed{x^y} \quad \boxed{4} \quad \boxed{=} \quad \boxed{+/-}$$

which displays the answer -625.

EXERCISE 1.2

Use your calculator to work out the following:

1   9^2

2   $9^2 \times 3$

3   3×9^2

4   $3 \times (-9)^2$

5   $3^2 \times 9^2$

6   $(3 \times 9)^2$

7   $\left(\dfrac{9+3}{2}\right)^2$

8   $7 + \left(\dfrac{9+3}{2}\right)^2$

9   $7^2 + \left(\dfrac{19+5}{2}\right)^2$

10   $4\left(\dfrac{19-3}{8}\right)^2$

11   $\sqrt{9}$

12   $\sqrt{81}$

13   $\sqrt{9^2}$

14   $\sqrt{3^2 \times 9^2}$

15   $\dfrac{\sqrt{19-3}}{4}$

16   $\sqrt{\dfrac{19-3}{4}}$

17   $14 + \sqrt{\dfrac{19-3}{4}}$

18   $14 - \dfrac{\sqrt{19-3}}{4}$

19   $2^2 - \sqrt{3^2 \times 9^2}$

20   $4 + 2^2 - \dfrac{\sqrt{19-3}}{4}$

21   6^4

22   $(-6)^4$

23   -6^4

24   $2 + 6^4$

25   $2 - 6^4$

26   2×3^3

27   $\dfrac{3 \times 4^5}{8}$

28   $21 - (2 \times 3^3)$

29   $21 + (3 \times 2^3)^3$

30   $\dfrac{21 + (3 \times 2^3)}{3}$

So far, we have looked at only some of the calculator keys to attempt our first Assignment, though we shall return to the same basic theme later in this chapter.

ASSIGNMENT

Now we are in a position to attempt the Assignment using only **whole numbers** (i.e. **integers**). Suppose that the vehicle is travelling at a speed of 30 ms^{-1} when the brakes are applied to produce a deceleration of 4 ms^{-2}. It is thought that once these brakes are applied the distance the vehicle travels is given by the formula

$$s = 30t - 2t^2.$$

> Deceleration and retardation, being negative acceleration, mean a slowing down.

Unfortunately there has been an error of judgment of 1 m in the measuring so that the formula requires correcting to

$$s = 1 + 30t - 2t^2$$

where s is the distance travelled
and t is the time from when the brakes are applied.

To many people this formula means very little. A diagram or graph of the distance plotted against time would be more appealing. Before attempting this it is useful to have a **table of values**. In later chapters we look at the graph. Suppose that the measurements are made at the moment the brakes are applied (i.e. $t=0$) and then after 1, 2, 4 and 6 seconds. We will go through the calculations stage by stage, ending up with the table at these times. We leave gaps so that you can work out the expected distance values at 3 and 5 seconds.

First we need a line of values of t, i.e.

t	0	1	2	3	4	5	6

Now the expected value of s is made up of three different terms on the right-hand side. We look at them in turn for each of these time values. The first term on that side is 1. It is a constant value and so is *unaffected* by each value of t, i.e.

1	1	1	1		1		1

The next term is $30t$. We look at each value of t in turn and multiply it by 30. For example when t is 4 the key operations are

$\boxed{30}\ \ \boxed{\times}\ \ \boxed{4}\ \ \boxed{=}$

to display an answer of 120. This line of the table is

$30t$	0	30	60		120		180

The final term on the right-hand side is $-2t^2$. Again we use each different value of t in the calculations. It is only the t that is squared. Once we have performed this operation, we multiply by -2. For example, when t is 6 the order of key operations is

$\boxed{6}\ \ \boxed{x^2}\ \ \boxed{\times}\ \ \boxed{2}\ \ \boxed{=}\ \ \boxed{^+/_-}$

to display an answer of -72. The line of the table is

$-2t^2$	0	-2	-8		-32		-72

After the first line of t values we have a series of rows from each of the three terms of $s = 1 + 30t - 2t^2$. Carefully drawn, we need to look closely at the columns formed once these rows have been put together. Because we have already taken care of the minus sign all that remains is to add each column of values to get the expected value for *s*. The last of these columns is given as

$$\frac{1}{180}$$

and -72.

These will add to 109 and it is this value that appears in the last column of the last row.

The complete table is given below:

t	0	1	2	3	4	5	6
1	1	1	1		1		1
$30t$	0	30	60		120		180
$-2t^2$	0	-2	-8		-32		-72
s	1	29	53		89		109

Row 1 shows the original values of *t*.
Rows 2–4 show all the workings for the various terms.
Row 5 is the row for the expected values of *s*, being the addition of row values 2, 3 and 4.

From this table we can check our practical distance measurements to see if they tally with our calculator values. The closer the tally the closer our formula is at predicting what will happen later to the vehicle, and eventually when it will stop.

Accuracy

A calculator will **not answer all your problems**. It is an aid to make the numerical calculations less tedious and speed up your work. Before you attempt to complete a calculation you need to understand the size and meaning of the figures you are working with. Often a value is quoted correct to a number of **decimal places**. This means it has been rounded or approximated to this figure. For example you
might read   53.46 correct to 2 decimal places,
perhaps shortened to 53.46 (2 dp),
because there are 2 **figures** (i.e. **digits**) to the right of the decimal point. This means that if we want accuracy to this number of decimal places then 53.46 is the closest figure. It is probable that we will have no idea of the true original figure.

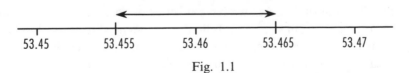

Fig. 1.1

Fig. 1.1 shows a magnified view of a number line. It is a simple scale measurement showing that any value between 53.455 and 53.465 is closer to 53.46 (2 dp) than any other value correct to 2 decimal places.

Examples 1.9

i) 53.458 could be rounded to 53.46 (2 dp).
ii) 53.454 could be rounded to 53.45 (2 dp).

If you have 53.455 you may round it up to 53.46 (2 dp) or down to 53.45 (2 dp). You could make a similar decision to round up 53.465 to either 53.47 (2 dp) or down to 53.46 (2 dp). Because numbers like 53.455 and 53.465 are on the borderline you may round either up or down. Whichever rounding decision you choose, be consistent in your calculations. For borderline decisions do *not* round up in one part of your calculation and then round down in a later part.

An alternative rounding to decimal places is a rounding to **significant figures**. For example 53.46 (2 dp) is the same as
                    53.46 correct to 4 significant figures,
perhaps shortened to 53.46 (4 sig fig),
or   53.46 (4 sf),
because there are 4 figures in significant positions within the overall number. You approximate significant figures by rounding up or down according to the same rules used for decimal places.

Examples 1.10

i) 1.07641 could be rounded to 1.0764 (5 sf).
ii) 1.07648 could be rounded to 1.0765 (5 sf).

Now consider rounding 1.07641 to only 3 significant figures. From the left we must consider the first 3 figures, i.e. 1.07, and ask ourselves whether 1.07641 is closer to 1.07 or 1.08. The borderline position is half-way between these values at 1.075. Fig. 1.2 shows that 1.07641 is slightly

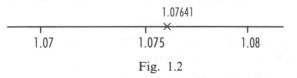

Fig. 1.2

bigger than the borderline value and therefore closer to 1.08 (3 sf) than any other number correct to 3 significant figures. Then 1.07641 may be rounded to 1.08 (3 sf).

If we are presented with a number correct to some significant figures it is unlikely that we would know of its true original value. Even then its 'true' value is only as accurate as the measuring device used. Suppose we are presented with a measurement of 76.2 mm. All we can say is that this is correct to 3 significant figures or 1 decimal place and that the true value lies somewhere between 76.15 and 76.25 mm.

Not all figures are significant. The general rule is that leading and trailing zeroes are not significant.

▄▄▄▄▄▄▄ **Examples 1.11** ▄▄▄▄▄▄▄▄▄▄▄▄▄▄▄▄▄

i)   0.00169 is correct to 3 significant figures because the zeroes at the beginning (leading) are not significant.
ii) 0.0016900 is still correct to only 3 significant figures because all the zeroes at the end (trailing) are not significant either.
iii) 0.0010900 is correct again to 3 significant figures. The zero sandwiched between the 1 and 9 is significant. It neither leads nor trails.

▄▄▄▄▄▄▄ **Example 1.12** ▄▄▄▄▄▄▄▄▄▄▄▄▄▄▄▄▄▄

A local engineering company made profits of £527 000 in the last financial year.

This profit is correct to 3 significant figures (the zeroes are trailing and so not significant) and has been rounded. It is very unlikely that the company can be exact when reporting its profit, and has said this is correct to the nearest thousand pounds.

Standard form

Numbers in standard form are all displayed in the same format: a value multiplied by 10 to a positive or negative whole number power. That value starts at 1 and almost, but not quite, reaches 10. We can use the numbers of Examples 1.11 and 1.12.

▄▄▄▄▄▄▄ **Examples 1.13** ▄▄▄▄▄▄▄▄▄▄▄▄▄▄▄▄▄

i) $0.00169 = 1.69 \times 10^{-3}$ in standard form. The decimal point of 0.00169 has leap-frogged 3 places (this is the figure in the power of 10) to the right (this is the negative in the power of 10) from the original number to create the standard form.

> The calculator display will miss out the 10. If you enter 0.00169 and press = the display will show 1.69^{-03}. It is taken for granted that you understand this to mean 1.69×10^{-3} in standard form.

ii) $0.0010900 = 1.09 \times 10^{-3}$ in standard form.

iii) $0.00023 = 2.3 \times 10^{-4}$ in standard form, having leap-frogged 4 places to the right.

iv) $527\,000 = 5.27 \times 10^{5}$ in standard form, having leap-frogged 5 places to the left from the original number to create the standard form.

v) $527 = 5.27 \times 10^{2}$ in standard form, having leap-frogged 2 places to the left.

vi)   5.27 would usually remain unchanged without any power of 10.

EXERCISE 1.3

Round the following values to the stated number of decimal places or significant figures.

1	0.992	(2 dp)		**12**	6.204	(1 dp)
2	0.38847	(4 dp)		**13**	20.3	(1 sf)
3	17.38847	(2 dp)		**14**	620.401	(4 sf)
4	5.69	(1 dp)		**15**	620.461	(1 dp)
5	21.399	(2 dp)		**16**	99.8	(1 sf)
6	4.32	(1 sf)		**17**	5488.97	(3 sf)
7	14.32	(1 sf)		**18**	1 400 000	(1 sf)
8	217 000	(2 sf)		**19**	4239.06	(2 sf)
9	0.0037602	(3 sf)		**20**	0.070629	(3 dp)
10	3165	(2 sf)		**21–40**	Using the original	
11	62.04	(1 dp)			questions **1–20** above re-write	
					each one in standard form.	

ASSIGNMENT

Let us return to our original assignment. This time, with improved accuracy, we have managed to measure the speed at 30.75 ms^{-1} just before the brakes are applied to give the deceleration at 4.40 ms^{-2}. The 1 m error measurement has been re-checked and found to be 0.86 m.

It is thought that once these brakes are applied the distance the vehicle travels is given by the formula

$$s = 0.86 + 30.75t - 2.2t^{2}$$

where s is the distance travelled
and t is the time from when the brakes are applied.

Just like last time, we will construct a table of values in stages. It will be based on times from when the brakes are applied (i.e. $t=0$) and then after 1, 2, 4 and 6 seconds. Again some gaps will be left in the table to give you some practice with your calculator. The calculation stages are as before, but this time using slightly different numbers in the formula for the expected value of s.

First we need a line of values of t, i.e.

t	0	1	2	3	4	5	6

Again s is made up of three different terms on the right-hand side which we will look at in turn for each of these time values. The first term on that side is 0.86. It is a constant value and so is unaffected by each value of t, i.e.

0.86	0.86	0.86	0.86		0.86		0.86

The next term is $30.75t$. We look at each value of t in turn and multiply by 30.75. For example, when t is 4 the key operations are

$\boxed{30.75}$ $\boxed{\times}$ $\boxed{4}$ $\boxed{=}$

to display an answer of 123. This line of the table is

$30.75t$	0	30.75	61.50		123.00		184.50

The final term on the right-hand side is $-2.2t^2$. Again we use each different value of t in the calculations. It is only the t that is squared. Once we have performed this operation, we multiply by -2.2. For example, when t is 6 the order of key operations is

$\boxed{6}$ $\boxed{x^2}$ $\boxed{\times}$ $\boxed{2.2}$ $\boxed{=}$ $\boxed{+/-}$

to display an answer of -79.2. The line of the table is

$-2.2t^2$	0	-2.20	-8.80		-35.20		-79.20

After the first line of t values we have a series of rows from each of the three terms of $s=0.86+30.75t-2.2t^2$. Carefully drawn we need to look closely at the columns formed once these rows have been put together. Because we have already taken care of the minus sign all that remains is to add each column of values to get the expected value for s. The last of these columns is given as

$$0.86$$
$$184.50$$
$$\text{and} -79.20.$$

These will add to 106.16 and it is this value that appears in the last column of the last row.

The complete table is given below:

t	0	1	2	3	4	5	6
0.86	0.86	0.86	0.86		0.86		0.86
30.75t	0	30.75	61.50		123.00		184.50
$-2.2t^2$	0	-2.20	-8.80		-35.20		-79.20
s	0.86	29.41	53.56		88.66		106.16

The original figures of 30.75, 4.40 and 0.86 are correct to 2 decimal places. There is an argument for approximating all these calculated values of *s* to only 1 decimal place. This is left as a simple exercise for you.

Row 1 shows the original values of *t*.

Rows 2–4 show all the workings for the various terms.

Row 5 is the row for *s*, being the addition of row values 2, 3 and 4.

From this table we can check our practical distance measurements to see if they tally with our calculator values. The closer the tally the closer our formula is at predicting what will happen later to the vehicle, and eventually when it will stop.

Truncation

Some calculators do *not* round numbers properly, but cut off according to how many figures (**digits**) can be displayed. When the rounding is a rounding down there is no problem as the calculator is deciding to round correctly. Unfortunately it is the **rounding up** that needs care. Suppose we have a calculator with an 8 digit display. If you try 14 divided by 3 it may display 4.6666666, showing the 6 repeated many times (called **recurring**). A bigger capacity display may just show more 6s. In fact 4.6666666 ... is closer to 4.6666667 than the figure shown on the display and it is this latter display that is shown on the better calculators. You need fewer decimal places than those actually displayed on the calculator. The last figure in the display should not be a problem.

To save space we have a notation to represent a number that is recurring. We place a dot or a line over the figures that recur; e.g
114.6666666 ... is shown by 114.$\dot{6}$ or   114.$\bar{6}$,
and   21.09090909 ... is shown by   21.$\dot{0}\dot{9}$ or   21.$\overline{09}$.

The true value of a number

If you have a £10 note you understand that it is worth £10. If you buy something costing less than this note, say something worth £9.60, then you will get some change. Also if your purchase is more than £10 you will need to offer more money. This accuracy is not true for everything. We can return to our engineering company profit of £527 000 which has been quoted correct to 3 significant figures. This means that the true (probably unknown) profit lies somewhere between £526 500 and £527 500 i.e. £527 000±£500. There is a variation hidden in that 4th significant figure we did not quote originally.

The same idea works for small values. 0.00169 is correct to 5 decimal places, but the exact value might be somewhere between 0.001685 and 0.001695. This means that 0.00169 is correct only to 0.000005, i.e. 0.00169±0.000005.

The true value of a number may be important when attempting simple arithmetic calculations.

Think about 2 values, 69 and 33.
Now 69 may vary between 68.5 and 69.5
and   33 may vary between 32.5 and 33.5.
Then 69 + 33 might lie between some limits as well.
Addition of the large extreme values gives   69.5 + 33.5   = 103,
                  of the quoted values gives   69   + 33   = 102
     and of the small extreme values gives   68.5 + 32.5   = 101.
Which answer is correct?

If we use the quoted values, which we believe to be true, then our answer is 102. It is possible that it may be slightly in error, perhaps as high as 103 or as low as 101 in the extreme cases. Overall we can quote our answer to be 102±1.

Similarly we may subtract our values.
                The biggest difference is   69.5 − 32.5   = 37,
    the difference using the quoted values is   69   − 33   = 36
            and the smallest difference is   68.5 − 33.5   = 35.
Again we can think about our answers and overall quote our answer to be 36 ± 1.

When attempting either addition or subtraction the general rule is that the maximum error is the sum of the original errors.

In these cases that error is ±(0.5 + 0.5) = ±1. In both our answers, 102 and 36, we were not confident of the last digits of 2 and 6 because this is where any change may occur.

What happens in multiplication?
            Multiplying the large values gives   69.5 × 33.5   = 2328.25,
                    the quoted values gives   69   × 33   = 2277
            and the small values gives   68.5 × 32.5   = 2226.25.
Again we must ask which of these answers is correct. If we know with absolute certainty that the original values of 69 and 33 are true then the

correct answer is 2277. If we are unsure about our values then in reality we may say only that the answer is 2000 correct to 1 significant figure. This is because all 3 answers will round to the same answer of 2000 correct to that 1 significant figure.

What happens in division?

Dividing extreme values gives $\dfrac{69.5}{32.5} = 2.138461$ (as displayed),

the quoted values gives $\dfrac{69}{33} = 2.090909$ (as displayed)

and other extreme values gives $\dfrac{68.5}{33.5} = 2.044776$ (as displayed).

Our original values were quoted to 2 significant figures, though our answers still differ, being 2.1, 2.1 and 2.0 respectively to this degree of accuracy. Only to 1 significant figure are all the answers the same.

For multiplication and division the general rule is that answers are correct to one significant figure less than the original values.

It is interesting to notice the arithmetic operations and the use of these extreme values. When adding and multiplying the pairing is both large extreme values together and both small extreme values together. When subtracting and dividing the choice is one extreme large and one extreme small value together. This is necessary to discover the maximum errors possible. Other combinations do not give us these wide variations.

The two general rules are important where a complete problem has a variety of part calculations before the final answer. Do not round your part answers too soon or you may find that the final answer is wildly inaccurate.

Rough checks

It is important to **have some idea of the size of your final answer** before attempting to use your calculator. This allows you to check roughly that you have pressed the correct keys.

We shall return to our values of 69 and 33. Correct to the nearest 10 we may think of them as 70 and 30. When multiplied together, $70 \times 30 = 2100$ which is roughly in the region of $69 \times 33 = 2277$.

During division it is easier to think of 70 in place of 69 and 35 in place of 33 because 70 is exactly divisible by 35,

i.e. $\dfrac{70}{35} = 2$, which again is close to our answer of $\dfrac{69}{33} = 2.090909$ as displayed.

███████ **Example 1.14** ████████████████████████████████████

i) 1.694×0.42 may be thought of as approximately 1.7×0.4.
    The approximate answer is 0.68 compared with the true displayed
    answer of 0.71148. Now 1.694 is quoted correct to 4 significant
    figures but 0.42 is quoted to only 2 significant figures. We pay
    attention to the lesser number of significant figures because this is
    where there is less accuracy. Then our answer must be to 1 significant
    figure less than this, i.e. $2 - 1 = 1$ significant figure,
    i.e. $1.694 \times 0.42 = 0.7$ (1 significant figure).

ii) $(2.143)^2$ may be thought of as only 2^2. This gives an approximate
    answer of 4 compared with 4.59 (3 significant figures).

iii) $(2.515)^2$ is closer to 3^2 than to 2^2. It would be safe to suggest that the
    answer lies somewhere between 9 and 4. In fact even though the
    original 2.515 lies closer to 3 than to 2 this value squared lies closer to
    4 than to 9. The reason for this is the effect of squaring a value.
    $(2.515)^2 = 6.33$ (3 significant figures).

████████ **EXERCISE 1.4** ██████████████████████████████████

For each question you should
  i) estimate (roughly work out) the size of your answer with a rough
      check on paper, not using your calculator;
  ii) work out the calculation based on those figures quoted being true,
      using your calculator.

1 3.12×7.90

2 4.35×1.6

3 $59.72 \div 4.11$

4 $3.672 \div 9.051$

5 $4.35^2 \times 1.6$

6 $(4.35 \times 1.6)^2$

7 $\dfrac{22.2 \times 3.9}{1.3}$

8 $\dfrac{14.72 \times 4.06}{1.28}$

9 $1.15 + \left(\dfrac{22.3 \times 3.9}{1.3}\right)$

10 $73.69 - \left(4.35 \times 1.6^2\right)^2$

11 $\dfrac{26.14 \times 958}{105.34 \times 9.26}$

12 $\dfrac{(153.4)^2 \times 7.85}{(295.7)^2 \times 6.91}$

13 $\dfrac{22.2^2 \times 3.9^2}{1.3^2}$

14 $\left(\dfrac{22.2 \times 3.9}{1.3}\right)^2$

15 $\dfrac{76.31 \times \sqrt{3.87}}{1.95 \div 15.03}$

16 $\dfrac{\sqrt{84.35} \times 12.62}{3.11 \times 25.09}$

17 $\dfrac{\sqrt{84.35 + 12.62}}{1.95 \div 15.03}$

18 $\dfrac{\sqrt{(84.35 \times 12.62)}}{(3.11 \times 25.09)}$

19 $\dfrac{(9.45)^2}{40.2} + \dfrac{(2.55)^2}{8.5}$

20 $\dfrac{12.35 \times 120.5 \times \sqrt{30.05}}{17.6 \times \sqrt{25.67} \times 3.45}$

Some other key operations

The majority of scientific calculators have many more function keys than we have discussed so far in this chapter. Because the models of calculators change quite often we cannot usefully consider all possibilities. In this section we will learn how to use some of them, but will leave more until later chapters.

$\sqrt[3]{}$ is the key that finds the cube root of a number.

▬▬▬▬▬ **Examples 1.15** ▬▬▬▬▬▬▬▬▬▬▬▬▬▬▬▬▬▬▬▬▬▬▬▬▬▬▬▬▬▬▬▬▬▬

i) $\sqrt[3]{64}$ is worked out using

$$64 | \quad \sqrt[3]{}|$$

to display the answer 4.

ii) $\sqrt[3]{-64}$ is worked out using

$$64 | \quad {}^{+}/_{-}|$$

in order to display -64 and then

$$\sqrt[3]{}|$$

to display the answer of -4.

───

$x^{1/y}|$ is the key that finds the yth root of a number x.
We may attempt the previous examples again, using this key, because the cube root has a power $1/3$.

▬▬▬▬▬ **Examples 1.16** ▬▬▬▬▬▬▬▬▬▬▬▬▬▬▬▬▬▬▬▬▬▬▬▬▬▬▬▬▬▬▬▬▬▬

i) $\sqrt[3]{64}$ may be thought of as $64^{1/3}$. It is understood that all of the number $(64)^{1/3}$ is raised to the power $1/3$.
    The order of key operations is

$$64| \quad x^{1/y}| \quad 3| \quad =|$$

to give the correct display of 4.
After pressing $x^{1/y}|$ it is the $3|$ that is the value used to replace y in the power.

ii) $\sqrt[3]{-64}$ is $(-64)^{1/3}$, emphasising the position of the minus (**negative**) sign.
    The order of key operations is

$$64| \quad {}^{+}/_{-}| \quad x^{1/y}| \quad 3| \quad =|$$

to give a display of -4.

───

When y is an **odd value** (i.e. 1, 3, 5, . . .) there is no problem finding the root of a positive or negative number. If the original number is positive then its root is positive as well. Similarly if the original number is negative then its root is negative.

A little more care is needed when y is an **even number** (i.e. 2, 4, 6, . . .). If the original number is positive there will be 2 real roots, one being positive and the other negative. The calculator will display the positive root and it is understood that you will know that there is a negative root as well. However, there are no real roots of a negative number.

Examples 1.17

i) $\sqrt[4]{625}$ may be thought of as $625^{1/4}$. In fact it will have 2 roots, one being positive that is displayed and one being negative that is understood.

The order of key operations is

$$625|\quad x^{1/y}|\quad 4|\quad =|$$

The display is 5 whilst the other answer of -5 is understood.

ii) $\sqrt[4]{-625}$ may be thought of as $(-625)^{1/4}$ which has no real roots. If you attempt the following order of

$$625|\quad ^+/_-|\quad x^{1/y}|\quad 4|\quad =|$$

the display will give the error message -E-.

$1/x|$ is the key that finds 1 divided by a number. It is called the **reciprocal**.

Examples 1.18

i) $\dfrac{1}{2}$ is the numerical way of writing 'one half'. It is also the reciprocal of 2. The usual way of finding its decimal value is to use the key operations of

$$1|\quad \div|\quad 2|\quad =|$$

An alternative method is

$$2|\quad 1/x|$$

Both of these methods will display 0.5.

ii) $\dfrac{1}{13 \times 5}$ may be calculated using the key operations

$$13|\quad \times|\quad 5|\quad =|$$

to display 65, i.e. $\dfrac{1}{13 \times 5} = \dfrac{1}{65}$.

Pressing the key $1/x$ displays the decimal value of 0.0153846.
The complete order of key operations is

$$13 \quad \times \quad 5 \quad = \quad 1/x$$

iii) $\dfrac{4}{13 \times 5}$ may be calculated in a few different ways. If we think of it as

$4 \times \left(\dfrac{1}{13 \times 5} \right)$ then we may use our previously displayed value of

0.0153846 multiplied by 4 to give a final display of 0.0615384.
The complete order of key operations is

$$13 \quad \times \quad 5 \quad = \quad 1/x \quad \times \quad 4 \quad =$$

π is the key representing the usual value of 3.14159... depending upon
the number of decimal places displayed on your calculator. It is a useful
key, saving you time and avoiding possible error inputting the value.

▰▰▰▰ Examples 1.19 ▰▰▰▰▰▰▰▰▰▰▰▰▰▰▰▰

i) 3π, understanding a multiplication sign between 3 and π, uses the key
operations

$$3 \quad \times \quad \pi \quad =$$

to display 9.424778.

ii) $\dfrac{1}{3\pi}$ may be worked out using the latest two key operations in the
following order because it is the reciprocal of the previous example.
The key operations

$$3 \quad \times \quad \pi \quad = \quad 1/x$$

display an answer of 0.1061033.

iii) $\sqrt{\dfrac{1}{3\pi}}$ extends the key operations a little further to use the $\sqrt{}$ key.
The complete order of operations is

$$3 \quad \times \quad \pi \quad = \quad 1/x \quad \sqrt{}$$

to display 0.325735.

▰▰▰▰ Example 1.20 ▰▰▰▰▰▰▰▰▰▰▰▰▰▰▰▰▰

In this example we will work out $\sqrt[3]{\left(\dfrac{1}{5\pi^2} \right)^4}$ using a combination of several

keys we have learned to use already. The order of key operations is
important, though there are several variations of the following order that
are possible. Think how you would write this down. One order might be

cube root

brackets to the power 4

one over

5

π squared.

The order of key operations is the reverse of this order of writing down, i.e. starting at the heart of the calculation and working outwards. We can take the complete order in several stages, working *up* the list above.

$$\boxed{\pi} \quad \boxed{x^2} \quad \boxed{\times} \quad \boxed{5} \quad \boxed{=}$$

to display 49.348022,

i.e. $\sqrt[3]{\left(\dfrac{1}{5\pi^2}\right)^4} = \sqrt[3]{\left(\dfrac{1}{49.348022}\right)^4}$

$\boxed{1/x}$ finds the reciprocal of 49.348022 to be 0.0202642.

This value can be raised to the power 4 using

$$\boxed{x^y} \quad \boxed{4} \quad \boxed{=}$$

The display is 1.6862463^{-07}, remembering the way the calculator represents standard form.

Finally the cube root can use

$$\boxed{x^{1/y}} \quad \boxed{3} \quad \boxed{=}$$

to display 5.5246784^{-03}.

$\boxed{EXP}$ is the key to input a number in standard form directly into the calculator.

▨▨▨ Examples 1.21 ▨▨▨▨▨▨▨▨▨▨▨▨▨▨

i) To input 1.25×10^3 the order of key operations is

$$\boxed{1.25} \quad \boxed{EXP} \quad \boxed{3}$$

and the display becomes 1.25^{03}.

ii) To input 6.089×10^{-2} the order of key operations is

$$\boxed{6.089} \quad \boxed{EXP} \quad \boxed{+/-} \quad \boxed{2}$$

giving the display 6.089^{-02}.

The memory

Different calculators can have slightly different memory operations (functions). **Whenever you have completely finished a calculation remember to cancel the contents of the memory.** We will mention the more usual keys common to many calculators.

MR⌋ recalls to the screen the contents of the memory.

M in⌋ transfers the screen display to the memory, deleting whatever was previously in the memory. When the transfer occurs, the display either may go blank or remain as before.

M+⌋ adds the screen display to the memory.

M–⌋ subtracts the screen display from the memory.

A calculator without the key M–⌋ presents no difficulties. +/–⌋ M+⌋ has the same effect because +/–⌋ changes the display to a negative value before adding that negative value to the memory using M+⌋. You should recall that subtraction is just the addition of a negative number, e.g. $5 - 3 = 5 + (-3) = 2$. Indeed, you should test this for yourself.

■■■■ EXERCISE 1.5 ■■■■

Use your calculator to work out the following:

1 2.479^4

2 4.3×2.96^3

3 $1.45^2 + 1.32^2$

4 $4.79(1.45^2 + 1.32^2)$

5 $4.79(1.45 + 1.32)^2$

6 $\sqrt[4]{74.19}$

7 $74.19^{1/4}$

8 $\sqrt[3]{33.249}$

9 $\sqrt[3]{(-33.249)}$

10 $74.19^{1/4} - \sqrt[3]{33.249}$

11 $74.19^{1/4} - \sqrt[3]{(-33.249)}$

12 $\sqrt{1.45^2 + 1.32^2}$

13 $\dfrac{1}{2\pi}$

14 $\dfrac{1}{2\pi + 4.9}$

15 $\sqrt{\dfrac{1}{2\pi - 4.9}}$

16 $\dfrac{\sqrt[3]{119.6}}{2.3}$

17 $\dfrac{2.3}{\sqrt[3]{119.6}}$

18 $\dfrac{1.32\pi + 2.3}{\sqrt[3]{119.6}}$

19 $1.32 + \dfrac{2.3}{\sqrt[3]{119.6}}$

20 $1.32\pi - \dfrac{2.3}{\sqrt[3]{119.6}}$

The final section of this chapter applies the calculator methods we have learned to some practical problems.

■■■■ Example 1.22 ■■■■

Suppose we have a copper rod of length 0.20000 m which we heat up, raising its temperature by 250°C. The rod will expand so that its new length is given by the formula $l = l_0 (1 + \alpha t)$. Before we attempt to calculate the newly expanded length we need to be able to understand the meaning of the formula.

l is the expanded length
l_0 is the original length
α is the coefficient of linear expansion
and t is the temperature rise.

All metals expand at different rates. The coefficient of linear expansion is a measure of that rate for every °C. In this problem we have $l_0 = 0.20000$, $\alpha = 17 \times 10^{-6}$ and $t = 250$ with all the units of measurement being consistent.

Using the methods of this chapter we can complete the calculation in a number of steps.

Step one	$\alpha t = 4.25 \times 10^{-3}$
Step two	$1 + \alpha t = 1.00425$
Step three	$l_0(1 + \alpha t) = (0.2000)(1.00425)$
	$= 0.20085.$

We need to interpret our answer bearing in mind the **original number of decimal places** when measuring the rod's length. This means that either we round up or down the last figure to give a new length of either 0.2009 m or 0.2008 m.

███████ **Example 1.23** ███████████████████████████████

In this example we will look at a body of mass 23 kg having kinetic energy of 1.8 kJ. We wish to find its speed.

The kinetic energy of a body is the energy it has by virtue of its motion, according to the formula $KE = \frac{1}{2}mv^2$.

In this formula for the body KE is the kinetic energy,
m is the mass
and v is the speed.

The energy is not given in units that are consistent with the formula. The 1.8 kJ needs to be written as 1800 J.

Firstly we will substitute the values into the formula so that

$$KE = \tfrac{1}{2}mv^2$$

becomes $1800 = \frac{1}{2}(23)v^2$

i.e. $1800 = 11.5v^2.$

Because v appears on the right we will concentrate on this side of the equation. If we had a value for v the calculation would be

input v
square
multiply by 11.5
value of 1800.

Remember that we read the list *upwards* and use the opposite operation to give a new list, i.e.

input 1800
divide by 11.5
square root
v.

Now our calculation is

$$\frac{1800}{11.5} = v^2$$

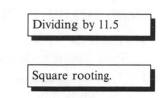

Dividing by 11.5

$$156.52 = v^2$$

$$\sqrt{156.52} = v$$

Square rooting.

$$12.5 = v.$$

The speed of the body is $12.5\,\text{ms}^{-1}$ correct to 1 decimal place.

Example 1.24

The circuit diagram shows an A.C. emf of 240 V and frequency 50 Hz with a resistance of 550 Ω and an inductance of 8 H in series. We are going to calculate the current, *i*, in the circuit given that $i = \dfrac{V}{\sqrt{(R^2 + \omega^2 L^2)}}$

where $\omega = 2\pi f$.

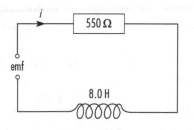

We need to check the meaning of all the letters in our formulae.

V is the voltage,	$V = 240$
R is the resistance,	$R = 550$
L is the inductance,	$L = 8$
f is the frequency,	$f = 50$
i is the current.	

The first stage of the calculation is to find ω so that

$$\omega = 2\pi f$$

becomes

$$\omega = 2(3.14159\ldots)(50)$$

$$\omega = 314.159$$

and then

$$\omega^2 = 98\,696.044.$$

In the denominator of the more complicated formula we have terms involving R^2, ω^2 and L^2 which we can use in stages.

$$\omega^2 L^2 = (98\,696.044)(64) \qquad = 6\,316\,546.8$$

and

$$R^2 + \omega^2 L^2 = 302\,500 + 6\,316\,546.8 \quad = 6\,619\,046.8$$

so

$$\sqrt{(R^2 + \omega^2 L^2)} = \sqrt{6\,619\,046.8} \qquad = 2\,572.751.$$

Finally

$$i = \frac{V}{\sqrt{(R^2 + \omega^2 L^2)}}$$

becomes

$$i = \frac{240}{2\,572.751} \qquad = 0.093.$$

This means that there is a current of 0.093 amps or 93 mA.

With a little care, you can find i using the calculator's memory to store the calculation in stages. You should try this for yourself. It will save you writing down answers and worrying about the number of decimal places to re-input into your calculator.

■■■■ EXERCISE 1.6 ■■■■

1   The area, A, of a trapezium is given by the formula $A = \frac{1}{2}(a + b)h$. Find the area when $a = 0.85\,\text{m}$, $b = 1.37\,\text{m}$ and $h = 0.2\,\text{m}$.

2   The surface area of a ball bearing, A, is given by $A = 4\pi r^2$ where r is the radius. For $r = 11\,\text{mm}$ calculate the surface area.

3   An increasing horizontal force is acting on a body on a rough horizontal plane. The body causes a normal reaction of $N = 50$ Newtons and the coefficient of friction between the body and the plane is $\mu = 0.24$. Calculate the amount of friction, F, acting if $\mu = \dfrac{F}{N}$. The total reaction, S, is given by $S = \sqrt{(F^2 + N^2)}$. Use your value of F to calculate S.

4   The initial velocity, u, constant acceleration, a, and time, t, are connected to the displacement, s, of a vehicle according to $s = ut + \frac{1}{2}at^2$. For $u = 1.56\,\text{ms}^{-1}$ and $a = 2.12\,\text{ms}^{-2}$ find the displacement after 7.50 seconds.

5   Given $l = l_0(1 + \alpha t)$ find the expanded length of a silver rod, l, if $l_0 = 0.3500\,\text{m}$, $\alpha = 19 \times 10^{-6}$ and $t = 850°\text{C}$.

6   The rate of heat energy transfer, Q, is related by $Q = \dfrac{kAT}{x}$ where A is the conducting area, T is the temperature difference between the two faces, x is the thickness of the material and k is the coefficient of thermal conductivity. If $k = 1.1$, $A = 6.5$, $T = 15$ and $x = 0.1$ find the value of Q.

7   Calculate the power, P, across a resistor, R, for a current, I, related by $P = I^2 R$ when $I = 1.25\,\text{A}$ and $R = 3\,\Omega$.

8   An object is placed at a distance (u) of 350 mm in front of a concave spherical mirror of radius (r) 200 mm. Find the distance of the image (v) from the mirror if $\dfrac{1}{v} + \dfrac{1}{u} = \dfrac{2}{r}$.

9   The effective resistance, R, of 2 resistors in parallel is $\dfrac{1}{R} = \dfrac{1}{r_1} + \dfrac{1}{r_2}$. Calculate R if $r_1 = 4.5\,\Omega$ and $r_2 = 3.6\,\Omega$.

10 Hooke's law relates the tension in an elastic spring, T, to the natural length of the spring, l, and its extension, x, according to $T = \dfrac{\lambda x}{l}$, where λ is the coefficient of elasticity. For $\lambda = 28\,\text{N}$ and $T = 9\,\text{N}$ find the extension from a natural length of 0.75 m.

11 A vehicle accelerates constantly from rest to $17.5\,\text{ms}^{-1}$ in 15 seconds. The displacement, s, is related to the initial velocity, u, the final velocity, v, and the time, t, by $s = \frac{1}{2}(u+v)t$. How far will it have travelled in this 15 seconds?

12 A van reaches a speed of $20.5\,\text{ms}^{-1}$ from rest over a distance of 200 m. The initial speed, u, the final speed, v, the acceleration, a, and the distance, s, are connected by the formula $v^2 = u^2 + 2as$. Calculate the acceleration of the van.

13 The area of an annulus (i.e. the area betwen two circles with the same centre and different radii), A, is given by $A = \pi(R^2 - r^2)$. Given $R = 250\,\text{mm}$ and $r = 150\,\text{mm}$ find the value of A.

14 For a simple pendulum the time period of an oscillation, in seconds, is given by $T = 2\pi\sqrt{(l/g)}$ where l is the length of the pendulum and g is the acceleration due to gravity. Find the value of T if $g = 9.81\,\text{ms}^{-2}$ and $l = 1.55\,\text{m}$.

15 The greatest height, h, of a body projected vertically upwards is given by $h = \dfrac{v^2}{2g}$ where v is the velocity of projection and g is the acceleration due to gravity. If the maximum height is 25 m and $g = 9.81\,\text{ms}^{-2}$ what is the velocity of projection?

16 The 3 sides of a triangle are a, b and c. For $a = 55\,\text{mm}$, $b = 34$ mm and $c = 46\,\text{mm}$ find s where $s = \dfrac{a+b+c}{2}$. Calculate the area of the triangle, A, where $A = \sqrt{s(s-a)(s-b)(s-c)}$.

17 The gas laws combine volume, V, pressure, P, and absolute temperature, T, according to $\dfrac{PV}{T} = \text{constant}$.
If $V = 19.2 \times 10^{-3}\,\text{m}^3$, $T = 273°\text{K}$ and $P = 1.013 \times 10^5\,\text{Nm}^{-2}$ find the value of the constant.
The pressure is increased by 10% due to a change in temperature whilst the other two values remain unchanged.
Calculate the new value of T.

18 The voltage magnification factor, Q, is related to the frequency, f, inductance, L, and resistance, R, by $Q = \dfrac{2\pi fL}{R}$. Given $f = 50\,\text{Hz}$, $L = 5\,\text{H}$ and $R = 200\,\Omega$ work out Q. For this value of Q, L is increased to 7.5 H. What is the new value of R?

19 The quantity of flow, $Q\,\text{m}^3\text{s}^{-1}$, along a pipe of cross-sectional area $0.005\,\text{m}^2$ is related to the velocity of the flow, $v\,\text{ms}^{-1}$ according to $Q = 0.005v$. Construct a table of values relating Q and v for values of v from 0 to 12 at intervals of $2\,\text{ms}^{-1}$.

20 The final velocity, v, of a van due to a constant acceleration of $2.45\,\text{ms}^{-2}$ over a period of time, t, is related to its initial velocity, $1.00\,\text{ms}^{-1}$, by $v = 1.00 + 2.45t$. From $t = 0$ to $t = 60$ seconds at intervals of 10 seconds construct a table of values relating t and v.

■■■■ MULTI-CHOICE TEST 1 ■■■■

1 The value of $\dfrac{325 \div \left(\dfrac{144}{9 \times 4}\right)}{5}$ is

A) 0.0125
B) 0.206
C) 16.25
D) 81.25

2 $4.36 - \dfrac{\sqrt{27.40 - 3.80}}{4.51}$ has a value of

A) 2.07
B) 2.85
C) 3.28
D) 4.04

3 The value of $\left(\dfrac{33.67 - 4.72^2}{11.31}\right)^2 + 6.90^2$ is

A) 18.42
B) 59.13
C 72.08
D) 17027.98

4 To 4 significant figures the value of $(62.18^2 - 47.39^2)^{1/2}$ is
A) 14.79
B) 40.26
C) 3819
D) 266 600

5 5.1578 correct to 3 significant figures is
A) 0.158
B) 5.15
C) 5.158
D) 5.16

6 0.00432 in standard form is
A) 4.32×10^{-3}
B) 4.32×10^3
C) 432×10^5
D) 432×10^0

7 The value of $\left(\dfrac{1}{0.3615} + \dfrac{1}{0.4709}\right)^2$ is

   A) 21.45
   B) 22.63
   C) 22.94
   D) 23.91

8 **Without using your calculator** the largest value is

   A) $\dfrac{1}{7 \times 9}$

   B) $\dfrac{1}{8 \times 8}$

   C) $\dfrac{1}{6 \times 11}$

   D) $\dfrac{1}{5 \times 13}$

9 **Without using your calculator** the approximate value of $\dfrac{48.8 \times 68.9 \times 0.00051}{0.0069}$ is

   A) 2.5
   B) 25
   C) 250
   D) 2500

10 $2.76 - \dfrac{4.35\pi}{\sqrt[3]{29.64}}$ has a value

   A) -1.66
   B) 0.25
   C) 1.35
   D) 2.76

11 A person quotes the value of π as 3.1415. Is this value
   A) truncated?
   B) correct to 4 decimal places?
   C) correct to 4 significant figures?
   D) correct to 5 significant figures?

12 A cone has a slant height, l, a base radius, r, and a curved surface area, A. They are connected by the formula $A = \pi r l$. Using the values from the diagram A (mm^2) is

   A) 980
   B) 1848
   C) 2310
   D) 3080

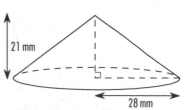

21 mm

28 mm

13 The volume V of a cylinder is given by $V = \pi r^2 h$. When the radius is $r = 70\,\text{mm}$ and the height is $h = 50\,\text{mm}$, $V\ (\text{mm}^3)$ is
A)   110 000
B)   220 000
C)   550 000
D)   770 000

14 For a simple pendulum the time period of an oscillation is given by

$T = 2\pi \sqrt{\dfrac{l}{g}}$. In this question $\pi = 3.142$, $l = 0.476\,\text{m}$ and $g = 9.81\,\text{ms}^{-2}$.

Hence T (s) is
A)   0.138
B)   1.38
C)   4.38
D)   13.8

Questions **15** and **16** refer to the following information.

$x = \dfrac{-b \pm \sqrt{b^2 - 4ac}}{2a}$ with $a = 3$, $b = -5$ and $c = -1$.

15 The value of $b^2 - 4ac$ is
A)   −11
B)   13
C)   20
D)   37

16 The values of x are
A)   −5.54 and 0.18
B)   −0.54 and 5.54
C)   −0.36 and 3.69
D)   −0.18 and 1.85

17 The volume V of a sphere is $66\,\text{mm}^3$. Its radius, r, is given by
$r^3 = \dfrac{66 \times 3}{4\pi}$. The value of r (mm) is
A)   2.3
B)   2.5
C)   4.0
D)   15.8

The effective resistance, R, of 2 resistors in parallel is given by $\dfrac{1}{R} = \dfrac{1}{R_1} + \dfrac{1}{R_2}$. Questions **18**, **19** and **20** use $R_1 = 25\,\Omega$ and $R_2 = 15\,\Omega$.

18 The value of R is

A) $\dfrac{1}{40}$ or $0.025\,\Omega$

B) $\dfrac{2}{40}$ or $\dfrac{1}{20}$ or $0.05\,\Omega$

C) $\dfrac{8}{75}$ or $0.10\bar{6}\,\Omega$

D) $\dfrac{75}{8}$ or $9.375\,\Omega$

19 If R_1 doubles in value then R
A)   decreases
B)   stays the same
C)   also doubles
D   generally increases

20 If both R_1 and R_2 double then R
A)   generally decreases
B)   halves
C)   stays the same
D)   also doubles

2 Solving Linear Equations

Introduction

The first task in this chapter is to explain the meaning of **linear** and **equation**. An **equation** is a relationship which generally contains numbers and letter(s). There are many sorts of equations, some of which can be

32

solved and some for which there is no real solution. In this book we will aim to solve equations, and in this particular chapter each equation will have just one solution. This means that only one value will be logically correct. That value is said to **satisfy** the equation. Whenever the word **linear** is mentioned it is understood to refer to straight lines. Combining both important words we will see that it is possible to represent a **linear equation** by a straight line graph.

ASSIGNMENT

The Assignment for this chapter involves the vehicle from Chapter 1 but this time paying attention to its speed. It is travelling at a speed of $30\,\text{ms}^{-1}$ when the brakes are applied to produce a deceleration of $4\,\text{ms}^{-2}$. During the motion the speed, v, is linked to the time, t, according to the formula

$$v = 30 - 4t.$$

The first use of this problem will be for one particular numerical calculation and the second will look more generally at the motion in graphical terms.

Simple equations

We could attempt to write some general algebra that is typical of an equation. As an introduction this just makes a simple problem needlessly complicated. Now $x - 12 = 8$ is an example of a simple equation. The $=$ sign can be thought of as a point about which the equation balances so that the left-hand side has the same overall value as the right-hand side. It is obvious from the right-hand side that in this case the value is 8. On the left-hand side reading $x - 12$ we can think of this as "some number minus twelve",

i.e. $$x - 12 = 8$$

is "some number minus twelve equals eight".

Easy arithmetic tells us that the number must be 20.

Suppose we have the simple equation

$$x + 5 = 11$$

i.e.   "some number plus 5 equals eleven".

Again easy arithmetic tells us that the number must be 6.

At this stage we have 2 important points to remember.

1. The equation must always balance about $=$.
2. Eventually only x must appear on one side of the equation.

Any one equation may be written in several ways. Our original example

$$x - 12 = 8$$

is equivalent to $8 = x - 12$

and $-12 + x = 8$

and $8 = -12 + x$

and yet other formats.

> It does *not* matter whether x, or any letter in use, lies on the left or right side.

Examples 2.1

i)   Solve $x - 12 = 8$.

We can see that we are subtracting 12 on the left. So that only x appears on the left we add 12 to that side. To keep the balance we must add 12 to the right as well,

i.e. $x - 12 + 12 = 8 + 12$.

Now $-12 + 12$ works out to 0, and 0 added to anything (x) leaves it unchanged,

i.e. $x = 8 + 12$.

On the right simple addition gives 20 so that

$$x = 20.$$

ii) Solve $x + 5 = 11$.

In this case we are adding 5 on the left. So that only x appears on the left we subtract 5 from that side. To keep the balance we must subtract 5 from the right as well,

i.e. $x + 5 - 5 = 11 - 5$.

Now $5 - 5$ works out to 0, and 0 added to anything (x) leaves it unchanged,

i.e. $x = 11 - 5$

and so $x = 6$.

To save time and space it is usual to simplify the left and right sides on the same line rather than in several repetitive stages.

iii) If α and β represent numbers solve $x - \alpha = \beta$.

To eliminate the subtraction of α on the left we must add α and do so to both sides to keep the balance,

i.e. $x - \alpha + \alpha = \beta + \alpha$

i.e. $x = \beta + \alpha$.

We have no actual values for α and β meaning that we cannot simplify the right to just a number. Instead we have to leave the answer alone.

Whenever letters are used to represent numbers in this way we term them
constants.

We have used x as the letter in these examples but could have used any
other letter in its place. This is shown in the previous examples again,
given in the same formats with different letters.

i) $y - 12 = 8$ to give $y = 20$;

ii) $t + 5 = 11$ to give $t = 6$;

iii) $m - \alpha = \beta$ to give $m = \beta + \alpha$ where α and β are constants.

EXERCISE 2.1

Solve the following simple linear equations, given that in Questions **9** and
10 α, β and γ are constants.

1	$x - 14 = 17$		**6**	$1 + t + 4 = 7$
2	$x + 19 = 3$		**7**	$-2 + t = -5$
3	$-14 + x = 17$		**8**	$4 + z - 3 = 13$
4	$5 = -2 + x$		**9**	$x + \alpha = \gamma$
5	$y + 6 = 0$		**10**	$x - \alpha + \beta = 0$

In this early section we have looked at the addition and subtraction of
numbers in simple equations. We have dealt with them together because
these arithmetic operations are the opposites of each other. Similarly
multiplication and division are opposites of each other, and so now we
will deal with them together.

Examples 2.2

i)   Solve the simple linear equation $4x = 3$.

We are multiplying x by 4 and the opposite of multiplication is
division. This means that we should divide by 4 and do it for both
sides to keep the balance of the equation,

i.e. $$\frac{4x}{4} = \frac{3}{4}$$

> 4 divides by 4 meaning that the 4s cancel to 1.

$$1x = 0.75$$
$$x = 0.75.$$

> 1 multiplied by any number leaves that number unchanged so that $1x = x$.

ii)   Solve the equation $\dfrac{x}{6} = 2.5$.

We are dividing x by 6 and the opposite of division is multiplication.
Again, attempting to leave x alone on the left we will multiply by 6
and do it for both sides to keep the balance,

i.e. $\dfrac{6x}{6} = 6(2.5)$

$1x = 15$

$x = 15.$

iii) We may combine both the needs of multiplication and division to solve $\dfrac{-2x}{5} = 3.42.$

The opposite of multiplying by -2 is dividing by -2 and the opposite of dividing by 5 is multiplying by 5. We apply these operations to both the left and right of the equation,

i.e. $\dfrac{5(-2x)}{-2(5)} = \dfrac{5(3.42)}{-2}$

> Both the -2s cancel to 1 and both the 5s cancel to 1 as well.

$1x = -8.55$

$x = -8.55.$

iv) If α, β and γ represent numbers solve $\dfrac{\alpha x}{\beta} = \gamma.$

The opposite of multiplying by α is dividing by α and the opposite of dividing by β is multiplying by β. We apply these operations to both the left and right of the equation,

i.e. $\dfrac{\beta(\alpha x)}{\alpha(\beta)} = \dfrac{\beta(\gamma)}{\alpha}$

$x = \dfrac{\beta\gamma}{\alpha}.$

■ EXERCISE 2.2

Solve the following equations.

1   $5x = 4$

2   $0.3y = 0.39$

3   $-2y = 3.6$

4   $2.14 = -0.2x$

5   $\dfrac{y}{4} = 7$

6   $\dfrac{x}{0.3} = 1.1$

7   $\dfrac{2x}{3} = 2.84$

8   $1.7 = \dfrac{3x}{4}$

9   $\dfrac{-12x}{5} = 6$

10   $\dfrac{13y}{4} = -1.09$

Now we are in a position to combine the effects of these pairs of opposite arithmetic operations. In Chapter 1 we concentrated on the heart of the calculation. When solving linear equations the heart of the problem is around x (or any other letter that is being used). We must remove other numbers from the same side as the letter (left or right) working our way towards the letter until it remains alone.

Example 2.3

i) Solve $3x + 4 = 20$.

Addition of 4 is less closely attached to x than anything else. This means that we will deal with 4 first.

$\therefore$ $3x + 4 - 4 = 20 - 4$

	Subtracting 4 from both sides.

$3x = 16$

$\dfrac{3x}{3} = \dfrac{16}{3}$

	Dividing both sides by 3.

$x = 5.\bar{3}$.

ii) Solve $2x - 5 = 12$.

Similarly, subtraction of 5 is less closely attached to x than any other part of the equation on the left.

$\therefore$ $2x - 5 + 5 = 12 + 5$

	Adding 5 to both sides.

$2x \qquad = 17$

$\dfrac{2x}{2} = \dfrac{17}{2}$

	Dividing both sides by 2.

$x = 8.5$.

iii) Solve $\dfrac{-x}{3} + 7 = 0$.

$\dfrac{-x}{3} + 7 - 7 = 0 - 7$

	Subtracting 7 from both sides.

$\dfrac{-x}{3} \qquad = -7$

$\dfrac{(-3)(-x)}{3} = -3(-7)$

	Multiplying both sides by −3.
	Remembering that $(-)(-) = +$.

$x = 21$.

iv) Solve $\dfrac{x}{3} + 31 = \dfrac{-2x}{5} + 36.5$.

Terms involving x appear on both sides of this equation. The first move is to gather these terms to one side. We will gather them together on the left, but the same final solution will be reached if we gather them on the right.

$\therefore$ $\dfrac{x}{3} + \dfrac{2x}{5} + 31 = \dfrac{2x}{5} - \dfrac{2x}{5} + 36.5$

$\dfrac{x}{3} + \dfrac{2x}{5} + 31 = 36.5$

	Adding $\dfrac{2x}{5}$.

$$\frac{x}{3} + \frac{2x}{5} + 31 - 31 = 36.5 - 31$$

> Subtracting 31.

$$\frac{x}{3} + \frac{2x}{5} = 5.5.$$

We can deal with the mixture of fractions, multiplying throughout by the lowest common multiple (LCM) of 3 and 5, i.e. multiplying throughout by 15,

i.e. $$15\left(\frac{x}{3} + \frac{2x}{5}\right) = 15\,(5.5)$$

> Multiplying each term by 15.

$$\frac{15x}{3} + \frac{30x}{5} = 82.5$$

$$5x + 6x = 82.5$$

$$11x = 82.5$$

$$\frac{11x}{11} = \frac{82.5}{11}$$

> Dividing by 11.

$$x = 7.5.$$

As equations involve more and more terms we may check that the answer is correct by separately substituting the answer into each side of the equation, e.g.

Left-hand side $= \dfrac{x}{3} + 31$ Right-hand side $= \dfrac{-2x}{5} + 36.5$

$$= \frac{7.5}{3} + 31 \qquad\qquad = \frac{-2(7.5)}{5} + 36.5$$

$$= 2.5 + 31 \qquad\qquad\quad = -3 + 36.5$$

$$= 33.5. \qquad\qquad\qquad = 33.5 \text{ also.}$$

ASSIGNMENT

Our problem involves a relationship between a vehicle's speed, v, and time, t, according to the formula $v = 30 - 4t$.

In this first discussion we are interested in the time taken for the speed to fall to 12.4 ms^{-1}. This means that we must substitute $v = 12.4$ into our formula,

i.e. $$12.4 = 30 - 4t$$

$$12.4 - 30 = -4t$$

> Subtracting 30 from both sides.

$$-17.6 = -4t$$

$$\frac{-17.6}{-4} = t$$

> Dividing both sides by -4.

$$t = 4.44$$

i.e. the speed falls to 12.4 ms^{-1} after 4.4 seconds.

Solve the following equations.

1 $6x + 2 = 11$

2 $2y - 3 = 7.6$

3 $x + \dfrac{x}{5} = 24$

4 $\dfrac{x}{5} = 1 + \dfrac{x}{4}$

5 $2y + 3 = 17 - 3y$

6 $\dfrac{5x}{6} - \dfrac{x}{2} = 3$

7 $\dfrac{5x}{6} + 3 = 18$

8 $\dfrac{3x}{5} - \dfrac{x}{3} = \dfrac{7}{2}$

9 $\dfrac{y-1}{4} = \dfrac{y+2}{3}$

10 $\dfrac{t}{2} + \dfrac{t}{3} + \dfrac{1}{7} = \dfrac{15}{7}$

■■■■ **Examples 2.4** ■■■■■■■■

i) Solve $5\left(\dfrac{x}{2} + 6\right) = 75$.

In this example the brackets affect the order of operations. x is at the heart of this problem on the left. Using our knowledge from Chapter 1, if x is replaced by a number the order of evaluation is

x (or the number)
divide by 2
plus 6
multiply by 5.

We use this list from the bottom working upwards with the opposite operation. This leads to the new list of operations we will carry out to find x, i.e.

divide by 5
subtract 6
multiply by 2
x (or the number).

Then

$$\dfrac{5}{5}\left(\dfrac{x}{2} + 6\right) = \dfrac{75}{5} \qquad \boxed{\text{Dividing by 5.}}$$

$$1\left(\dfrac{x}{2} + 6\right) = 15 \qquad \boxed{\text{Cancelling the 5s.}}$$

$$\dfrac{x}{2} + 6 = 15 \qquad \boxed{\begin{array}{l}\text{Multiplication by 1 –}\\ \text{no change.}\end{array}}$$

$$\dfrac{x}{2} + 6 - 6 = 15 - 6 \qquad \boxed{\text{Subtracting 6.}}$$

$$\dfrac{x}{2} = 9$$

$$\dfrac{2x}{2} = 2\,(9) \qquad \boxed{\text{Multiplying by 2.}}$$

$$x = 18.$$

Again we may check our answer by substitution, this time on the left to see that it gives the value of 75 on the right,

i.e. $5\left(\dfrac{x}{2}+6\right)=5\left(\dfrac{18}{2}+6\right)=5(9+6)=5(15)=75.$

ii) Solve $5(x-2)+3(2x-1)=7.$

As the brackets become more complicated it is useful to multiply them out,

i.e. $5x-10+6x-3=7$

> 5 times each value in the first bracket and 3 times each value in the second bracket.

$$11x-13=7$$
$$11x-13+13=7+13$$
$$11x=20$$
$$\dfrac{11x}{11}=\dfrac{20}{11}$$
$$x=1.8\overline{1}.$$

iii) Solve $5(x-2)-3(2x-1)=7$.

This example is similar to the previous one, again needing the multiplication of the brackets as a first step. We need to be careful with the second bracket because each value is multiplied by -3,

i.e. $5x-10-6x+3=7$
$$-1x-7=7$$
$$-x-7+7=7+7$$
$$-x=14$$
$$(-1)(-x)=(-1)(14)$$

> Multiplying by -1.

$$x=-14.$$

iv) Solve $\dfrac{(2x+7)}{4}-\dfrac{2(x+1)}{5}=\dfrac{1}{4}.$

There are a number of possible first moves in the solution of this equation. One possibility is to remove the fractions, multiplying throughout by the LCM of 4, 5 and 4, i.e. multiplying by 20,

i.e. $20\dfrac{(2x+7)}{4}-20\dfrac{2(x+1)}{5}=20\dfrac{(1)}{4}$

> Cancelling down the 20s.

$$5(2x+7)-8(x+1)=5$$

> Multiplying brackets

$$10x+35-8x-8=5$$
$$2x+27=5$$
$$2x+27-27=5-27$$
$$2x=-22$$

$$\frac{2x}{2} = \frac{-22}{2}$$

$$x = -11.$$

> Dividing both sides by 2.

■■■■ EXERCISE 2.4 ■■■■

Solve the following linear equations.

1 $5(x + 2) + 3x - 3 = 2$

2 $4(x - 1) - 5(1 - x) - 2 = 3x$

3 $\frac{1}{5}(2x - 4) + 1 = \frac{1}{4}x$

4 $2(y + 1) - 5(y - 4) = 0$

5 $\frac{2}{5}x + 4 = \frac{1}{4}x + 7$

6 $\frac{1}{2}y + 1 - \frac{1}{3}(y - 1) + 2 = 0$

7 $2\left(\frac{4}{3}x - 1\right) = x + 9$

8 $\frac{2}{3}(x - 6) = \frac{x + 9}{4}$

9 $\left(\frac{3y + 1}{2}\right) + \left(\frac{y + 2}{3}\right) = 1$

10 $\frac{2}{5}(x + 5) - \frac{10}{3}(x + 1) = 2$

All the examples so far have included the reasons and operations needed to reach a final solution. It is usual for some of the method to be omitted as you become more experienced with the algebra. We can repeat Example 2.3i) to show a simplified version,

i.e. $3x + 4 = 20$

i.e. $3x = 20 - 4$

> Subtracting 4 from both sides.

$3x = 16$

$x = \frac{16}{3}$ or $5.\bar{3}$.

> Dividing both sides by 3.

Graphs and equations

A graph is a straight line or curve that represents a relationship, often between y and x. The graph of a linear equation is always a straight line. We will draw the line with reference to a pair of **Cartesian** (i.e. **rectangular**) **axes** (plural of **axis**) and attempt to relate y to x. Fig. 2.1 shows a typical set, or pair, of axes together with some points marked O, P, Q and R.

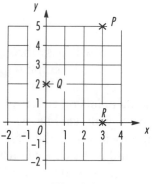

Fig. 2.1

The **horizontal** axis is the **independent** axis and is labelled in this case with *x*, the **independent** variable. This means that the original choice of values is a choice of values of *x*. The **vertical** axis is the **dependent** axis and is labelled in this case with *y*, the **dependent** variable. This means that the values of *y* depend on the choice of *x* values and the particular way *y* is related to *x*. *y* is said to be given in terms of *x*.

The number scale on each axis is regular. It is easier to choose scales involving 1s, 2s, 5s, 10s, 20s, 50s, 100s, . . . or 0.1s, 0.01s, ... This allows you to find parts of graph paper squares and interpret your answers accurately. Fig. 2.1 shows both axes with the same scales, but often they are different, as in Fig. 2.2.

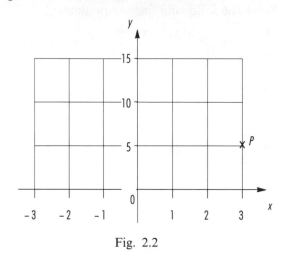

Fig. 2.2

The point *O* is called the **origin**. It is where the 0 values for *y* and for *x* **intersect** (i.e. **cross**). The origin is shown in Fig. 2.1. To use the graph paper efficiently it is sometimes omitted, as in Figs. 2.3.

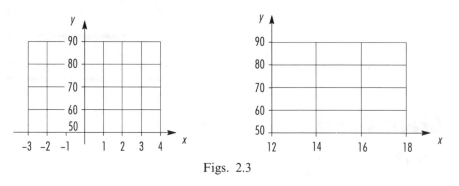

Figs. 2.3

A graph will pass through many points, some of which you will have marked on the graph paper. Each point may be marked by either a cross, x, or by a dot within a circle, ⊙. Due to error or inaccuracy you may see some points on either side of the line. We can label these points or refer to them with **coordinates**. A pair of coordinates is an ordered pair of numbers where the first is the independent variable, usually *x*, and the second is the dependent variable, usually *y*,

i.e.   (independent variable, dependent variable),

i.e. (x, y).

By way of example (3, 5) are the coordinates representing P in both Fig. 2.1 and Fig. 2.2 . The position of P only looks different because of the different scales in the two figures.

The origin has the coordinates $(0, 0)$, which can be written $O(0, 0)$.

If the first coordinate value is 0 (i.e. $x = 0$) then the point lies on the **vertical axis**. Q, which may be written $Q(0, 2)$, is such a point in Fig. 2.1. If the second coordinate value is 0 (i.e. $y = 0$) then the point lies on the **horizontal axis**. Again in Fig. 2.1 we have $R(3, 0)$.

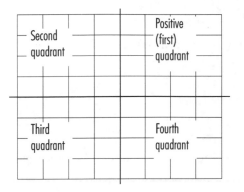

Fig. 2.4

P is said to lie in the **positive quadrant**, i.e. the quarter of the graph paper where both x and y are positive. The "quarter" is not precise because the axes may not be in the middle of the paper to divide it equally. The **positive quadrant** is sometimes called the first quadrant, the other quadrants then being taken in the anticlockwise direction shown in Fig. 2.4.

■■■■■ EXERCISE 2.5 ■■■■■■■

On one sheet of graph paper draw and label a pair of axes x and y. Mark on your paper the following coordinates.

1	(0, 0)	**6**	(−1, 6)
2	(2, 0)	**7**	(7, −1)
3	(0, 4)	**8**	$(-\frac{1}{2}, -\frac{1}{2})$
4	(2, 2)	**9**	(3, 5)
5	(−1, −1)	**10**	(5, 3)

On a separate sheet of graph paper with fully labelled axes mark the following coordinates for each question. Is there any pattern in each case?

11 (1, 1), (6, 6), (3.5, 3.5), (−5, −5), (−2.5, −2.5)

12 (−1, 1), (−6, 6), (−3.5, 3.5), (2.5, −2.5), (5, −5)

13 (2, 1), (3, 1), (−4.5, 1), (−0.5, 1)

14 (3, 1), (3, −3), (3, −0.5), (3, −4.5)

15 (1, 4), (6, 9), (3.5, 6.5), (−5, 2), (−2.5, 0.5)

The straight line law

The 2 important features of a straight line graph are the **gradient** (i.e. **slope**) and the **intercept** on the vertical axis (the y axis) where the graph crosses that axis. In the standard equation the gradient is represented by m and the intercept by c.

There are 4 possible types of gradient.

1. Positive gradients

They slope from **bottom left** to **top right**.

2. Negative gradients

They slope from **top left** to **bottom right**.

3. Zero gradients

The lines are **horizontal**, i.e. they go neither up nor down and so have **no slope**.

4. Infinite gradients

The lines are **vertical**.

Gradient, m is defined as

$$m = \frac{\text{vertical change}}{\text{horizontal change}}$$

The horizontal change is always for x increasing, i.e. moving to the right. During this movement if the vertical change is an increase then the gradient is positive, whilst if the vertical change is a decrease then the gradient is negative.

Before we look at the next set of examples we need to mention something about drawing graphs. An accurate drawing, generally on graph paper, is called a **plot**. A less accurate drawing, usually on ordinary paper, is called a **sketch**. A sketch is used as a general guide to show the basic line or curve.

Example 2.5

i) Find the gradient of the line joining the points (2, 1) and (5, 3).
    A sketch of the coordinates and this line on a pair of labelled axes is a
    help. As soon as it is drawn it will be obvious whether the gradient is
    positive or negative. In this example you see that we have a positive
    gradient.

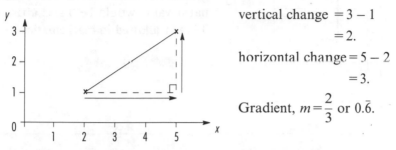

vertical change $= 3 - 1$
$$= 2.$$
horizontal change $= 5 - 2$
$$= 3.$$

Gradient, $m = \dfrac{2}{3}$ or $0.\bar{6}$.

ii) Find the gradient of the line joining the points $(-4, -1.5)$ and $(2, -3)$.
    Again we can mark the coordinates and sketch the line.

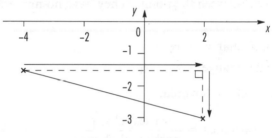

vertical change $= -3 - (-1.5) = -3 + 1.5 = -1.5$.

horizontal change $= 2 - (-4) = 2 + 4 = 6$.

Gradient, $m = \dfrac{-1.5}{6} = -0.25$.

iii) Find the gradient of the line joining the points $(-4, -1.5)$ and
     $(2, -1.5)$.

     The y coordinate, -1.5, remains the same for both pairs of
     coordinates meaning that we have a horizontal line and so no
     vertical change. This means that there is no gradient,

i.e. $m = \dfrac{0}{6} = 0$.

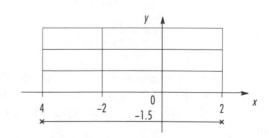

iv)   Find the gradient of the line joining the points $(1, 2)$ and $(1, 5)$.

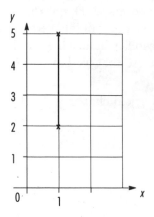

The x coordinate remains the same for both pairs of coordinates, meaning that we have a vertical line and so no horizontal change. We cannot use the gradient formula because the denominator value would be 0 and **division by 0 is not allowed in mathematics**.

There is another version of the gradient formula that saves you drawing the line. Suppose we have 2 points (x_1, y_1) and (x_2, y_2). The positions of the 1s and 2s are called **subscript** positions. They only label the xs and ys to distinguish between the points. They have no numerical meaning in the calculation.

$$\text{Vertical change} = y_2 - y_1.$$
$$\text{Horizontal change} = x_2 - x_1.$$

Using our gradient formula,

$$\text{Gradient} = \frac{\text{vertical change}}{\text{horizontal change}}$$

becomes

$$m = \frac{y_2 - y_1}{x_2 - x_1}.$$

There is another version of this formula,

i.e.

$$m = (1)\frac{(y_2 - y_1)}{(x_2 - x_1)}$$

Multiplication by 1 – no change.

$$= \frac{(-1)(y_2 - y_1)}{(-1)(x_2 - x_1)}$$

$\frac{-1}{-1}$ cancels to 1.

$$= \frac{-y_2 + y_1}{-x_2 + x_1}$$

Multiplying out the brackets.

$$m = \frac{y_1 - y_2}{x_1 - x_2}.$$

Re-writing with "–" centrally.

The formula takes into account whether the line has a positive or negative gradient. Sometimes a quick sketch can boost your confidence and confirm your answer. We will repeat Examples 2.5 i) and ii) using the formula.

 **Example 2.6**

i) Find the gradient of the line joining the points (2, 1) and (5, 3). Compare these values with the general (x_1, y_1) and (x_2, y_2) so that $x_1 = 2$, $y_1 = 1$ and $x_2 = 5$, $y_2 = 3$. Now the gradient is given by

$$m = \frac{y_2 - y_1}{x_2 - x_1}$$

$$= \frac{3 - 1}{5 - 2}$$

> Substituting the values.

$$= \frac{2}{3} \text{ or } 0.\bar{6}.$$

ii) Find the gradient of the line joining the points $(-4, -1.5)$ and $(2, -3)$. Again we can compare these coordinates with the general (x_1, y_1) and (x_2, y_2) so that $x_1 = -4$, $y_1 = -1.5$ and $x_2 = 2$, $y_2 = -3$. The gradient is given by

$$m = \frac{y_2 - y_1}{x_2 - x_1}$$

$$= \frac{-3 - (-1.5)}{2 - (-4)}$$

> Substituting the values.

$$= \frac{-3 + 1.5}{2 + 4}$$

$$= \frac{-1.5}{6}$$

$$= -0.25.$$

In both these examples the previous sketches confirm the slopes being positive in (i) and negative in (ii).

The second important feature of the straight line is the intercept, c, on the vertical axis (generally the y axis). At this point on the y axis the value of x is 0. Using the axes of Fig. 2.3 we may draw some straight lines.

Examples 2.7

i)

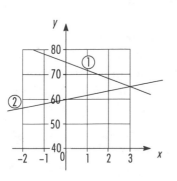

Line ① has an intercept of 75, i.e. $c = 75$. Also we know from the previous section of work that the gradient is negative.

Line ② has an intercept of 60, i.e. $c = 60$. Again we can see that for this line the gradient is positive.

ii)

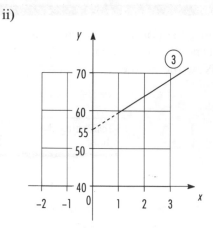

Line ③ does not actually cut the y axis, but we can extend it backwards with the same slope (----). We see that it intercepts at $c = 55$.

iii)

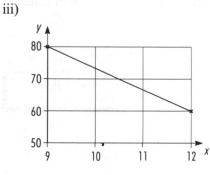

We cannot read off the intercept in this example because in our sketch the x axis does not include where $x = 0$. Finding the intercept on the vertical axis needs an extra calculation which is demonstrated in Example 2.12.

So far in this section we have learned about the 2 important features of a straight line, the gradient and the intercept. Now we are able to link them together because the general equation of the straight line is given by

$y = mx + c$

where m is the gradient

and c is the intercept on the vertical axis.

Algebraically m, being a number just before x, is said to be the **coefficient** of x.

The format of the equation may not always be exactly the same as this, but often a little algebraic manipulation will re-create the standard format. Whatever the original form and the final, standard, format you must have both x and y to the power 1. As usual we do not write the power 1, usually treating it as being understood. In practical examples it is possible that letters other than x, y, m and c will be used. Each practical example will need to be compared with the standard $y = mx + c$.

We need to be able to interpret a known straight line to give its gradient and intercept. In addition, knowing the 2 important features we need to be able to write down the equation.

████████ **Examples 2.8** ████████

This set of examples looks to compare each one against the standard form $y = mx + c$ in order to find the values of m and c.

  i) $y = 2x + 1$ $m = 2,$ $c = 1$

 ii) $y = 2x - 1$ $m = 2,$ $c = -1$

iii) $y = \dfrac{1}{3}x + 4$ $m = \dfrac{1}{3},$ $c = 4$

iv) $y = 1 - 2x$ $m = -2, c = 1$

████████ **Examples 2.9** ████████

Our next set of examples is just as easy when combining with some of the algebraic rearrangement we learned earlier in the chapter. This is to re-create the form of $y = mx + c$.

  i) $y + 3x - 5 = 0$ is not in the correct format because y must stand alone on one side of the equation.

      i.e. $y = -3x + 5$ | Subtracting $3x$ from and adding 5 to both sides.

      or $y = 5 - 3x$

      are preferred forms.

$$m = -3, c = 5.$$

 ii) $2y + 3x - 5 = 0$ also needs some attention to re-create the standard format,

      i.e. $2y = -3x + 5$

$$y = \frac{-3}{2}x + \frac{5}{2}$$

      | Dividing by 2.

$$m = \frac{-3}{2}, c = \frac{5}{2}.$$

iii) $\dfrac{1}{3}y - 4x + 10 = 0$ is an equation that also needs some attention.

      ie. $\dfrac{1}{3}y = 4x - 10$ | Adding $4x$ to and subtracting 10 from both sides.

             $y = 12x - 30$ | Multiplying by 3.

$$m = 12, c = -30.$$

iv)   You may find the relationship $y = x$ easier to deal with in the form

$$y = x + 0$$

      | Coefficient of x is 1.

$$m = 1, c = 0.$$

 v) $y = 2$ may be thought of as

$$y = 0x + 2$$

$$m = 0, c = 2.$$

vi) $x = 3$ does not involve any letter other than x and so it is impossible to create the standard form $y = mx + c$. In fact if you quickly sketch the graph you will see that it is a vertical line passing through the value 3 on the horizontal axis. This means that its gradient is infinite and it does not intercept the vertical axis.

███████ **Examples 2.10** ███████

In this group of examples we are going to construct the equation representing the straight line for the given gradient and intercept on the vertical axis. Each equation will be created by substituting the numerical values into the standard format $y = mx + c$.

i)   A gradient of 3 and an intercept of 2,
     i.e. $m = 3$ and $c = 2$.
     $\therefore \qquad y = 3x + 2.$ Alternative forms of this
                                           are $y - 3x = 2$
                                           and $y - 3x - 2 = 0.$

ii)   A gradient of -3 and an intercept of 2,
      i.e. $m = -3$ and $c = 2$.
      $\therefore \qquad y = -3x + 2.$ Alternative forms of this
                                            are $y + 3x = 2$
                                            and $y + 3x - 2 = 0.$

iii)   An intercept of -2 and a zero gradient,
       i.e. $m = 0$ and $c = -2$.
       $\therefore \qquad y = 0x - 2$
       i.e. $\qquad y = -2.$ An alternative form of this
                                             is $y + 2 = 0.$

iv)   A gradient of $\dfrac{-1}{4}$ and an intercept of $\dfrac{3}{5}$,

      i.e. $m = \dfrac{-1}{4}$ and $c = \dfrac{3}{5}$.

      $\therefore \qquad y = \dfrac{-1}{4}x + \dfrac{3}{5}.$

This form of the equation is perfectly all right, though alternative versions can be created to avoid the fractions,

      i.e. $20y = 20\dfrac{(-1x)}{4} + 20\dfrac{(3)}{5}$ | Multiplying by 20 and cancelling. |

         $20y = -5x + 12.$ Alternative forms of this
                                            are $20y + 5x = 12$
                                            and $20y + 5x - 12 = 0.$

The recent groups of examples have involved numbers given at the beginning of each one. Suppose we have no written numbers, just a graph drawn on a pair of labelled axes.

Example 2.11

From the points marked on this graph we see that it passes through the points $(1, 60)$ and $(5, 76)$. We need to find the gradient and the intercept so that we may write down the equation of the line.

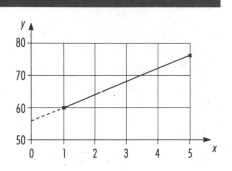

By comparing our coordinates with (x_1, y_1) and (x_2, y_2) we see that $x_1 = 1$, $y_1 = 60$ and $x_2 = 5$, $y_2 = 76$.

$$\therefore \qquad m = \frac{y_2 - y_1}{x_2 - x_1}$$

$$\text{becomes} \qquad m = \frac{76 - 60}{5 - 1}$$

$$= \frac{16}{4} = 4.$$

To find the intercept we project the line backwards and see that it intercepts at 56.

$$\therefore \qquad y = mx + c$$

$$\text{becomes} \qquad y = 4x + 56 \qquad \text{in this case.}$$

We can go further with the calculations and avoid drawing a graph altogether, though you will often find one very helpful. This is relevant too when we are unable to read off the intercept value because the origin does not appear on our axes.

Example 2.12

Suppose we know that a straight line passes through the points with coordinates $(9, 80)$ and $(12, 60)$, and that we wish to find its equation.

Comparing these coordinates with our general ones, (x_1, y_1) and (x_2, y_2) we see that $x_1 = 9$, $y_1 = 80$ and $x_2 = 12$, $y_2 = 60$.

$$\therefore \qquad m = \frac{y_2 - y_1}{x_2 - x_1}$$

$$\text{becomes} \qquad m = \frac{60 - 80}{12 - 9}$$

$$= \frac{-20}{3}.$$

This value for the gradient could be given as a recurring decimal of $6.\bar{6}$, but the fraction is preferred for accuracy.

$$\therefore \qquad y = mx + c$$

becomes $\qquad y = \dfrac{-20}{3}x + c.$

Now every point that lies on the line (i.e. that the line passes through) satisifies the equation representing the line. This means that we can substitute the values for x and y from any such point to leave the only unknown, c. We can easily use either of the original points and do so below.

Using $(9, 80)$

and $\qquad y = \dfrac{-20}{3}x + c$

then $\qquad 80 = \dfrac{-20\,(9)}{3} + c$

$$80 = -60 + c$$

$$80 + 60 = c$$

i.e. $\qquad c = 140.$

or using $(12, 60)$

and $\qquad y = \dfrac{-20}{3}x + c$

then $\qquad 60 = \dfrac{-20\,(12)}{3} + c$

$$60 = -80 + c$$

$$60 + 80 = c$$

i.e. $\qquad c = 140.$

Either of these calculations may be used to complete the equation

$$y = \dfrac{-20}{3}x + 140.$$

We can re-write this in any format, perhaps multiplying through by 3 to remove the fraction and moving some or all terms to one side,

i.e. $\qquad\qquad 3y = -20x + 420$

or $\qquad\qquad 3y + 20x = 420$

or $\qquad 3y + 20x - 420 = 0$

are some of these possible alternatives, though there are others.

▰▰▰ Example 2.13 ▰▰▰

We may extend the previous example and attempt to either sketch or plot the straight line whose equation we have found. The first piece of information we need is that we have a straight line. This means that we must have 3 points to mark – the third one being a check on the accuracy of the other 2. Already we know of our 2 original points $(9, 80)$ and $(12, 60)$. Also we know that the intercept on the vertical axis is 140 and that all the way along the y axis the value of x is 0. This gives us the third point of $(0, 140)$.

As an exercise you should check for yourself with a graph on a pair of labelled axes that all the information connecting these points, the gradient and the equation is true.

■ EXERCISE 2.6 ■

The first 5 questions show a straight line graph on a pair of labelled axes.
In each case find the   i)   gradient,
                ii)   intercept,
              iii)   equation of the line.

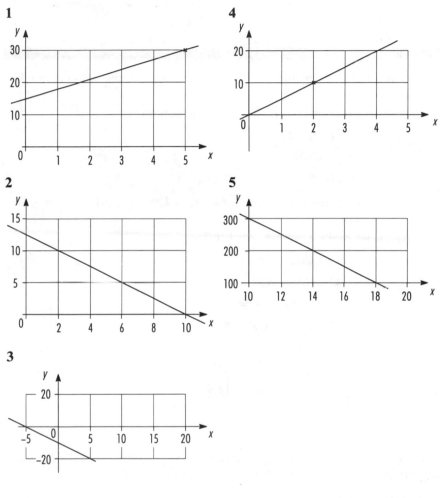

In this second set of 5 questions you have no sketch of the straight line
graph. Using the pairs of coordinates in each case find

                i)   gradient,
              ii)   intercept,
             iii)   equation of the line.

6 $(0, 4)$ and $(2, 2)$ **9** $(1, 4)$ and $(6, 9)$

7 $(-1, -1)$ and $(5, 3)$ **10** $(7, 1)$ and $(\frac{1}{2}, -\frac{1}{2})$

8 $(3, 5)$ and $(5, 3)$

We have spent many pages and quite some time finding out about the straight line law. What use is an equation once we know of its features? We can use it to **predict** a value. Given a value of x we can find the associated value of y, or given a value of y we can find the value of x leading to it. The next 2 examples use equations we calculated in earlier work.

▬▬▬ **Example 2.14** ▬▬▬▬▬▬▬▬▬▬▬▬▬▬▬▬▬▬▬▬▬▬▬

i) Given the straight line law $y = 4x + 55$ what is the value of y when $x = -7$?
   We substitute this value of x into the equation,

   i.e.　　$y = 4(-7) + 55$

   　　　　$= -28 + 55$

   　　　　$= 27.$

ii) Given the straight line law $3y + 20x - 420 = 0$ find the value of x when $y = 14$.
   Again we make a simple substitution,

   i.e.　　$3(14) + 20x - 420 = 0$

   　　　　$42 + 20x - 420 = 0$

   　　　　$20x - 378 = 0$

   　　　　$20x = 378$

   　　　　$x = \dfrac{378}{20}$

   　　　　$x = 18.9.$

We may take our theory a stage further, linking together some of the work we did solving equations with this more recent graphical work.

▬▬▬ **Example 2.15** ▬▬▬▬▬▬▬▬▬▬▬▬▬▬▬▬▬▬▬▬▬▬▬

We will use the simple equation $3x + 4 = 20$ from Examples 2.3i).

i) First we notice that there is no y in this equation, only x, but the left-hand side looks like an application of $mx + c$ with $m = 3$ and $c = 4$.

   $\therefore$ let $y = 3x + 4$ which means that, linking this with the right-hand side, we have $y = 20$.

   We can sketch both of these lines on one pair of labelled axes.

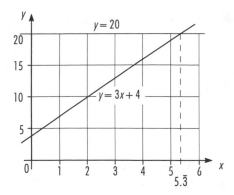

Because the original question was in terms of x our answer must give x as a number, i.e. $x = 5.\overline{3}$ where the two lines cross.

ii) We may re-write our original equation

$$3x + 4 = 20$$

to become $$3x + 4 - 20 = 0$$

| Subtracting 20. |

$$3x - 16 = 0.$$

Again we may compare this equation with the standard format $y = mx + c$. Comparing $mx + c$ with $3x - 16$ we have $m = 3$ and $c = -16$. We will sketch $y = 3x - 16$.

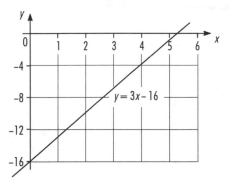

Comparing $3x - 16 = 0$ with $y = 3x - 16$ we have $y = 0$. This means we will see where the graph crosses the line $y = 0$, i.e. where it crosses the x axis. In fact $x = 5.\overline{3}$ is that point, meaning that this is the solution to our original equation.

This method is most useful for simple linear equations. When the linear equations become complicated,

e.g. $\dfrac{2}{5}(x + 5) - \dfrac{10}{3}(x + 1) = 2$

with x appearing more than once, then a purely algebraic solution is easier.

■ ASSIGNMENT ■

Let us return to our original Assignment. Can we draw a straight line graph of our equation $v = 30 - 4t$? The answer is yes!

We can compare our equation with the standard format,

i.e. $v = 30 - 4t$

with $y = mx + c$.

v and y look to be in similar positions and on the other side of each equation are the letters t and x. This means that we should label the horizontal axis with the independent variable t and the vertical axis with the dependent variable v, i.e. the velocity, v, depends upon the time, t, of the motion. m is the gradient, shown as the coefficient of x. This means that in our case -4 is the gradient, -4 is the coefficient of t. All that remain in the comparison are the intercept c and 30,

i.e. $c = 30$.

We have enough information to sketch the graph. However, with a little more care and accuracy we might plot the straight line on graph paper. Because it is a known straight line we need just 3 points, 2 and the other 1 as a check. The intercept on the vertical axis is 30 so already we have the first pair of coordinates $(0, 30)$. We may choose any other values of t and calculate the values of v by substitution.

e.g. $t = 2$, $\therefore v = 30 - 4\,(2) = 22$

and $t = 5$, $\therefore v = 30 - 4\,(5) = 10$.

These give the 3 pairs of coordinates $(0, 30)$, $(2, 22)$ and $(5, 10)$, through which we can draw the straight line shown in Fig. 2.5.

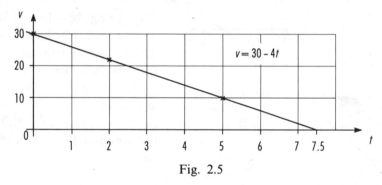

Fig. 2.5

We can use our graph to estimate one value when given the other variable.

i)   Suppose we want to know when the vehicle stops, i.e. when it is at rest. This occurs when $v = 0$, the point where the graph crosses the horizontal axis. We can read off that this occurs at 7.5, i.e. after 7.5 seconds from the start of the motion the vehicle is at rest.

ii) Suppose we want to know the initial velocity of the vehicle. "Initially" means "when $t = 0$" which is where the graph crosses the vertical axis, i.e. the vertical intercept is 30 meaning that the initial velocity is $30\,\text{ms}^{-1}$.

iii) Suppose we want to know the time taken for the speed to fall to $12.4\,\text{ms}^{-1}$. Fig. 2.6 shows this line drawn on the same pair of labelled axes as our original graph $v = 30 - 4t$. Where the two lines cross we can read off the value of t as 4.4, i.e. it has taken 4.4 seconds for the speed to fall to $12.4\,\text{ms}^{-1}$.

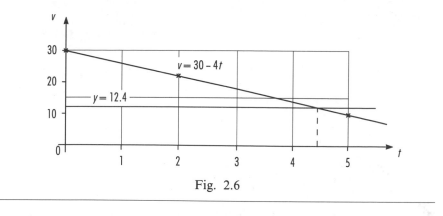

Fig. 2.6

Experimental data

Experiments and the use of test rigs often involve taking numerical readings. Even when the readings are neatly tabulated they can appear as a jumble of numbers. We need to see if there is any **pattern**. There are many ways of displaying the results in a variety of charts, diagrams and graphs. As a first step, in this chapter, we are going to see if sets of results obey a linear law, i.e. if when they are plotted accurately on a pair of labelled axes on graph paper a straight line may be drawn through them. It is highly unlikely in a practical situation that all the points will lie on the line. The aim when drawing the line is to have as many points as possible lying on the line with any others evenly scattered on either side. This will be our attempt to draw a **line of best fit**.

Example 2.16

In this example we have some mechanical test figures relating load, W Newtons, and effort, E Newtons. We wish to find out if they are related. If they are related is that relationship $E = mW + c$?

W	5	10	15	20	25	30
E	2.91	6.20	9.00	11.76	14.90	18.00

The first thing to notice is how the load figures increase by an equal amount each time. This is because they have been selected independently, meaning that the horizontal axis will be labelled with W. The effort figures are not as obviously regular because they depend on the values of W and the test rig. We will label the vertical axis E.

Because the first figures for both W and E are close to 0 we will include the origin on our graph. If this was not so we should use the graph paper more efficiently by omitting the origin.

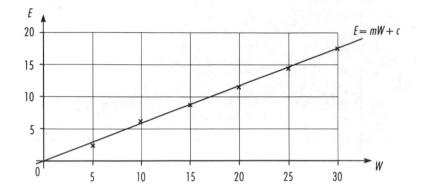

We can see from the graph that, within experimental error, the figures show that W and E are related by a straight line. In the next part of the example we can look at the form of this relationship. We need to compare our example,

$$E = mW + c$$

with the standard straight line law

$$y = mx + c.$$

In the standard format y is plotted vertically and x is plotted horizontally. Our example has E plotted vertically and W plotted horizontally. We can see that these letters have the same relative positions in their equations. Comparing further we have m as the gradient and c as the intercept in both cases.

By extending the line backwards to the vertical axis we intercept at the origin. This means that $c = 0$.

We can create a right-angled triangle to find the vertical and horizontal changes so that we can calculate the gradient. If you were doing this for yourself you should aim for a large triangle. This gives a better chance of accuracy when drawing the changes and reading off the scales. In this example we have chosen the points $(15, 9)$ and $(30, 18)$ lying on the line and apply the gradient formula

$$m = \frac{y_2 - y_1}{x_2 - x_1}$$

i.e. $$m = \frac{18 - 9}{30 - 15}$$

$$= \frac{9}{15}$$

$$= 0.6.$$

At the end of this example we are in a position to bring together our results, stating that W and E are related according to the straight line law $E = 0.6W$.

We might use this relationship to estimate the effort requirèd to move, for example, a load of 21.5 N,

i.e. $W = 21.5$ in the formula $E = 0.6W$

so that
$$E = (0.6)(21.5)$$
$$E = 12.9$$

i.e. an effort of 12.9 N is needed to move 21.5 N.

Example 2.17

The table in this example relates the resistance, $R\,\Omega$, to the temperature, $t\,°C$, of a length of wire.

t	20	40	60	80	100
R	5.30	5.68	6.11	6.50	7.00

We think they are related according to the straight line law $R = R_0 + kt$. This thought suggests some more questions.
  i)   If the relationship is true what are the values of R_0 and k?
  ii) How does our equation match the usual equation $R = R_0(1 + \alpha t)$?
  iii) What is the resistance at 150°C?
  iv) At what temperature will the resistance be 6.25 Ω?

First we look at the table. The regularity of the values of t suggest that it is the independent variable to be plotted horizontally. Therefore R is the dependent variable to be plotted vertically.

We omit the origin to use the graph paper more efficiently. The graph shows that the table values are related according to a straight line. We need to see how our equation compares with the standard format,

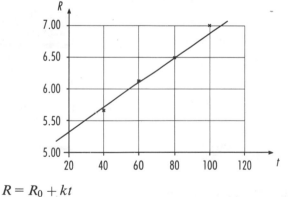

i.e. $R = R_0 + kt$

and $y = mx + c$.

Already we have compared R and y vertically, and t and x horizontally. Earlier in the chapter we mentioned that the gradient, m, is the coefficient of x. In our equation k is the coefficient of t, meaning that it must be the gradient. This leaves R_0 to be the intercept on the vertical axis.

i) We can choose two points from the graph, $(20, 5.30)$ and $(80, 6.50)$, to calculate the gradient using the formula

$$m = \frac{y_2 - y_1}{x_2 - x_1}$$

i.e.
$$k = \frac{6.50 - 5.30}{80 - 20}$$

$$= \frac{1.20}{60}$$

$$= 0.02.$$

Therefore in our equation we can substitute for k so that

$$R = R_0 + 0.02t.$$

Because we are unable to use our graph to read off the intercept we must use our equation. We may choose any point that actually lies on our line, e.g. $(20, 5.30)$, and substitute the coordinate values into our equation. This means

$$R = R_0 + 0.02t$$

becomes
$$5.30 = R_0 + 0.02\,(20)$$

$$5.30 = R_0 + 0.40$$

$$5.30 - 0.40 = R_0 \qquad \boxed{\text{Subtracting } 0.40.}$$

$$4.90 = R_0.$$

Our completed equation may be written

$$R = 4.90 + 0.02t.$$

ii) This version may be compared to the usual one

$$R = R_0\,(1 + \alpha t) \qquad \boxed{\begin{array}{l}\text{Expanding the}\\\text{brackets.}\end{array}}$$

i.e.
$$R = R_0 + R_0\,\alpha t.$$

By comparing terms we see that

$$R_0\,\alpha = 0.02$$

i.e.
$$4.90\,\alpha = 0.02 \qquad \boxed{\text{Substituting for } R_0.}$$

$$\alpha = \frac{0.02}{4.90} \qquad \boxed{\text{Dividing by } 4.90.}$$

$$= 4.08 \times 10^{-3}.$$

iii) We will return to our version of the equation from the straight line graph and find the resistance by substituting for $t = 150$,

i.e.
$$R = 4.90 + 0.02\,(150)$$

$$= 4.90 + 3.00$$

$$= 7.9$$

i.e. the resistance of the wire at $150\,°C$ is $7.9\,\Omega$.

iv) Using the same equation again, substituting for $R = 6.25$, we may calculate the temperature at which this occurs,

i.e. $\qquad\qquad 6.25 = 4.90 + 0.02t$

> Subtracting 4.90.
> Dividing by 0.02.

$\qquad\qquad\qquad 1.35 = 0.02t$

$\qquad\qquad\qquad 67.5 = t$

i.e. the temperature of the wire is $67.5\,°C$ when its resistance is $6.25\,\Omega$.

EXERCISE 2.7

1 For the table of values plot y vertically against x horizontally.

x	-3	-2	-1	0	1	2	3
y	21	15	9	3	-3	-9	-15

With the aid of your graph check that the relationship is $y = 3 - 6x$.

2 y is thought to be related to x by the formula $y = mx + c$.
Using the table of values plot a graph of y vertically against x horizontally to check that it is correct. Use your graph to find the values of m and c.

x	0	1	4	9	16	25
y	-5	1	7	17	31	47

Use the equation to estimate the values of   i) y when $x = 8$
and ii) x when $y = 4$.

3 A lifting machine raises loads, W kg, with effort E N according to the equation $E = mW + c$. A selection of test results are

W	10	20	40	60	100	150
E	75	102	168	225	340	500

Draw a graph to help you check that this relationship is true and use it to find values for the gradient and vertical intercept. Write down your equation using the values you have found. Estimate the effort needed to raise 130 kg.

4 The following table shows the distance travelled, d m, in time, t s, measured initially from 0 m.

d	95	190	290	385	480	575
t	30	60	90	120	150	180

By plotting a graph of d against t show that they are related by $d = kt$. Use your graph to estimate the time taken to travel 400 m. Find the value of k and use your equation to check your graphical time estimate for $d = 400$.

5 There is a temperature connection between Fahrenheit, °F, and Celsius, °C. Suppose this may be written $F = aC + b$.
   i)   Which letter represents the gradient and which one represents the intercept on the vertical axis?
   ii) Plot F vertically against C horizontally using the 3 pairs of coordinates $(0, 32)$, $(50, 122)$ and $(100, 212)$.
   iii) From your graph find the values of a and b and substitute them into the equation $F = aC + b$.
   iv) If your workplace temperature is 68°F what is this in °C?

6 A loaded beam has a shear force given by $S = mx + c$ where x is the distance from one end. Using the following table of values plot a graph of S against x to check that it is a straight line.

x (m)	1	2	3.5	4	5
S (N)	−800	−400	200	400	800

   From your graph work out values for m and c so that you can use your new equation to find
   i)   the shear force when $x = 1.75$ m,
   ii) the distance from the end where the shear force disappears (i.e. where the shear force is zero).

7 Hooke's law states that the tension, T N, in an elastic spring is related to the extension, x m, of the spring.

x	1.5	2.5	4.0	5.0	7.5	10.0
T	20.2	33.8	53.6	67.9	101.3	135.1

   Use the table of values to decide whether this is true according to the relation $T = kx$ where k is a value to be found from your graph. Is k the gradient or the intercept?

8 The velocity of a van, v ms^{-1}, is connected to time, t s, by the formula $v - mt - c = 0$ where m and c are the gradient and intercept as usual. Re-write this formula in the standard format and use the graph obtained from the following table to work out values for m and c.

t	10	14	20	26	31	35
v	3.90	5.15	6.50	7.82	9.32	10.20

   Substitute your newly found values for m and c into the formula and use it to work out the velocity after a minute. Your answer will be in ms^{-1}. Convert it to kmh^{-1} and then mph to see if it is reasonable.

9 According to Ohm's law $V = IR$ where V is the voltage, I is the current (amp) and R is the resistance (ohm). The table shows some test results for a particular resistor.

V	0	3	7	10	14	20	25
I	0	0.077	0.179	0.256	0.359	0.513	0.641

Plot V vertically and I horizontally so that you can use your straight line graph to find the value of the resistor.

10 The length of a copper rod, l m, expands as its temperature, $t\,°C$ is raised probably according to the formula $l = l_0 + mt$. Use the following table of values to draw a graph and check the truth of this theory.

t	0	100	250	500	750
l	0.2000	0.2017	0.2042	0.2085	0.2128

Find the values of l_0 and m from your graph. Substitute these values into the original equation and use it to estimate the temperature for a length of 0.2100 m.

▬▬ MULTI-CHOICE TEST 2 ▬▬

1 Given $x + 4 = 12$ the value of x is
A) $\frac{1}{3}$
B) 3
C) 8
D) 16

2 If $2 + x - 5 = 7$ then x is
A) 0
B) 4
C) 10
D) 14

3 Where $\dfrac{4x}{5} = 20$ then x is
A) 3.75
B) 6.25
C) 16
D) 25

4 Given $\dfrac{2x}{5} + 3 = 38$ the value of x is
A) 11.5
B) 31.6̇
C) 87.5
D) 102.5

5 If $\dfrac{x+1}{6} = \dfrac{x-2}{5}$ then x equals

A) -16
B) -1.75
C) 7
D) 17

6 Where $2x + \dfrac{x}{4} = 36$ then x is

A) 16
B) 18
C) 48
D) 72

7 The value of x when $4(x-2) + 5(x+1) = 7$ is

A) $-6.\overline{66}$
B) $0.\overline{88}$
C) $1.\overline{11}$
D) $1.\overline{55}$

8 The axes in the diagram relate the temperatures in °F and °C. The points are labelled $(32,0)$, $(86,30)$, $(100,212)$ and $(131,55)$. How many points are correctly labelled?

A)   one
B)   two
C)   three
D)   four

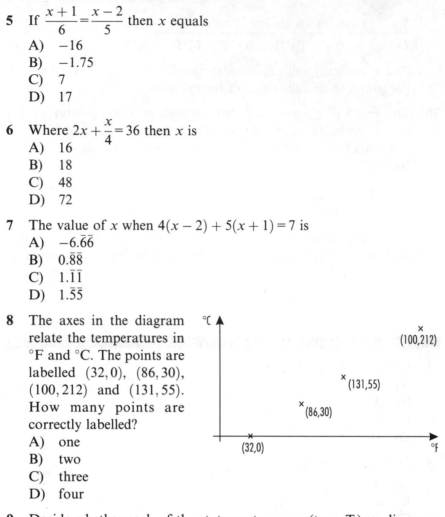

9 Decide whether each of the statements agrees (true, T) or disagrees (false, F) with its accompanying diagram.

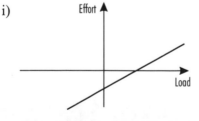

i) The gradient of this first straight line is positive.

ii) The gradient of this second straight line is negative.

A)   i) T   ii) T
B)   i) T   ii) F
C)   i) F   ii) T
D)   i) F   ii) F

10 The graph shows displacement, s, plotted against time, t. The gradient of this straight line, representing velocity, is

A) $\dfrac{s_1 - s_2}{t_1 - t_2}$

B) $\dfrac{t_1 - t_2}{s_1 - s_2}$

C) $\dfrac{t_2 - t_1}{s_2 - s_1}$

D) $\dfrac{s_1 - t_1}{s_2 - t_2}$

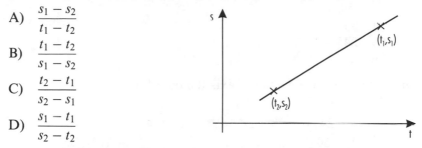

11 The resistance, R, of a length of wire is related to temperature, t, by the straight line equation $R = c + mt$. The letter m represents the

A) intercept on the t axis

B) intercept on the R axis

C) R coordinate of a point

D) gradient of the line

12

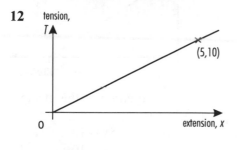

According to Hooke's law tension, T, in an elastic string is related to extension, x. The gradient of the straight line is

A) $\frac{1}{2}$

B) 2

C) 5

D) 10

13 A straight line passes through the points $(-11, 2)$ and $(1, -1)$. The gradient is

A) -4

B) $-\frac{1}{4}$

C) $\frac{1}{4}$

D) 4

Questions **14**, **15** and **16** refer to the following information. The resistance $R\,\Omega$, of a length of wire is related to temperature, $t\,^\circ\text{C}$, by $50R = t + 245$. This is the equation of a straight line.

14 Its gradient is

A) $\dfrac{1}{245}$

B) $\dfrac{1}{50}$

C) 1

D) 245

15   The intercept on the vertical axis is

A) $\dfrac{245}{50}$

B)   50

C)   245

D) 245×50

16   Where the line cuts the *t*-axis the coordinates are

A) $(0, -245)$

B) $\left(0, \dfrac{245}{50}\right)$

C) $(-50, 0)$

D) $(-245, 0)$

17   *v* is the velocity and *t* is the time of motion for a vehicle. Its motion is represented by the equation of this straight line

A) $v = -0.4t + 10$

B) $v = 2.5t + 25$

C) $v = 25 - 0.4t$

D) $v = 25 - 2.5t$

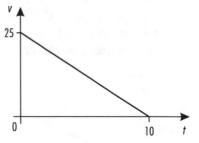

18

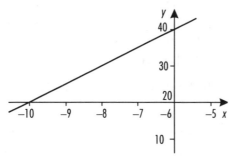

The equation of this straight line is

A) $y = -5x - 10$

B) $y = -5x + 10$

C) $y = 5x + 40$

D) $y = 5x + 70$

19   Which of the following points lies on the straight line $2x - 3y = 18$?

A) $(6, 2)$

B) $(18, 0)$

C) $(6, -2)$

D) $(0, 6)$

20   A vehicle is decelerating at $4\,\text{ms}^{-2}$. Its velocity, *v*, is related to the time of motion, *t*, by $v = 7 - 4t$. The equation of the straight line passing through the origin parallel to the given straight line is

A) $v + 4t = 0$

B) $v = 7 + 4t$

C) $v = 4t$

D) $v = 7t - 4$

3 Using Formulae

Element: Use algebra to solve engineering problems.

PERFORMANCE CRITERIA
- Formulae appropriate to the engineering problem are selected.
- The algebraic manipulation of formulae is carried out.

RANGE
Engineering problems: electrical, electronic; mechanical.
Formulae: linear equations, quadratic equations.
Algebraic manipulation: transposition.

Introduction

Engineers and scientists tend to remember formulae in a standard format. This may not always be convenient when attempting to solve a problem. **Transposition** of formulae tells us from the title that we will **move (trans)** the **position** of various letters and operations. **Transposition** is sometimes called **transformation**. Already we know many of the transposition rules because we used them to solve simple equations in Chapter 2. We will use some of the examples again. In fact Example 2.1 iii) was a case of simple transposition.

ASSIGNMENT

We use an electrical problem to show how we might apply the knowledge of this chapter.

Let us consider a current, I, flowing through a resistor, R. The power, P, associated with the resistor is given by the usual formula $P = I^2 R$.

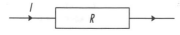

This format for the formula is useful if we know the current and the resistor values. However in our Assignment we are going to look at finding current values for a given resistance at various powers. This will involve transposing the formula into a more useful format that will help with the calculations.

Proportionality

We can link together any 2 variables by some relation. One of them is an independent variable (e.g. x) and one of them is a dependent variable (e.g. y). In Chapter 2 we saw the independent variable plotted horizontally and the dependent variable plotted vertically. This gave us a straight line. In this section we will see that only some straight lines show 2 variables in proportion. It is possible for curves to show proportion as well.

We can write "y is proportional to x". This means that whatever happens to x in turn affects y.
$\propto$ is the symbol for **is proportional to**,
e.g. $y \propto x$ means y **is proportional to** x.
The question now arises: how are they in proportion? Without some numbers we do not know anything about the proportionality. The mention of numbers encourages us to reach for a calculator. Our new $\propto$ symbol is not there. In fact we replace $\propto$ with $= k$. k is the **constant of proportionality** (or **coefficient of proportionality**). $y \propto x$ and $y = kx$ are the same relationship, but $y = kx$ is more useful because of the "$=$" symbol.

▬▬▬▬ Example 3.1 ▬▬▬▬

Generally when running a car the volume of petrol is related to the miles driven. Suppose a car uses 7.5 gallons of petrol whilst travelling 240 miles. How far will it travel on 5.0 gallons?

In this example the distance we can drive the car depends on the volume of petrol in the tank. (We know it depends on many other things too.) The volume of petrol is the independent variable (x) so the distance travelled is the dependent variable (y);

i.e.   distance travelled $\propto$ volume of petrol

i.e. $y \propto x$

i.e. $y = kx$

We are given $x = 7.5$ together with $y = 240$ so that

$$240 = k\,(7.5)$$

$$\frac{240}{7.5} = k$$

$$\boxed{\frac{\text{miles}}{\text{gallons}}}.$$

i.e. $k = 32.$

The units for k are miles per gallon (mpg), though knowing this is not vital to the example.

For this particular car we know now that

$$y = kx$$

is $\qquad y = 32x.$

How far it travels on 5.0 gallons of petrol means we can write

$$y = 32 \times 5.0 = 160$$

i.e. on 5 gallons of petrol the car travels 160 miles.

We also know that if there is no petrol in the fuel tank we cannot drive the car anywhere, i.e. $x = 0$, $y = 0$. Now we have 3 pairs of values for petrol and distance. We can write these as coordinates (x, y); $(0, 0)$, $(5.0, 160)$ and $(7.5, 240)$. In Fig. 3.1 we see them plotted on a pair of axes. They all lie on a straight line passing through the origin. We see that the more petrol we have the greater the distance generally we are able to travel.

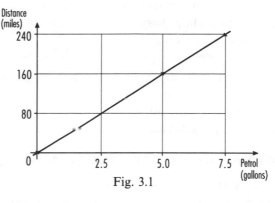

Fig. 3.1

Many short distances around towns are likely to affect our calculation. How somebody drives and in which particular car are some other effects. We can cope with these changes. All we would need to do is re-work our calculation with revised numbers.

Example 3.2

At home the more units of electricity we use the greater the electricity bill. Is the cost of electricity proportional to the number of units used?

In the previous example distance and volume of petrol were in proportion. Also "We see that the more petrol we have the greater the distance generally we are able to travel". At first glance we might expect the cost of electricity to be proportional to the number of units used. However, if you look at a domestic electricity bill there is a quarterly standing charge. If no electricity is used the charge will still appear on the bill. We can look at this as a graph of costs against units of electricity used in Fig. 3.2. This time you can see the straight line graph does *not* pass through the origin. This means the cost of

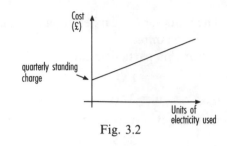

Fig. 3.2

electricity is *not* directly proportional to the number of units used. In fact the intercept on the vertical axis is the cost of the quarterly charge.

We could extend Example 3.1 to include all a person's motoring costs. Included in the total costs are insurance and the Road Fund Licence, two types of standing charge. These must be paid no matter how great or small the distance driven.

$y \propto x$, or $y = kx$, says that y is directly proportional to x. The powers of x and y are 1.

There are many other types of proportionality. The next set of examples looks at a few types.

Example 3.3

One version of Ohm's law says that the current, I, is inversely proportional to the resistance, R. When $R = 5\,\Omega$, $I = 3\,\text{A}$. Find the constant of proportionality. What is the value of I when $R = 9\,\Omega$?

We write this as I is inversely proportional to R

i.e. $$I \propto \frac{1}{R}$$

i.e. $$I = k\left(\frac{1}{R}\right) = \frac{k}{R}$$

> k is the constant of proportionality.

Substituting for $R = 5$ and $I = 3$ gives

$$3 = \frac{k}{5}$$

i.e. $$3 \times 5 = k$$

i.e. $$k = 15.$$

This means $$I = \frac{15}{R}$$ so that when $R = 9$ we have

$$I = \frac{15}{9}$$

i.e. $$I = \frac{5}{3} \text{ or } 1.\bar{6},$$

i.e. the current for this resistance is $1.\bar{6}\,\text{A}$.

In Chapter 12 we will look at the type of graph of $I=\dfrac{k}{R}$. $I=\dfrac{k}{R}$ may be re-written as $IR=k$, i.e. $IR=$ constant. For any general inversely proportional relation, $y=\dfrac{k}{x}$, we may write $xy=$ constant. This means that for an inversely proportional relationship the product of the variables is constant.

Example 3.4

y is proportional to x^2. If $y=135$ when $x=4$ find the constant of proportionality. What is the value of x when $y=76.5$?

$$y \text{ is proportional to } x^2$$

is written as $y \propto x^2$

i.e. $y = kx^2.$

> k is the constant of proportionality.

Substituting for $x=4$ and $y=135$ gives

$$135 = k \times 4^2$$

i.e. $\dfrac{135}{4^2} = k$

i.e. $k = 8.4375.$

This means $y = 8.4375x^2$ so that when $y=76.5$ we have

$$76.5 = 8.4375x^2$$

i.e. $\dfrac{76.5}{8.4375} = x^2$

$$9.0\bar{6} = x^2$$

i.e. $x = \pm\sqrt{9.0\bar{6}}$

$$x = \pm3.01.$$

In Chapter 12 we will look at a graph of the type $y=kx^2$.

Example 3.5

Without skidding a vehicle negotiates a curved horizontal track. Its velocity, v, is proportional to the (positive) square root of the radius of the curve, $\sqrt{r}$. Write this statement in mathematical terms.

If a vehicle's velocity is $14\,\text{ms}^{-1}$ around a curve of radius $45\,\text{m}$ find the coefficient of proportionality.

On the same type of road surface if the radius is $60\,\text{m}$ what is the maximum permitted velocity? What must be the curve's radius for no skidding at $10\,\text{ms}^{-1}$?

We have $\qquad v \propto \sqrt{r}$

i.e. $\qquad v = k\sqrt{r}.$

> k is the constant of proportionality.

Substituting for $v = 14$ and $r = 45$ gives

$$14 = k\sqrt{45}$$

i.e. $\qquad \dfrac{14}{6.708} = k$

i.e. $\qquad k = 2.087 \quad$ is the coefficient of proportionality.

Now to find v given $r = 60$ we use the calculator value of k so

$$v = 2.086\ldots \times \sqrt{60}$$

i.e. $\qquad v = 16.17 \text{ ms}^{-1}$ is the velocity.

Then to find r given $v = 10$ we have

$$10 = 2.086\ldots \times \sqrt{r}$$

i.e. $\qquad \dfrac{10}{2.087} = \sqrt{r}$

i.e. $\qquad (4.792)^2 = r$

to give $\qquad r = 22.96$ m as the curve's radius.

In Chapter 12 we will look at a graph of the type $v = k\sqrt{r}$ (or $y = k\sqrt{x}$).

■ EXERCISE 3.1 ■

1   We know that $y \propto x$ and $y = 4$ when $x = 10$. Find the coefficient of proportionality. What is the value of y when $x = 1.5$?

2   $y = 3$ when $x = 4$ according to $y \propto x^2$. When $x = 32$ what is the value of y?

3   Given that $y \propto x^3$ and $y = 80$ when $x = 2$ find the constant of proportionality. What is the value of x when $y = 16$?

4   $y \propto 1/x$. If $y = 8$ when $x = 5$ what is the value of y when $x = 5.5$?

5   If $y \propto \sqrt{x}$ and $x = 6$ when $y = 8$ find the coefficient of proportionality. Find a value for y when $x = 1$. What would change in this question if $y \propto -\sqrt{x}$?

6   Given that $y \propto 1/x^2$ and $y = 8$ when $x = 9$ find the values of x when $y = 16$.

7   We know that $y \propto \pm\sqrt{x}$. If $y = 45$ when $x = 9$ calculate the coefficient of proportionality. What can you deduce when $x = -3$?

8   If $y \propto 1/x^2$ find the value of y when $x = 2$ given that $y = 8$ when $x = 5$.

9   If $y \propto \pm 1/\sqrt{x}$ and $x = 25$ when $y = 30$ find the constant of proportionality. What is the value of x when $y = 36$?

10 In a particular motion the distance travelled, dm, is proportional to the time for the motion, ts. We know that $d = 480$m when $t = 150$s. What distance has been travelled in 210 seconds?

11 Hooke's law states that the tension, TN, in an elastic spring is proportional to the extension, xm. Given that $T = 20$N when $x = 0.15$m find the tension for an extension of 0.25m.

12 The volume, Vm^3, of a sphere is proportional to its radius, rm, cubed. If $r = 0.35$m and $V = 0.18$m^3 are related what is the volume for $r = 0.48$m?

13 For a simple pendulum the time, Ts, for a beat is proportional to the square root of its length, lm. If $l = 0.65$m and $T = 1.62$s find the time for a beat when $l = 0.95$m.

14 If a body is dropped, the distance it has fallen is proportional to its velocity squared. Having fallen 5m the velocity of the body is $9.90\,\text{ms}^{-1}$. What is its velocity after 15 m?

15 In simple harmonic motion the time period, Ts, for an oscillation is inversely proportional to the angular velocity, ω. If $\omega = 8$ and $T = 0.785$ what is the value of T when $\omega = 11$?

Transposition of formulae

Let us look at a basic formula involving the initial velocity of a vehicle, u, and its final velocity, v, over a time, t, due to an acceleration a. The formula is $v = u + at$. Because v is alone on one side of the formula (generally it will be the left side) v is said to be the **subject** of the formula. Alternatively v is given in terms of u, a and t. It is possible that we might not be interested in the final velocity, but in the time taken to reach this velocity. It is possible to re-arrange (transpose) the letters so that t appears on its own while all the other letters appear on the opposite side. This means that t will be given in terms of u, v and a. This re-arrangement is useful when we need to repeat a type of calculation several times for different numerical values of u, v and a.

One of the difficult things is to give all our attention to getting one particular letter in terms of all the others. Often the letters seem to be a jumble. When numbers are involved with just one letter then that letter stands out. To help in early examples we will use Greek letters in the same way as numbers so that you can concentrate on the other letter, usually x. Don't worry about understanding what the Greek letters represent; just think of them as numbers that cannot be simplified.

As a starting point we will look again at some early examples from Chapter 2.

Examples 3.6

i) Solve the equation $x - 12 = 8$.

i.e. $x = 8 + 12$.

> Addition and subtraction are opposites.

It is not important in this chapter that the actual solution of the equation is $x = 20$.

ii) Make x the subject of the formula $x - \alpha = \beta$.

i.e. $x = \beta + \alpha$

> Adding α to both sides.

or $x = \alpha + \beta$.

Examples 3.7

i) Solve the equation $4x = 3$.

i.e. $x = \dfrac{3}{4}$.

> Multiplication and division are opposites.

The actual numerical answer is not important in this example. We need to pay attention to the operation used.

ii) Make x the subject of the formula $\alpha x + \beta = \gamma$.
Addition of β is less closely attached than α to x. This means we will deal with β first.

i.e. $\alpha x = \gamma - \beta$

> Subtracting β.

                                      $x = \dfrac{\gamma - \beta}{\alpha}$.

> Dividing by α.

Example 3.8

Make x the subject of the formula

$$\frac{x}{\alpha} - \theta = -\beta x + \phi.$$

Terms involving x appear on both sides of this formula. The first move is to gather these terms on one side. We will gather them together on the left, but the same final solution will be reached if we gather them on the right.

$\therefore$ $\dfrac{x}{\alpha} + \beta x - \theta = \phi$

> Adding βx.

                                      $\dfrac{x}{\alpha} + \beta x = \phi + \theta$.

> Adding θ.

The α in the denominator position needs some attention to ease the algebra later in the problem. We can move it from this position by multiplying throughout by α.

$$\frac{\alpha x}{\alpha} + \alpha \beta x = \alpha(\phi + \theta)$$

> Multiplying by α.

$$x + \alpha \beta x = \alpha(\phi + \theta)$$

> Cancelling the αs.

$$x(1 + \alpha \beta) = \alpha(\phi + \theta)$$

> x is a common factor.

$$x = \frac{\alpha(\phi + \theta)}{1 + \alpha \beta}.$$

> Dividing by $1 + \alpha \beta$.

██████ **Example 3.9** ██

Make x the subject of the formula $\alpha\left(\dfrac{x}{\beta} + \gamma\right) = \epsilon$.

In this example the brackets affect the order of operations. x is at the heart of this formula on the left. Using our earlier knowledge, if the letters are replaced by numbers the order of evaluation is

x (or the number)

divide by β

plus γ

multiply by α.

We use this list from the bottom working upwards with the opposite operation. This leads to the new list of operations we perform to find x, i.e.

divide by α

subtract γ

multiply by β

x (or the number).

Then
$$\frac{x}{\beta} + \gamma = \frac{\epsilon}{\alpha} \qquad \boxed{\text{Dividing by } \alpha.}$$

$$\frac{x}{\beta} = \frac{\epsilon}{\alpha} - \gamma \qquad \boxed{\text{Subtracting } \gamma.}$$

$$x = \beta\left(\frac{\epsilon}{\alpha} - \gamma\right). \qquad \boxed{\text{Multiplying by } \beta.}$$

██████ **Example 3.10** ███

Make t the subject of the formula $v = u + at$.

t appears on the right. Suppose that we had actual numbers for u, a and t on the right. Using our calculator we would

input t

multiply by a

add u.

This order of operations gives the value for v. We use this list of operations from the bottom working upwards with the opposite operation. This leads to a new list

value of v

subtract u

divide by a.

This gives t as the subject of the formula,

i.e. $\qquad\qquad v = u + at$

becomes $\qquad\qquad v - u = at \qquad \boxed{\text{Subtracting } u.}$

$$\frac{v - u}{a} = t. \qquad \boxed{\text{Dividing by } a.}$$

It is more usual to write this as $t = \dfrac{v - u}{a}$ so that the subject, t, appears on the left.

■■■■■ **Example 3.11** ■■■■■

Make l the subject of the formula $W = \dfrac{\frac{1}{2}\lambda x^2}{l}$.

We can re-write the $\frac{1}{2}$ in this formula to give $W = \dfrac{\lambda x^2}{2l}$. Both versions of the formula have exactly the same meaning. On the right we are dividing by l. The opposite of division is multiplication. We write

$$Wl = \frac{l\lambda x^2}{2l}$$
> Multiplying by l.

i.e.
$$Wl = \frac{\lambda x^2}{2}.$$
> Cancelling ls on the right.

We need l alone on the left, rather than being multiplied by W. The opposite of multiplication is division. We write

$$\frac{Wl}{W} = \frac{\lambda x^2}{2W}$$
> Dividing by W.

i.e.
$$l = \frac{\lambda x^2}{2W}.$$
> Cancelling Ws on the left.

■■■■■ **Example 3.12** ■■■■■

Make r the subject of the formula $\dfrac{r}{1-r} = S$.

r already appears on the left in this formula; but it appears more than one. For ease we remove the fraction by multiplying by $(1 - r)$,

i.e.
$$\frac{r(1-r)}{1-r} = S(1-r)$$
> Multiplying by $(1 - r)$.

i.e.
$$r = S(1-r)$$
> Cancelling on the left.

This version of the formula is simpler than before because it has no fractional algebra. However we do need to re-gather the r terms.

$$r = S - Sr$$
> Multiplying the bracket.

Then $\qquad r + Sr = S - Sr + Sr$
> Adding Sr to both sides.

i.e. $\qquad r + Sr = S$

i.e. $\qquad r(1 + S) = S$
> r is a factor on the left.

We leave r alone on the left dividing through by $(1 + S)$,

i.e.
$$\frac{r(1+S)}{1+S} = \frac{S}{1+S}$$
> Dividing by $(1 + S)$.

i.e.
$$r = \frac{S}{1+S}$$
> Cancelling on the left.

■ **Example 3.13** ■

Make r_1 the subject of the formula $\dfrac{1}{R} = \dfrac{1}{r_1} + \dfrac{1}{r_2}$

where r_1 and r_2 are resistors in parallel.

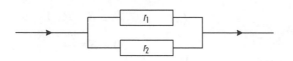

There are several ways of making r_1 the subject of the formula. In earlier examples we attempted to get rid of fractions, multiplying through by the lowest common multiple (LCM) of the denominators.

$$\frac{1}{R} - \frac{1}{r_2} = \frac{1}{r_1}.$$

> Subtracting $\dfrac{1}{r^2}$.

The LCM of R, r_1 and r_2 is Rr_1r_2 and so

$$Rr_1r_2\left(\frac{1}{R} - \frac{1}{r_2}\right) = Rr_1r_2\left(\frac{1}{r_1}\right)$$

> Multiplying by Rr_1r_2.

$$Rr_1r_2\left(\frac{1}{R} - \frac{1}{r_2}\right) = Rr_2$$

> Cancelling r_1 on the right.

$$r_1\left(\frac{Rr_2}{R} - \frac{Rr_2}{r_2}\right) = Rr_2.$$

> Multiplying the brackets apart from r_1.

We will keep r_1 outside the brackets because it appears only once at this point. In a few lines we will be able to move all the other terms away from r_1 to leave it alone as the subject of the equation.

$$r_1(r_2 - R) = Rr_2$$

> Simplifying the left.

$$r_1 = \frac{Rr_2}{r_2 - R}.$$

> Dividing by $r_2 - R$.

■ **EXERCISE 3.2** ■

Each question is a relationship followed by a letter in brackets. In each case transpose the relationship to make that letter the subject.

1   $x + \alpha - \beta = 0$ (x)
    i.e. make x the subject of this formula or give x in terms of α and β.

2   $x + \alpha - \beta = \gamma$ (x)

3   $x + \alpha = \beta + \gamma$ (x)

4   $x\alpha - \beta = \gamma$ (x)

5   $\dfrac{x}{\alpha} + \beta = \delta$ (x)

6   $\left(\dfrac{x}{\alpha} + \beta\right)\gamma = \delta$ (x)

7   $\alpha x + \gamma = \beta x$ (x)

8   $\dfrac{x}{a} - c = bx$ (x)

9   $\dfrac{x}{a} + d - c = \dfrac{x}{b}$ (x)

10   $\dfrac{1}{x} + a = b$ (x)

11 $\dfrac{a}{x} - b = c$ $\qquad$ (x)

12 $ax + c = \dfrac{x}{b}$ $\qquad$ (x)

13 $\dfrac{a}{x} - d = \dfrac{b}{x}$ $\qquad$ (x)

14 $I = \dfrac{PRT}{100}$ $\qquad$ (P)

15 $C = \pi d$ $\qquad$ (d)

16 $A = \frac{1}{2}bh$ $\qquad$ (b)

17 $C = \dfrac{5}{9}(F - 32)$ $\qquad$ (F)

18 $A = \frac{1}{2}(a + b)h$ $\qquad$ (h)

19 $S = Wx - 2$ $\qquad$ (x)

20 $s = \frac{1}{2}(u + v)t$ $\qquad$ (t)

21 $I_1 + I_2 + I_3 + I_4 = 0$ $\qquad$ (I_1)

22 $C = 2\pi r$ $\qquad$ (r)

23 $I = \dfrac{PRT}{100}$ $\qquad$ (R)

24 $A = \frac{1}{2}bh$ $\qquad$ (h)

25 $\dfrac{1}{C} = \dfrac{1}{c_1} + \dfrac{1}{c_2}$ $\qquad$ (c_1)

26 $\dfrac{1}{v} + \dfrac{1}{u} = \dfrac{1}{f}$ $\qquad$ (v)

27 $pV = mRT$ $\qquad$ (T)

28 $p_1V_1 - p_2V_2 = 0$ $\qquad$ (p_2)

29 $R = R_0(1 + \alpha t)$ $\qquad$ (t)

30 $S = \dfrac{N - R}{N}$ $\qquad$ (R)

31 $A = \frac{1}{2}(a + b)h$ $\qquad$ (a)

32 $I = \dfrac{PRT}{100}$ $\qquad$ (T)

33 $s = \frac{1}{2}(u + v)t$ $\qquad$ (u)

34 $V = \dfrac{\pi D^2 f N}{4}$ $\qquad$ (N)

35 $A = \frac{1}{2}(a + b)h$ $\qquad$ (b)

36 $Ft = m(v - u)$ $\qquad$ (v)

37 $S = \dfrac{N - R}{N}$ $\qquad$ (N)

38 $\Omega = \omega + \alpha t$ $\qquad$ (α)

39 $l = l_0(1 + \alpha t)$ $\qquad$ (t)

40 $\dfrac{R_1}{R_2} = \dfrac{1 + \alpha t_1}{1 + \alpha t_2}$ $\qquad$ (α)

Square and square roots

If we have a number which we square, this is the opposite operation to finding the square root of a number. We may try this with any number first of all using a calculator.

Suppose we choose the number 81.

Now $\qquad 81^2 = 6561$

and $\qquad \sqrt{6561} = 81.$

We will use the positive square root but remember that there is a negative one as well. The calculator displays only the positive root.

We may combine both these operations and write

$$\sqrt{81^2} = \sqrt{6561} = 81$$

We treat the $\sqrt{}$ symbol in a similar way to brackets. Usually we calculate first whatever appears in the brackets and so should calculate first whatever appears under the square root, $\sqrt{}$.

$\sqrt{}$ has another form. It is the power $\frac{1}{2}$.

e.g. $(81^2)^{\frac{1}{2}} = (6561)^{\frac{1}{2}} = 81.$

We can omit the middle stage to reveal

$(81^2)^{\frac{1}{2}} = 81^1.$

> Usually we do not write the power 1.

Looking at the powers,

$2(\frac{1}{2}) = 1.$

> Simple multiplication.

Using letters makes no difference to the operations so that with x in place of 81 we have

$$\sqrt{x^2} \text{ or } (x^2)^{\frac{1}{2}} = x.$$

We can interchange the order of the square and square root operations so that

$$(\sqrt{x})^2 \text{ or } (x^{\frac{1}{2}})^2 = x.$$

Example 3.14

The volume of a cone is given by $V = \frac{1}{3}\pi r^2 h.$

We may make r the subject of the formula. The aims are to make r^2 the subject of the formula and then, by taking the square root of both sides, obtain r. On the right r^2 is multiplied by $\frac{1}{3}$, π and h. To leave r^2 on the right we must apply the opposite operation of division,

i.e. $\dfrac{V}{\frac{1}{3}\pi h} = r^2.$

This may be simplified because dividing by $\frac{1}{3}$ is the same as multiplying by 3, i.e. "invert the fraction and multiply", so that

$$r^2 = \frac{3V}{\pi h}$$

$$r = \sqrt{\frac{3V}{\pi h}}.$$

> Square root of both sides.
>
> $\sqrt{r^2} = r.$

Example 3.15

$\frac{1}{2}mu^2 = mgh + \frac{1}{2}mv^2$ represents a simple system in which mechanical energy (kinetic and potential) is conserved. We wish to make v the subject of the formula.

We may look at the formula in 3 sections (**terms**): $\frac{1}{2}mu^2,$

$$mgh$$

and $\frac{1}{2}mv^2.$

Because we need v let us give our attention to this last section. mgh is added in our formula and the opposite of addition is subtraction,

i.e. $\frac{1}{2}mu^2 - mgh = \frac{1}{2}mv^2.$ Subtracting mgh from both sides.

v^2 is multiplied by $\frac{1}{2}$ and m. In fact all the terms are multiplied by m so we may cancel this letter throughout the formula,

i.e. $\frac{1}{2}u^2 - gh = \frac{1}{2}v^2.$ Dividing each term by m.

v^2 is still multiplied by $\frac{1}{2}$, so

$$\frac{\frac{1}{2}u^2}{\frac{1}{2}} - \frac{gh}{\frac{1}{2}} = \frac{\frac{1}{2}v^2}{\frac{1}{2}}$$

Dividing each term by $\frac{1}{2}$.

Division by $\frac{1}{2}$ is the same as multiplication by 2.

$$u^2 - 2gh = v^2.$$

Finally $\sqrt{u^2 - 2gh} = v.$ Square root of both sides.

This is usually written as

$$v = \sqrt{u^2 - 2gh}$$

with the subject, v, on the left.

Example 3.16

Suppose that $a - 3\sqrt{2x} = b$ and that we want to make x the subject of the formula.

We need to look at the term $-3\sqrt{2x}$.

Now $-3\sqrt{2x} = b - a$ Subtracting a from both sides.

$3\sqrt{2x} = -(b - a)$ Multiplying both sides by -1.

$3\sqrt{2x} = -b + a$

i.e. $3\sqrt{2x} = a - b$

$\sqrt{2x} = \dfrac{a - b}{3}$ Dividing by 3.

$2x = \left(\dfrac{a - b}{3}\right)^2$ Squaring both sides.

$x = \dfrac{1}{2}\left(\dfrac{a - b}{3}\right)^2.$ Dividing by 2.

Multiplying by $\frac{1}{2}$ has the same effect as dividing by 2.

ASSIGNMENT

Our electrical problem involves the power relationship $P = I^2 R$ for the current I flowing through the resistor R. We are going to make I the subject of this formula.

$$P = I^2 R$$

becomes $\quad\dfrac{P}{R} = I^2$

Dividing by R.

so that $\quad I = \sqrt{\dfrac{P}{R}}.$

Square rooting both sides.

Our particular resistor is 470 Ω.

$\therefore \qquad\qquad I = \sqrt{\dfrac{P}{470}}.$

Substituting for $R=470$.

Suppose we wish to know the current needed for a power of $5\times10^{-4}\,\text{W}$,

i.e. $\qquad\qquad I = \sqrt{\dfrac{5 \times 10^{-4}}{470}}$

Substituting for $P=5 \times 10^{-4}$.

$$= \sqrt{1.06\ldots \times 10^{-6}}$$
$$= \sqrt{1.06\ldots} \times \sqrt{10^{-6}}$$
$$I = 1.03 \times 10^{-3}\,\text{A or } 1.03\,\text{mA}.$$

We might go further with this Assignment taking into account a design variation allowing us to vary the power between $4.6 \times 10^{-4}\,\text{W}$ and $5.4 \times 10^{-4}\,\text{W}$. The table below shows the calculations leading to the current needed for each value of P.

$P \qquad \times 10^{-4}$	4.6	4.8	5.0	5.2	5.4
$\dfrac{P}{470} \quad \times 10^{-6}$	0.979	1.021	1.064	1.106	1.149
$I \qquad \times 10^{-3}$	0.99	1.01	1.03	1.05	1.07

EXERCISE 3.3

Each question is a relationship followed by a letter in brackets. In each case transpose the relationship to make that letter the subject.

1 $A = \pi r^2$ (r)
i.e. make r the subject of this formula or give r in terms of A and π.

2 $P = \dfrac{Wv^2}{32r}$ (v)

3 $a^2 = b^2 + c^2$ (b)

4 $V = \pi r^2 h$ (r)

5 $W = \dfrac{\lambda x^2}{2a}$ (x)

6 $\dfrac{v^2}{r} = \omega^2 r$ (ω)

7 $C = 1 - \dfrac{x^2}{2}$ (x)

8 $y = ax^2 + b$ (x)

9 $v = \sqrt{2gh}$ (h)

10 $y = a\sqrt{x} + b$ (x)

11 $Ch = 1 + \dfrac{x^2}{2}$ (x)

12 $A = P\left(1 - \dfrac{r}{100}\right)^2$ (r)

13 $c = \sqrt{a^2 - b^2}$ (b)

14 $A = \dfrac{\pi d^2}{4} + 320$ (d)

15 $V = \sqrt{V_R{}^2 + (V_L b - V_C)^2}$

    (V_R)

16 $A = P\left(1 + \dfrac{r}{200}\right)^2$ (r)

17 $Ck = \tfrac{1}{2}m(v^2 - u^2)$ (u)

18 $v = \sqrt{\dfrac{gr(v + t)}{(1 - vt)}}$ (t)

19 $Lf = \dfrac{1}{(s - c)^2}$ (s)

20 $C = \dfrac{1 - t^2}{1 + t^2}$ (t)

Powers and roots

This is a more general version of what we did with squares and square roots.

Raising a number to some power n is the opposite operation to finding the nth root of a number. The nth root may be written as $\sqrt[n]{\ }$ or $(\ )^{1/n}$. Thus $\sqrt[n]{81^n}$ or $(81^n)^{1/n} = 81^1$.

Looking at the powers only we have

$$(n)\left(\frac{1}{n}\right) = 1.$$

> Simple multiplication.

Again using a letter, x, in place of 81 we have

$$\sqrt[n]{x^n} \quad \text{or} \quad (x^n)^{1/n} = x^1 = x.$$

> Usually we do not write the power 1.

Interchanging the order of operations has no effect upon the final answer,

i.e. $(\sqrt[n]{x})^n$ or $(x^{1/n})^n$ $= x$.

Example 3.17

The volume of a sphere is given by $V = \dfrac{4\pi r^3}{3}$.

Suppose we wish to make the radius, r, the subject of this formula i.e. give r in terms of V.

The style of solution is similar to the one used in Example 3.9. On the right r^3 is multiplied by $\dfrac{4}{3}$ and π. To leave r^3 on the right we must apply the opposite operation, division,

i.e. $\dfrac{V}{\dfrac{4}{3}\pi} = r^3.$

This may be simplified because dividing by $\dfrac{4}{3}$ is the same as multiplying by $\dfrac{3}{4}$, i.e. "invert the fraction and multiply", so that

$$r^3 = \frac{3V}{4\pi}$$

$$r = \left(\frac{3V}{4\pi}\right)^{\frac{1}{3}}.$$

> Cube root of both sides.
> $\sqrt[3]{r^3} = r$.

Example 3.18

If $a\sqrt[4]{bx} + c = d$ suppose we wish to make x the subject of the relationship. The heart of the problem is x. If we consider numbers in place of all the letters our list of calculator operations is

> input x
> multiply by b
> fourth root
> multiply by a
> add c.

This list will give a calculated value for d.

We can use this list from the bottom working upwards with the opposite operation. The new list to make x the subject of the formula is

> value for d
> subtract c
> divide by a
> raise to the power 4
> divide by b.

x is now the subject of the formula.

i.e. $\qquad a\sqrt[4]{bx} + c = d$

> Subtracting c.

becomes $\qquad a\sqrt[4]{bx} = d - c$

> Dividing by a.

$$\sqrt[4]{bx} = \frac{d - c}{a}$$

> Raising both sides to the power 4.

$$bx = \left(\frac{d - c}{a}\right)^4$$

> Dividing by b.

$$x = \frac{1}{b}\left(\frac{d - c}{a}\right)^4.$$

Multiplying by $\dfrac{1}{b}$ has the same effect as dividing by b. Because of the complicated bracket, multiplying by $\dfrac{1}{b}$ is the preferred format.

■■■■ EXERCISE 3.4 ■■■■

Each question is a relationship followed by a letter in brackets. In each case transpose the relationship to make that letter the subject.

1 $y = ax^3 + b$ $\qquad$ (x)
i.e. make x the subject
of this formula or give
x in terms of a and b.

2 $y = mx^5 + c$ $\qquad$ (x)

3 $y + Mx^4 + c = 0$ $\qquad$ (x)

4 $y = a\sqrt[3]{x} + b$ $\qquad$ (x)

5 $y = m\sqrt[4]{x} + c$ $\qquad$ (x)

6 $y + m\sqrt[5]{x} + c = 0$ $\qquad$ (x)

7 $(c + ts)^4 = z$ $\qquad$ (c)

8 $A = P\left(1 + \dfrac{R}{100}\right)^{10}$ $\qquad$ (R)

9 $ST^{\frac{1}{7}} = c$ $\qquad$ (T)

10 $z = \dfrac{\pi D^3}{32}$ $\qquad$ (D)

11 $pV^\gamma = c$ $\qquad$ (V)

12 $I = \dfrac{bd^3}{12}$ $\qquad$ (d)

13 $ST^{0.1} = c$ $\qquad$ (T)

14 $(c - st)^3 = z$ $\qquad$ (s)

15 $A = p\left(1 - \dfrac{R}{100}\right)^7$ $\qquad$ (R)

16 $T = kf^{0.75}D^{1.8}$ $\qquad$ (D)

17 $I = \dfrac{\pi D^4}{32}$ $\qquad$ (D)

18 $T = kf^{0.75}D^{1.8}$ $\qquad$ (f)

19 $I = \dfrac{BD^3 - bd^3}{12}$ $\qquad$ (D)

20 $z = \dfrac{\pi}{32}\left(\dfrac{D^4 - d^4}{D}\right)$ $\qquad$ (d)

■■■■ MULTI-CHOICE TEST 3 ■■■■

A vehicle has an initial velocity, u. Its acceleration is f to reach a final velocity, v. This takes place during a time, t, over a displacement, s.

Questions **1**, **2** and **3** refer to the formula $v = u + ft$.

1 Transposing we get $u =$
  A) $v + ft$

  B) $\dfrac{v}{-ft}$

  C) $v - ft$

  D) $\dfrac{v}{ft}$

2 Transposing we get $f=$

A) $\dfrac{v}{u}-t$

B) $\dfrac{v}{ut}$

C) $v-u-t$

D) $\dfrac{v-u}{t}$

3 Transposing we get $t=$

A) $\dfrac{v}{u-f}$

B) $\dfrac{v-u}{f}$

C) $\dfrac{\frac{v}{u}}{f}$

D) $\dfrac{v}{fu}$

Questions **4** and **5** refer to the following information. S is the sum of a geometric series. The series' first term is a and the common ratio is r. They are connected by $S=\dfrac{a}{1-r}$.

4 As the subject of this formula $a=$
A) $S+1-r$
B) $S(1-r)$
C) $S-r$
D) $S-1+r$

5 As the subject of this formula $r=$

A) $1+\dfrac{a}{S}$

B) $a-S$

C) $\dfrac{S-a}{S}$

D) $\dfrac{1+a}{S}$

6 You are given $\dfrac{2x}{1+x}=T$. Re-arranging this formula $x=$

A) $\dfrac{T}{2-T}$

B) 1

C) T

D) $T(2-T)$

A vehicle has an initial velocity, u. Its acceleration is f to reach a final velocity, v. This takes place during time, t, over a displacement, s.

Questions **7**, **8** and **9** refer to the formula $s = \dfrac{(u+v)}{2}t$.

7   Re-arranging we have $u =$

A) $\dfrac{st}{2} - v$

B) $\dfrac{2s}{t} - v$

C) $\dfrac{st}{2} + v$

D) $\dfrac{2v}{t} + s$

8   Re-arranging we have $t =$

A) $s - \dfrac{(u+v)}{2}$

B) $2s - (u+v)$

C) $\dfrac{s}{2}(u+v)$

D) $\dfrac{2s}{u+v}$

9   Re-arranging we have $v =$

A) $\dfrac{st}{2} - u$

B) $\dfrac{2s}{t} - u$

C) $\dfrac{st}{2} + u$

D) $s + \dfrac{2u}{t}$

10   A system has total energy E made up of kinetic energy, $\frac{1}{2}mv^2$, and potential energy, mgh. Given $E = \frac{1}{2}mv^2 + mgh$ then $m =$

A) $\dfrac{2E}{v^2 + 2gh}$

B) $\dfrac{2E - 2gh}{v^2}$

C) $\pm\sqrt{\dfrac{2E}{v^2 + 2gh}}$

D) $\dfrac{E}{-v^2 - 2gh}$

11 A solid cylinder has radius r and height h. Its total surface area A, is given by $A = 2\pi r^2 + \pi r h$. From this formula $h =$

A) $A - 2r$

B) $\dfrac{A}{2\pi r^2} - \pi r$

C) $\dfrac{A - 2\pi r^2}{\pi r}$

D) $\dfrac{\dfrac{A}{2\pi r^2}}{\pi r}$

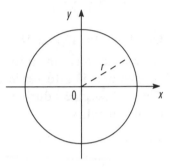

12 The equation of a circle with centre $(0,0)$ and radius r is given by $x^2 + y^2 = r^2$. From this equation $y =$

A) $r - x$

B) $\pm\sqrt{r^2 - x^2}$

C) $\pm\sqrt{r - x}$

D) $\pm\sqrt{r^2 + x^2}$

13 In a system the mechanical energy is conserved. The initial kinetic energy is $\frac{1}{2}mu^2$ and the final kinetic energy is $\frac{1}{2}mv^2$. The change in potential energy is mgh. They are connected according to $\frac{1}{2}mv^2 = mgh + \frac{1}{2}mu^2$. Hence $v =$

A) $\pm\sqrt{2gh + u^2}$

B) $\pm\sqrt{2gh + u}$

C) $\pm\sqrt{\frac{1}{2}gh + u^2}$

D) $\pm\sqrt{gh + \frac{1}{2}u^2}$

14 Given $y = ax^2 - b$ then $x =$

A) $\pm\sqrt{\dfrac{y + b}{a}}$

B) $\pm\sqrt{\dfrac{y - b}{a}}$

C) $\pm\sqrt{\dfrac{y}{b} - a}$

D) $\pm\sqrt{\dfrac{y - a}{b}}$

15   Where $y = \dfrac{1+x^2}{1-x^2}$ we can re-arrange to get $x =$

A) $\dfrac{y^2-1}{y^2+1}$

B) $\pm\sqrt{\dfrac{y+1}{y-1}}$

C) $\pm\sqrt{\dfrac{y-1}{2}}$

D) $\pm\sqrt{\dfrac{y-1}{y+1}}$

Questions **16**, **17** and **18** are based on the following information.

A vehicle has an initial velocity, u. Its acceleration is a to reach a final velocity, v. This takes place during a time, t, over a displacement, s.

16   If $s = ut + \frac{1}{2}at^2$ then $a =$

A) $\dfrac{2(s-u)}{t}$

B) $\dfrac{2s-u}{t}$

C) $\dfrac{2s-ut}{t^2}$

D) $\dfrac{2(s-ut)}{t^2}$

17   Where $s = vt - \frac{1}{2}at^2$ we can re-arrange to get $v =$

A) $\dfrac{s-at^2}{2t}$

B) $\dfrac{s-\frac{1}{2}at^2}{t}$

C) $s + \frac{1}{2}at^2$

D) $\dfrac{s+\frac{1}{2}at^2}{t}$

18   For $v^2 = u^2 + 2as$ with u as the subject we have $u =$
A) $\pm\sqrt{v^2-2as}$
B) $\pm\sqrt{v^2+2as}$
C) $v - \sqrt{2as}$
D) $v + \sqrt{2as}$

19 g is the acceleration due to gravity. The time period, T, of a simple pendulum is related to the length of the pendulum, l, by $T = 2\pi\sqrt{\dfrac{l}{g}}$. Where l is the subject of this formula we have $l =$

A) $(T - 2\pi)g$

B) $g\left(\dfrac{T}{2\pi}\right)^2$

C) $\dfrac{Tg}{2\pi}$

D) $(T - 2\pi)^2 g$

20 A circle has radius r. Its area is given by $A = \pi r^2$ and its circumference by $C = 2\pi r$. If we eliminate r we get $A =$

A) $\dfrac{C^2}{2\pi}$

B) $\pi^2 C$

C) $\dfrac{Cr}{2\pi}$

D) $\dfrac{C^2}{4\pi}$

4 Introducing Trigonometry

Introduction

We start this chapter with the measurement of angles using degrees and radians. We look at the conversion formulae between them. In the remainder of the chapter we deal with **trigonometric ratios**. They may be called **trigonometric operations** or **trigonometric functions**. In this chapter

90

we look at all 6 of them: sine, cosine, tangent, secant, cosecant and cotangent. The sine and cosine are the most important ones in engineering and science. We study their graphs and the tangent graph. By way of important revision we apply trigonometry to right-angled triangles, including 2 special triangles.

▬▬ ASSIGNMENT ▬▬

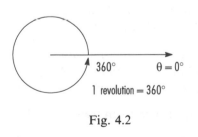

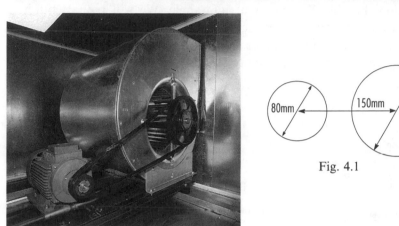

Fig. 4.1

The Assignment for this chapter looks at the 2 pulleys in Fig. 4.1. One of them has a diameter of 120 mm and the other of 80 mm. Their centres are 150 mm apart. Firstly we will find the length of the belt around them.

Degrees

The usual way to measure an angle is by using degrees, often with a protractor. A protractor is marked off (**graduated**) in degrees.

In general we draw a horizontal line towards the right to act as a base line, or line of reference. We call this the line $\theta = 0°$ and from it rotate either clockwise or anticlockwise. In Fig. 4.2 the rotation is the more usual anticlockwise one, with a complete revolution being 360°.

360° $\theta = 0°$

1 revolution = 360°

Fig. 4.2

180° is said to be "the angle on a straight line". This means that from the base line $\theta = 0°$ if you turn through 180° your position is the same as if the straight line had been projected backwards. Figs. 4.3 show we can use any starting line as well as the usual horizontal $\theta = 0°$.

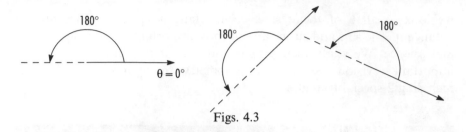

Figs. 4.3

A rotation through 90° creates a right-angle. We will pay attention to right-angled triangles (triangles containing one right-angle and 2 other acute angles) later in this chapter. Again Figs. 4.4 show that the base line does not have to be horizontally to the right.

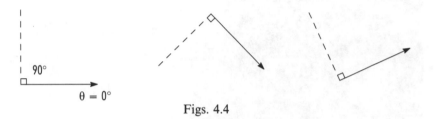

Figs. 4.4

We may relate these angles to a revolution, i.e.

$$1 \text{ rev} = 360°$$

$$\tfrac{1}{2} \text{ rev} = 180°$$

$$\tfrac{1}{4} \text{ rev} = 90°.$$

Radians

The definition of a radian is based on a circle. Draw a circle with centre O and radius r. Mark a point A on the circumference and measure an arc equal in length to the radius. If this arc length ends at B, the angle AOB, $\angle AOB$, is defined to be of size 1 radian. This is shown in Fig. 4.5.

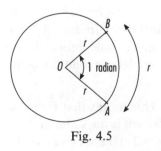

Fig. 4.5

A formal definition is that one radian is the angle **subtended** at the centre of a circle by an arc equal in length to the radius of the circle.

We can use the short form **rad** instead of radian.

There is a connection between degrees and radians. The circumference of the circle in Fig. 4.5 is given as $2\pi r$. To draw a circle we need to turn through 1 revolution (360°).

When constructing (drawing) a circle we can see that the greater the angle we turn for a given radius the greater the arc formed until we have completed a full circle. An arc of length r is linked with an angle of 1 radian, an arc of length $2r$ is linked with an angle of 2 radians and so an arc of length $2\pi r$ is linked with an angle of 2π radians, which is the circumference.

We have $\boxed{1 \text{ rev} = 360° = 2\pi \text{ rad.}}$

Degrees/radians conversions

Already we have $360° = 2\pi$

so that $\dfrac{360°}{360} = \dfrac{2\pi}{360}$

i.e. $\boxed{1° = \dfrac{\pi}{180} \text{ rad.}}$

This is the **conversion factor** to move from degrees to radians.

▰▰▰▰ **Examples 4.1** ▰▰▰▰▰▰▰▰▰▰▰▰▰▰▰▰▰

Convert i) 30°, ii) 90°, iii) 135°, iv) 290°, v) θ° into radians correct to 3 decimal places.

i) 30° is $30 \times 1°$ $= \dfrac{30 \times \pi}{180}$

$= \dfrac{\pi}{6}$ or 0.524 rad.

ii) 90° is $90 \times 1°$ $= \dfrac{90 \times \pi}{180}$

$= \dfrac{\pi}{2}$ or 1.571 rad.

iii) 135° is $135 \times 1°$ $= \dfrac{135 \times \pi}{180}$

$= \dfrac{3\pi}{4}$ or 2.356 rad.

iv) 290° is $290 \times 1°$ $= \dfrac{290 \times \pi}{180}$

$= 5.061$ rad

v) θ° is $\theta \times 1°$ $= \theta \times \dfrac{\pi}{180}$ or 0.017θ rad.

Using $\qquad 2\pi = 360°$

so that $\qquad \dfrac{2\pi}{2\pi} = \dfrac{360°}{2\pi}$

i.e. $\qquad \boxed{1 \text{ rad} = \dfrac{180°}{\pi}.}$

This is the **conversion factor** to move from radians to degrees. A calculator check finds that 1 radian is slightly less than 60°, approximately 57.3° (3 sf).

■■■■■■■ **Examples 4.2** ■■■■■■■

Convert   i) 1.5 rad,   ii) 2.6 rad,   iii) $\dfrac{7\pi}{4}$ rad,   iv) θ rad to degrees.

i)   1.5 rad is $1.5 \times 1 = \dfrac{1.5 \times 180°}{\pi}$

$\qquad\qquad\qquad\qquad = 85.94°$ (4 sf).

ii)   2.6 rad is $2.6 \times 1 = \dfrac{2.6 \times 180°}{\pi}$

$\qquad\qquad\qquad\qquad = 148.97°$ (2 dp)

$\qquad\qquad\qquad\text{or } = 149.0°$ (3 sf).

iii) $\dfrac{7\pi}{4}$ rad is $\dfrac{7\pi}{4} \times 1 = \dfrac{7\pi}{4} \times \dfrac{180°}{\pi}$

$\qquad\qquad\qquad\qquad = 315°$

iv) θ is $\theta \times 1 = \theta \times \dfrac{180°}{\pi}$

$\qquad\qquad\qquad\qquad = 57.3\theta°$ (3 sf).

■■■■■■■ **EXERCISE 4.1** ■■■■■■■

Convert the following angles from degrees to radians, leaving each answer as a multiple of π.

1	10°	6	210°
2	45°	7	315°
3	60°	8	360°
4	150°	9	540°
5	180°	10	1080°

Convert the following angles from degrees to radians, giving your answer as a decimal (2 dp) using the usual calculator value of π.

11	25°	14	225.1°
12	73.2°	15	255.2°
13	167.4°	16	296.5°

17 307.75°

18 345°

19 500°

20 610°

Convert the following radian measures into degrees, giving your answer correct to 2 decimal places.

21 7.00

22 4.55

23 5.20

24 $\dfrac{2\pi}{3}$

25 1.25

26 13.6

27 $\dfrac{5\pi}{4}$

28 1.9

29 $\dfrac{5\pi}{3}$

30 6.15

Rotations

Most of our examples and exercises have been between 0° and 360°. There is no reason to stop there. We can keep going round and round by further revolutions using our earlier relationship that 1 rev = 360° = 2π rad.

All revolutions will be multiples of this relationship,

e.g.   2 rev = 2 × 2π = 4π rad

$5\frac{1}{2}$ rev = $5\frac{1}{2} \times 2\pi$ = 11π rad.

Commonly used angles

Some angles in degrees or radians occur more often than others. The following lists show these common ones in both degrees and radians.

0°		180°		360°
0		π		2π

	90°		270°	
	$\dfrac{\pi}{2}$		$\dfrac{3\pi}{2}$	

45°	135°	225°	315°
$\dfrac{\pi}{4}$	$\dfrac{3\pi}{4}$	$\dfrac{5\pi}{4}$	$\dfrac{7\pi}{4}$

60°	120°	240°	300°
$\dfrac{\pi}{3}$	$\dfrac{2\pi}{3}$	$\dfrac{4\pi}{3}$	$\dfrac{5\pi}{3}$

30°	150°	210°	330°
$\dfrac{\pi}{6}$	$\dfrac{5\pi}{6}$	$\dfrac{7\pi}{6}$	$\dfrac{11\pi}{6}$

Using the calculator

The shortened forms of sine, cosine and tangent appear on your scientific calculator as **sin**, **cos** and **tan**. We may use either degrees or radians if there is a facility to change the angular measurement.

███████ **Examples 4.3** ███████████████████████████

Use your calculator to find the values of
i) sin 53°,   ii) sin 147°,   iii) − sin 147°.

i)   The trigonometric operation of sin acts on the angle 53°. This affects how we use the calculator as the following order of operations shows:

$$\underline{53|} \quad \underline{\sin|}$$

to display 0.7986 correct to 4 decimal places.

ii)   sin 147° uses the order

$$\underline{147|} \quad \underline{\sin|}$$

to display 0.5446.

iii)   − sin 147° just extends the previous list of operations with a final use of the $\underline{+/_-|}$ button to give

$$\underline{147|} \quad \underline{\sin|} \quad \underline{+/_-|}$$

to display −0.5446.

███████ **Examples 4.4** ███████████████████████████

Use your calculator to find the values of
i) cos 31°,   ii) cos 209°,   iii) cos 335.5°.

i)   For cos 31° the order of operations is

$$\underline{31|} \quad \underline{\cos|}$$

to display 0.8572 correct to 4 decimal places.

ii)   cos 209° uses the order

$$\underline{209|} \quad \underline{\cos|}$$

to display −0.8746.

iii)   cos 335.5° uses the order

$$\underline{335.5|} \quad \underline{\cos|}$$

to display 0.9100.

███████ **Examples 4.5** ███████████████████████████

Use your calculator to find the values of
i) tan 82.75°,   ii) tan 115°,   iii) tan 297°.

i)   For tan 82.75° the order of operations is

$$\underline{82.75|} \quad \underline{\tan|}$$

to display 7.8606 correct to 4 decimal places.

ii) tan 115° uses the order

  115⌋ tan⌋

to display −2.1445.

iii) tan 297° uses the order

  297⌋ tan⌋

to display −1.9626.

If you use your calculator to find tan 90° or tan 270° you will find that it gives an error display. We will come back to these strange values when we look more closely at the tangent function.

███ **Examples 4.6** ███████████████████████████████

Use your calculator to find the values of
i) cos (−127°), ii) − cos 127°, iii) − cos (−127°).

The orders of operations for all these examples are similar, only differing by the positions of − signs.

 i) For cos (−127°) the order of operations is

  127⌋ ⁺/₋⌋ cos⌋

to display −0.6018 correct to 4 decimal places.

ii) − cos 127° puts the − sign in a different position after the cosine operation using the order

  127⌋ cos⌋ ⁺/₋⌋

to display 0.6018.

iii) − cos (−127°) extends the previous example, using the order

  127⌋ ⁺/₋⌋ cos⌋ ⁺/₋⌋

to display 0.6018.

███ **Examples 4.7** ███████████████████████████████

Use your calculator to find the values of
i) sin (−94°), ii) − sin (−94°), iii) tan (−315°).

 i) For sin (−94°) the order of operations is

  94⌋ ⁺/₋⌋ sin⌋

to display −0.9976.

ii) − sin (−94°) extends the previous example, using the order

  94⌋ ⁺/₋⌋ sin⌋ ⁺/₋⌋

to display 0.9976.

iii) tan (−315°) uses the order

  315⌋ ⁺/₋⌋ tan⌋

to display 1.

■■■■■ EXERCISE 4.2 ■■■■■

Use your calculator to find the values of

1	$\sin 33°$	16	$\sin 295°$
2	$\sin 127°$	17	$\tan 26°$
3	$\cos 149°$	18	$\cos(-151°)$
4	$\cos 37°$	19	$\cos 246°$
5	$\tan 98°$	20	$\tan 46°$
6	$\tan 65°$	21	$\tan 103°$
7	$\sin(-59°)$	22	$\sin 310.25°$
8	$\cos 173°$	23	$\tan(-213°)$
9	$\tan 115°$	24	$\cos 217°$
10	$\sin 222°$	25	$\sin 199°$
11	$\sin 180°$	26	$\tan 415°$
12	$\cos 293.75°$	27	$\cos(-340°)$
13	$\tan 59.5°$	28	$\sin 161°$
14	$\tan 302°$	29	$\sin 393°$
15	$\cos 341°$	30	$\cos 509°$

■■■■■ Examples 4.8 ■■■■■

Use your calculator to find values of

i) $\sin 3.025$,   ii) $\cos \dfrac{5\pi}{12}$,   iii) $\tan(-0.975)$

where the angles are in radians.

Before you start working out these trigonometric ratios make sure that your calculator is in radian mode. If your calculator only uses degrees you must use the conversion from p. 94.

i)   For $\sin 3.025$ the order of operations is

$$\boxed{3.025}\quad \boxed{\sin}$$

to display 0.1163 correct to 4 decimal places.

ii) $\cos \dfrac{5\pi}{12}$ uses the order

$$\boxed{5}\quad \boxed{\times}\quad \boxed{\pi}\quad \boxed{\div}\quad \boxed{12}\quad \boxed{=}\quad \boxed{\cos}$$

to display 0.2588.

iii) $\tan(-0.975)$ uses the order

$$\boxed{0.975}\quad \boxed{^{+}/_{-}}\quad \boxed{\tan}$$

to display -1.475.

■■■ EXERCISE 4.3 ■■■

This set of questions use radians in place of degrees.

Use your calculator to find the values of

1 $\sin 2$

2 $\cos 0.76$

3 $\sin 1.23$

4 $\tan 6.50$

5 $\tan \dfrac{3\pi}{4}$

6 $\sin(-1.54)$

7 $\cos 4.25$

8 $\sin \dfrac{5\pi}{3}$

9 $\cos\left(\dfrac{-5\pi}{6}\right)$

10 $\tan(-0.9)$

Sine and cosine curves

We may use the calculator to construct a table of values of sine from $0°$ to $360°$. The chosen angles are specimen ones because they generate a particular pattern of values for the sine.

θ	$0°$	$30°$	$60°$	$90°$	$120°$	$150°$	
$\sin\theta$	0	0.500	0.866	1.000	0.866	0.500	

θ	$180°$	$210°$	$240°$	$270°$	$300°$	$330°$	$360°$
$\sin\theta$	0	-0.500	-0.866	-1.000	-0.866	-0.500	0

There is an alternative method for finding sine that uses a circle in place of a calculator (see Fig. 4.6). We draw a circle with centre O and a unit radius. If you do this for yourself draw it accurately to a large scale so that it is easy to read the measurements.

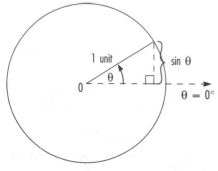

Fig. 4.6

The usual axis is $\theta = 0°$ horizontally to the right, in the same place as the x-axis. Values of θ are measured from this axis in an anticlockwise direction. For each value of θ the unit radius touches the circumference of the circle at a different point. The distance of each point above or below the axis gives the value of $\sin\theta$.

A circle has 4 **quadrants** (i.e. quarters) shown in Fig. 4.7.

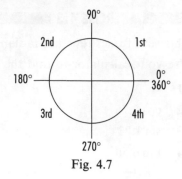

Fig. 4.7

Figs. 4.8 shows examples of sine, one from each quadrant in turn.

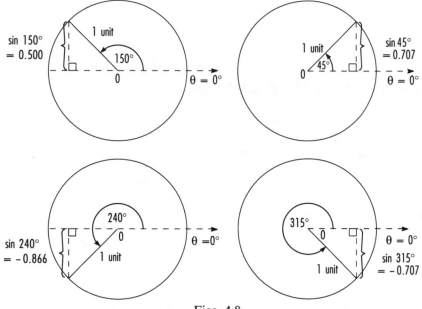

Figs. 4.8

The calculator method is quicker than the circle method. Whichever method you choose the values can be used to draw the graph of $y = \sin\theta$ shown in Fig. 4.9.

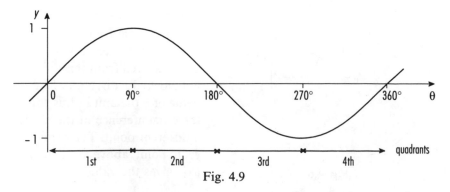

Fig. 4.9

The graph continues this pattern to the left and to the right of where we have drawn it, repeating itself every 360°. Each repetition is a **cycle**. The sine graph is important. You should know how to sketch it over one cycle.

In the first quadrant as the values of θ increase from 0° to 90° the values of sin θ increase from 0 to a maximum of 1. Moving into the second quadrant (90° to 180°) the values of sin θ decrease from 1 to 0. In the third quadrant (180° to 270°) that change continues from 0 down to −1. Finally in the fourth quadrant (270° to 360°) sin θ turns, increasing from −1 back to its starting value of 0. The shape of the curve in each quadrant is just a twisted repetition of the one from the first quadrant.

You can check that a tangent drawn to the curve at these maximum and minimum values is horizontal. Tangents in the first and fourth quadrants have positive gradients. Those in the second and third quadrants have negative ones.

We may repeat these techniques with small alterations to draw the cosine curve. Firstly we can construct a table of values of cosine from 0° to 360° using a calculator. The table below again shows some specimen values at intervals of 30°. They were chosen to create a pattern.

θ	0°	30°	60°	90°	120°	150°	
cos θ	1.000	0.866	0.500	0	−0.500	−0.866	

θ	180°	210°	240°	270°	300°	330°	360°
cos θ	−1.000	−0.866	−0.500	0	0.500	0.866	1.000

Again we can consider an alternative method for finding the cosine. It uses a circle with centre O and a unit radius.

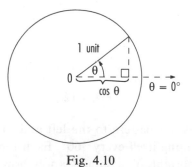

Fig. 4.10

The same axis θ = 0° is used and the points where the radius touches the circumference are important again. This time different distances are needed. We are interested in the horizontal distances along the axis θ = 0°, either to the left or right of O. These distances give the values of cos θ.

Figs. 4.11 shows specimen values of $\cos\theta$ using the circle method, one from each quadrant.

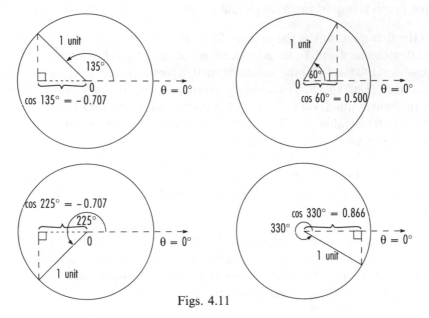

Figs. 4.11

You can choose either the calculator or circle method to work out the cosines. Again, the calculator will work out the cosine values quickly. Now we are able to draw the graph of $y=\cos\theta$ in Fig. 4.12.

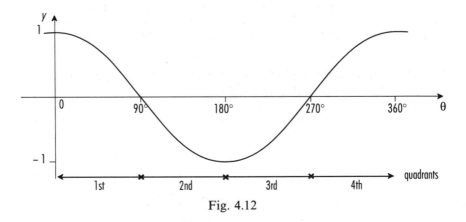

Fig. 4.12

The graph continues this pattern to the left and to the right of where we have drawn it, repeating itself every 360°. Each repetition is a **cycle**. The cosine graph is important. You should know how to sketch it over one cycle.

In the first quadrant as the values of θ increase from 0° to 90° the values of $\cos\theta$ decrease from a maximum of 1 to 0. Moving into the second quadrant (90° to 180°) the values of $\cos\theta$ continue to decrease from 0 to a minimum of -1. In the third quadrant ($180°$ to $270°$) the curve turns, increasing from -1 to 0. Finally in the fourth quadrant ($270°$ to $360°$) $\cos \theta$ returns to its starting value of 1. The shape of the curve in each quadrant is just a twisted repetition of the one from the first quadrant.

You can check that a tangent drawn to the curve at these maximum and minimum values is horizontal. Tangents in the first and second quadrants have negative gradients. Those in the third and fourth quadrants have positive ones.

If you study the graphs of $\sin \theta$ and $\cos \theta$ and attempt to sketch them you will get a "feel" for their shapes. There are a number of connections between them. We can mention a simple relationship here. **cos** stands for **co**mplementary **s**ine. Two angles are complementary if they add up to $90°$. Thus there is a connection between sine, cosine and $90°$. We may write the relationship as

$$\sin \theta = \cos (90° - \theta).$$

The next set of examples uses numerical substitution to demonstrate our new relation.

Examples 4.9

i)   If $\theta = 30°$ then $\sin 30° = 0.5000$
               and $\cos (90° - 30°) = \cos 60° \quad = 0.5000.$

ii)   If $\theta = 165°$ then $\sin 165° = 0.259$
               and $\cos (90° - 165°) = \cos (-75°) = 0.259.$

Also sine and cosine are related by

$$\cos \theta = \sin (90° - \theta).$$

The next set of examples uses substitution to demonstrate this variation of our new relation.

Examples 4.10

i)   If $\theta = 55°$ then $\cos 55° = 0.5736$
              and $\sin (90° - 55°) = \sin 35° \quad = 0.5736.$

ii)   If $\theta = 201°$ then $\cos 201° = -0.9336$
              and $\sin (90° - 201°) = \sin (-111°) = -0.9336.$

iii)   If $\theta = -24°$ then $\cos(-24°) = 0.9135$
             and $\sin (90° - -24°) = \sin (90° + 24°)$
                            $= \sin 114°$
                            $= 0.9135.$

We have mentioned already that each graph repeats itself every 360°. This means that the sine curve has the same shape from −360° to 0° as from 0° to 360°. Then that shape is repeated from 360° to 720°, and from 720° to 1080° and so on. The same rule applies to the cosine curve. In fact Figs. 4.13 show each graph over 3 cycles. A cycle can start at any point on the curve. It stops when the next similar position on the curve is reached. Two examples of cycles are drawn in Figs. 4.13.

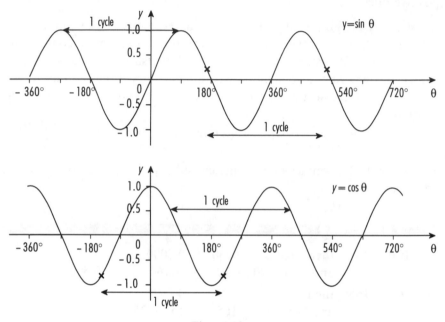

Figs. 4.13

Inverse sine and cosine

So far in this chapter we have always worked from the angle and found either the sine or cosine. What happens if we start with $\sin \theta$ and attempt to find the angle θ? What happens if we start with $\cos \theta$ and attempt to find θ? In other words we are using the opposite or inverse operations to those we have used already. $\sin^{-1}$ represents **inverse sin** and $\cos^{-1}$ represents **inverse cos**. $\sin^{-1}$ is the opposite operation to sin and $\cos^{-1}$ is the opposite operation to cos. Alternatives to $\sin^{-1}$ and $\cos^{-1}$ are **arcsin** and **arccos**.

First we will return to the calculator. Most calculators use the keys inv| and sin| or inv| and cos| with the shortened form $\sin^{-1}$ or $\cos^{-1}$ above the sin| and cos| buttons.

With sin| we input an **angle** and found the value of the **sine** of that angle.

With inv| sin| we input the **sine value** and find the **angle**. Similarly with inv| cos| we input the **cosine value** and find the **angle**. The next set of examples show the techniques.

▄▄▄▄▄▄ **Examples 4.11** ▄▄▄▄▄▄▄▄▄▄▄▄▄▄▄▄▄▄▄▄▄▄▄▄▄▄▄▄▄▄▄▄▄▄

Use your calculator to find the values (angles) of
i) $\sin^{-1} 0.5$,   ii) $\sin^{-1}(-0.6)$,
iii) $\cos^{-1} 0.75$,   iv) $\cos^{-1}(-0.866)$.

i)   For $\sin^{-1} 0.5$ the order of operations is

.5| inv| sin|

to display 30°.

ii) For $\sin^{-1}(-0.6)$ the order is

.6| +/_| inv| sin|

to display −36.87°.

iii) For $\cos^{-1} 0.75$ the order is

.75| inv| cos|

to display 41.41°.

iv) For $\cos^{-1}(-0.866)$ the order is

.866| +/_| inv| cos|

to display 150°.

We can check these answers by looking at the graphs of sine and cosine. An accurate check depends upon the accuracy and scale of your drawing. When we wanted to find the sine of 30° we started at 30° on the horizontal axis. We moved directly to the curve and then towards the vertical axis. The reading of 0.5000 from that vertical allowed us to write sin 30° = 0.5000. The next set of examples, for inverse sine, reverses the process. It starts with the value from the vertical axis and ends with angles on the horizontal axis.

▄▄▄▄▄▄ **Examples 4.12** ▄▄▄▄▄▄▄▄▄▄▄▄▄▄▄▄▄▄▄▄▄▄▄▄▄▄▄▄▄▄▄▄▄▄

Use your calculator and the sine graph between 0° and 360° to find the values (angles) of
i) $\sin^{-1} 0.5$,   ii) $\sin^{-1}(-0.6)$.

i)   We can look at the sine curve and draw a horizontal line through 0.5 on the vertical axis (Fig. 4.14). The line cuts the curve in 2 places, the first one being at $\theta = 30°$. You can see from the graph that it cuts the curve at 30° before the curve crosses the horizontal axis again, i.e. $180° - 30° = 150°$.

This means that there are two angles with a sine value of 0.5, i.e. $\sin^{-1} 0.5 = 30°$ and $150°$.

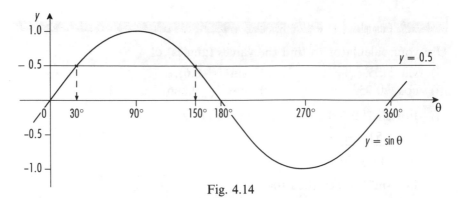

Fig. 4.14

ii) This time we draw a horizontal line through -0.6 on the vertical axis. Fig. 4.15 shows that it cuts the curve in 2 places between $0°$ and $360°$.

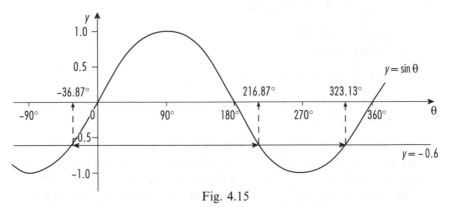

Fig. 4.15

The calculator's answer of $-36.87°$ is before our cycle of interest. By comparing positions with our cycle we see these are $36.87°$ after (addition) $180°$ and before (subtraction) $360°$,

i.e. $180° + 36.87° = 216.87°$

and $360° - 36.87° = 323.13°$

so that $\sin^{-1}(-0.6) = 216.87°$ and $323.13°$.

We can turn these into 2 rules. You should always use the calculator **and** a sketch of the sine curve rather than blindly following them.

$\sin^{-1}(\text{POSITIVE VALUE})$ = CALCULATOR DISPLAY and

$180°-$ CALCULATOR DISPLAY

$\sin^{-1}(\text{NEGATIVE VALUE})$ gives a negative angle.

Press the $+/-$ calculator button to get a new positive angle on the CALCULATOR DISPLAY so that the answers are

180° + CALCULATOR DISPLAY and

360° − CALCULATOR DISPLAY.

These ideas can be extended to other cycles. Because each cycle is 360° you just add 360° or 720° or . . . for further cycles. For earlier cycles you subtract the multiples of 360°.

Examples 4.13

Use your calculator and the cosine graph between 0° and 360° to find the values (angles) of

i) $\cos^{-1} 0.75$,   ii) $\cos^{-1}(-0.866)$.

i) We can look at the cosine curve and draw a horizontal line through 0.75 on the vertical axis (Fig. 4.16). The line cuts the curve in 2 places, the first one being at $\theta = 41.41°$. You can see from the graph that it cuts the curve at 41.41° before it reaches another maximum value, i.e. $360° - 41.41° = 318.59°$. This means that there are 2 angles with a cosine value of 0.75, i.e. $\cos^{-1} 0.75 = 41.41°$ and $318.59°$.

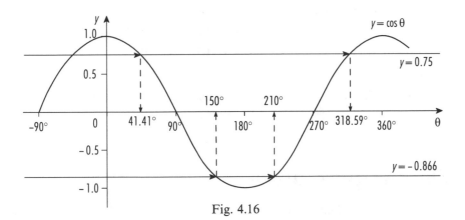

Fig. 4.16

ii) This time we draw the horizontal line through −0.866 on the vertical axis. Again it cuts the curve in 2 places, the first one being at $\theta = 150°$ after the first maximum value. You can see that it cuts the curve again at 150° before the next maximum value,

i.e. at $360° - 150° = 210°$,

so that $\cos^{-1}(-0.866) = 150°$ and $210°$.

We can turn these into 2 rules. You should always use the calculator **and** a sketch of the cosine curve rather than blindly following them.

$\cos^{-1}(\text{POSITIVE VALUE})$ = CALCULATOR DISPLAY and

360°− CALCULATOR DISPLAY

$\cos^{-1}(\text{NEGATIVE VALUE})$ = CALCULATOR DISPLAY and

360°− CALCULATOR DISPLAY.

These ideas can be extended to other cycles. Because each cycle is 360° you just add 360° or 720° or . . . for further cycles. For earlier cycles you subtract the multiples of 360°.

■■■■ EXERCISE 4.4 ■■■■■■■■■■■■■■■■■■

Find the angles, in the range 0° to 360°, for the following inverse trigonometric functions. (Use your graphs to check that your answers lie in the correct quadrants.)

1	$\sin^{-1} 0.866$	**11**	$\cos^{-1} 0.1326$
2	$\sin^{-1} 0.7071$	**12**	$\sin^{-1} 1$
3	$\cos^{-1} 0.5$	**13**	$\sin^{-1} 0$
4	$\cos^{-1} 0.236$	**14**	$\cos^{-1}(-1.372)$
5	$\cos^{-1} 0.7071$	**15**	$\cos^{-1}(-0.5)$
6	$\sin^{-1}(-0.5)$	**16**	$\sin^{-1}(-1)$
7	$\sin^{-1}(-0.866)$	**17**	$\sin^{-1}(4.09)$
8	$\sin^{-1} 0.5936$	**18**	$\cos^{-1} 0$
9	$\cos^{-1} 1$	**19**	$\cos^{-1}(-1)$
10	$\cos^{-1}(-0.893)$	**20**	$\sin^{-1}(-0.7071)$

The tangent curve

We may use the calculator to construct a table of values for tangent from 0° to 360°. Strange things happen to the tangent curve around 90° and 270°. Because of this more table values are included.

θ	0°	15°	30°	45°	60°	75°	90°
$\tan\theta$	0	0.268	0.577	1.000	1.732	3.732	∞

θ		105°	120°	135°	150°	165°	180°
$\tan\theta$		−3.732	−1.732	−1.000	−0.577	−0.268	0

θ	195°	210°	225°	240°	255°	270°
tan θ	0.268	0.577	1.000	1.732	3.732	∞

θ	285°	300°	315°	330°	345°	360°
tan θ	−3.732	−1.732	−1.000	−0.577	−0.268	0

∞ is the symbol for **infinity**, i.e. a very large number. It is possible that your calculator may display -E- instead of the symbol for tan 90° and tan 270°. This is because it cannot cope with the calculation. The problem arises because $\tan 90° = \dfrac{\sin 90°}{\cos 90°}$ and $\tan 270° = \dfrac{\sin 270°}{\cos 270°}$. Remember that both cos 90° and cos 270° are 0, and division by 0 is not allowed in Mathematics.

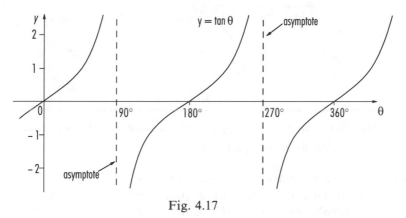

Fig. 4.17

The graph of $y = \tan θ$ is drawn in Fig. 4.17. It continues this pattern to the left and to the right of where we have drawn it. It repeats itself every 180°. You can see the same basic shape from the first quadrant (0° to 90°) is re-drawn in a different position for the second quadrant. The pattern is then repeated. The ¦ lines are not part of the tangent graph, but are included as a guide. They are **asymptotes**. A graph approaches an asymptote but never quite touches it. Unlike the graphs of sine and cosine, tangent is discontinuous at 90° and 270°, shown by the breaks in the curve.

Inverse tangent

We start with tan θ and attempt to find the angle θ using the inv⌋ and tan⌋ calculator keys. The shortened form $\tan^{-1}$ appears above the tan⌋ key. The alternative to **tan⁻¹** is **arctan**. Finding the angle from the tangent value is similar to the method used for $\sin^{-1}$ and $\cos^{-1}$.

███████ **Examples 4.14** ███████

Use your calculator and the graph of $\tan\theta$ between $0°$ and $360°$ to find the values of   i) $\tan^{-1}0.65$ ii) $\tan^{-1}(-1.29)$.

i)   We can look at the tangent curve and draw a horizontal line through 0.65 on the vertical axis. The line cuts the curve in 2 places, the first one being at $33°$ (2 sf). You can see from Fig 4.18 that it cuts the curve at $33°$ after the curve next crosses the horizontal axis,

i.e. at $180° + 33° = 213°$.

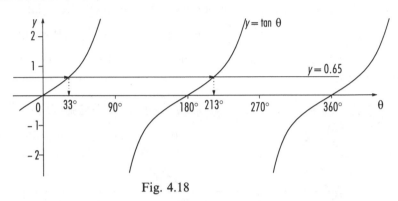

Fig. 4.18

This means there are 2 angles with a tangent value of 0.65,

i.e. $\tan^{-1}0.65 = 33°$ and $213°$.

ii)   This time we draw a horizontal line through -1.29 on the vertical axis. It cuts the curve in 2 places between $0°$ and $360°$. The calculator's angle of $-52.22°$ is not in this range. By comparing the position of $-52.22°$, in Fig. 4.19, with those of our cycle we have $52.22°$ before $180°$ and before $360°$,

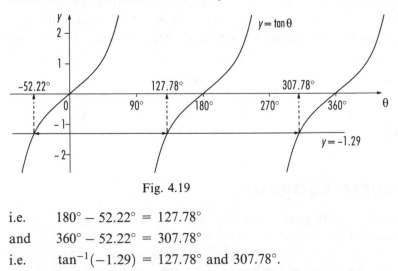

Fig. 4.19

i.e. $180° - 52.22° = 127.78°$

and $360° - 52.22° = 307.78°$

i.e. $\tan^{-1}(-1.29) = 127.78°$ and $307.78°$.

We can turn these into 2 rules. You should always use the calculator **and** a sketch of the tangent curve rather than blindly following them.

$\tan^{-1}$(POSITIVE VALUE) = CALCULATOR DISPLAY and

180°+ CALCULATOR DISPLAY

$\tan^{-1}$(NEGATIVE VALUE) gives a negative angle on the display.

Press $^{+/-}$ to get a new, positive angle, on the CALCULATOR DISPLAY. Now the answers are

180°− CALCULATOR DISPLAY and

360°− CALCULATOR DISPLAY.

These ideas can be extended to other cycles. Because each cycle is 360° you just add 360° or 720° or . . . for further cycles. For earlier cycles you subtract the multiples of 360°.

■■■ EXERCISE 4.5 ■■■■■■■■■■■■■■■■■

Find the angles for the following inverse tangents. Use your calculator and graph in the range 0° to 360°.

1 $\tan^{-1} 0.5$

2 $\tan^{-1} 1.5$

3 $\tan^{-1}(-2.45)$

4 $\tan^{-1} 6.28$

5 $\tan^{-1} 1$

6 $\tan^{-1}(-1.732)$

7 $\tan^{-1}(-1.5)$

8 $\tan^{-1}(3.25)$

9 $\tan^{-1}(-3.25)$

10 $\tan^{-1}(-8.35)$

Summary of sine, cosine and tangent values

We can bring together sine, cosine and tangent values in a 4 quadrant diagram between 0° and 360° (Fig. 4.20). We can compare the axes with the usual graphical ones of

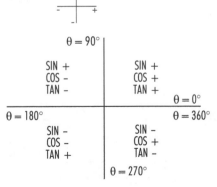

Fig. 4.20

Remember the circle has a rotating positive radius.

All the trigonometric ratios are positive in the first quadrant. In turn they are each positive in another quadrant. We can highlight this in Fig. 4.21.

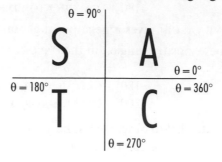

Fig. 4.21

Right-angled triangles

Every triangle has 3 sides and 3 angles. The sum of those 3 angles is always 180°. In a right-angled triangle (Fig 4.22) one of the angles is 90°.

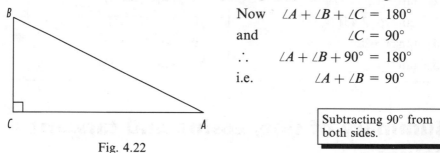

Now $\angle A + \angle B + \angle C = 180°$

and $\angle C = 90°$

∴ $\angle A + \angle B + 90° = 180°$

i.e. $\angle A + \angle B = 90°$

Fig. 4.22

> Subtracting 90° from both sides.

Let us choose 1 of the angles other than the right-angle, say $\angle A$, in Fig. 4.23. The shortest side is always opposite the smallest angle. The longest side is always opposite the largest angle. In this case the largest angle is $\angle C = 90°$. In a right-angled triangle the longest side, opposite the right-angle, is the **hypotenuse**. The side opposite $\angle A$ (i.e. length BC) is called the **opposite**. The remaining side is next to $\angle A$ (i.e. length AC) and is called the **adjacent**.

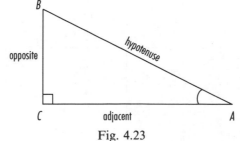

Fig. 4.23

Suppose we had chosen $\angle B$ instead of $\angle A$. The hypotenuse remains as before. The opposite and adjacent sides are interchanged because they refer to $\angle B$ rather than $\angle A$.

For any right-angled triangle (Fig. 4.23) we define

sine by $\qquad \sin A = \dfrac{\text{length of the \textbf{opposite} side}}{\text{length of the \textbf{hypotenuse}}}$,

cosine $\qquad \cos A = \dfrac{\text{length of the \textbf{adjacent} side}}{\text{length of the \textbf{hypotenuse}}}$,

tangent $\qquad \tan A = \dfrac{\text{length of the \textbf{opposite} side}}{\text{length of the \textbf{adjacent}side}}$.

We simplify these formulae to read

$$\sin A = \frac{\textbf{opp}}{\textbf{hyp}}, \qquad \cos A = \frac{\textbf{adj}}{\textbf{hyp}}, \qquad \tan A = \frac{\textbf{opp}}{\textbf{adj}}.$$

These are the ratios of 2 chosen sides. Suppose we have a right-angled triangle, large or small with, say, an angle of $43°$. The ratios for $\sin 43°$ will be the same for both triangles. Exactly the same ideas apply to cosine and tangent.

Earlier in this chapter we introduced the radian as an alternative angular measure. You may not have quite adjusted to this alternative yet. For this section we concentrate upon using degrees.

The following set of examples and exercises are based upon the 3 formulae. We find angles and sides in turn.

▰▰▰ Examples 4.15 ▰▰▰

In each case we find the angle with the marked letter. Notice the method is similar based upon either sine or cosine or tangent.

i) In Fig. 4.24 we have the hypotenuse and the side opposite $\angle A$.

Using $\qquad \sin = \dfrac{\text{opp}}{\text{hyp}}$

we substitute to get

$$\sin A = \frac{3.6}{4.95} = 0.7\overline{2}$$

i.e. $\qquad \angle A = 46.7°$ (1 dp).

The calculator sequence is

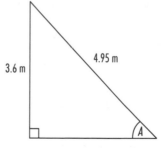

Fig. 4.24

$\underline{3.6|} \quad \underline{\div|} \quad \underline{4.95|} \quad \underline{=|} \quad$ to display $0.7\overline{2}$

$\underline{\text{inv}|} \quad \underline{\sin|} \qquad\qquad$ to display $46.658\ldots$

ii)   In Fig. 4.25 we have the hypotenuse and the side adjacent to $\angle B$.

Using $\cos = \dfrac{\text{adj}}{\text{hyp}}$

we substitute to get

$$\cos B = \dfrac{45}{97} = 0.4639\ldots$$

i.e. $\angle B = 62.4°$ (1 dp).

The calculator sequence is

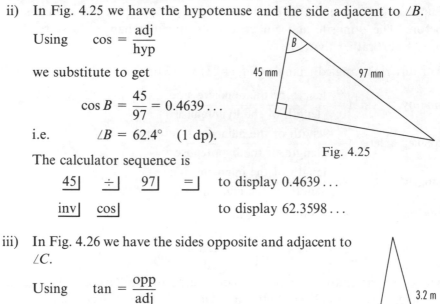

Fig. 4.25

$\underline{45|}$ $\underline{\div|}$ $\underline{97|}$ $\underline{=|}$ to display 0.4639...

$\underline{\text{inv}|}$ $\underline{\cos|}$ to display 62.3598...

iii) In Fig. 4.26 we have the sides opposite and adjacent to $\angle C$.

Using $\tan = \dfrac{\text{opp}}{\text{adj}}$

we substitute to get

$$\tan C = \dfrac{3.2}{1.5} = 2.1\bar{3}$$

$$\angle C = 64.9°\quad\text{(1 dp)}.$$

The calculator sequence is

Fig. 4.26

$\underline{3.2|}$ $\underline{\div|}$ $\underline{1.5|}$ $\underline{=|}$ to display $2.1\bar{3}$

$\underline{\text{inv}|}$ $\underline{\tan|}$ to display 64.885...

EXERCISE 4.6

In each question use sine or cosine or tangent to find the unknown, labelled, angle.

1

3

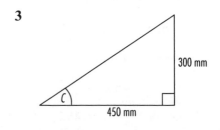

2

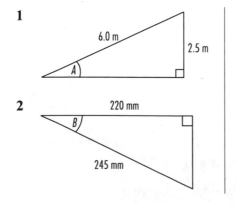

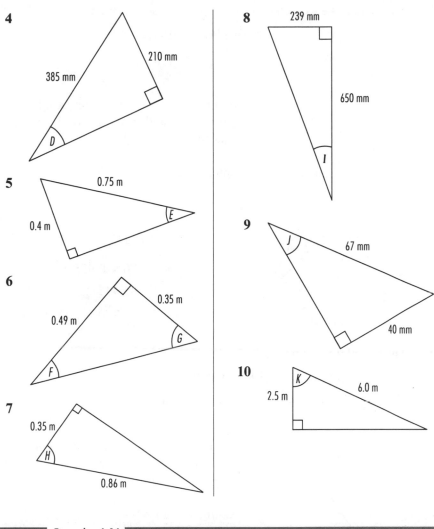

4 210 mm, 385 mm, D

5 0.75 m, 0.4 m, E

6 0.35 m, 0.49 m, G, F

7 0.35 m, H, 0.86 m

8 239 mm, 650 mm, I

9 J, 67 mm, 40 mm

10 K, 6.0 m, 2.5 m

▓▓▓ **Examples 4.16** ▓▓▓

In each case we find the side with the marked letter. Notice the method is similar based upon either sine or cosine or tangent.

i) In Fig. 4.27 we have an angle, the hypotenuse and need to find the side opposite the angle.

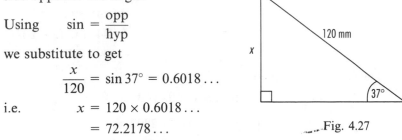

Using $\quad \sin = \dfrac{\text{opp}}{\text{hyp}}$

we substitute to get

$$\frac{x}{120} = \sin 37° = 0.6018\ldots$$

i.e. $\qquad x = 120 \times 0.6018\ldots$

$\qquad\qquad = 72.2178\ldots$

$\qquad\qquad = 72.2\,\text{mm (3 sf) is the opposite side.}$

Fig. 4.27

The calculator sequence is

37°| sin| to display 0.6018 ...

×| 120| =| to display 72.2178 ...

ii)

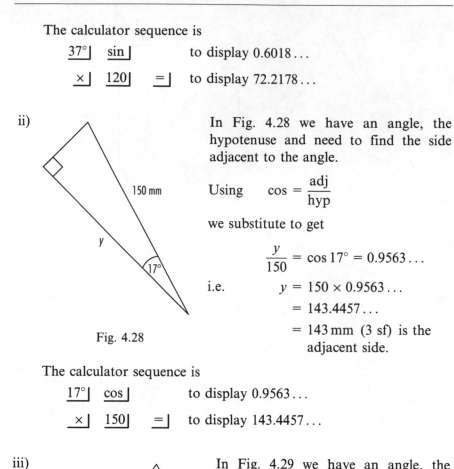

Fig. 4.28

In Fig. 4.28 we have an angle, the hypotenuse and need to find the side adjacent to the angle.

Using $\cos = \dfrac{\text{adj}}{\text{hyp}}$

we substitute to get

$$\frac{y}{150} = \cos 17° = 0.9563\ldots$$

i.e. $y = 150 \times 0.9563\ldots$

$$= 143.4457\ldots$$

$$= 143\,\text{mm (3 sf) is the}$$
adjacent side.

The calculator sequence is

17°| cos| to display 0.9563 ...

×| 150| =| to display 143.4457 ...

iii)

Fig. 4.29

In Fig. 4.29 we have an angle, the adjacent side and need to find the side opposite the angle.

Using $\tan = \dfrac{\text{opp}}{\text{adj}}$

we substitute to get

$$\frac{z}{95} = \tan 25° = 0.4663\ldots$$

i.e. $z = 95 \times 0.4663\ldots$

$$= 44.2992\ldots$$

$$= 44\,\text{mm (2 sf) is the}$$
adjacent side.

The calculator sequence is

25°| tan| to display 0.4663 ...

×| 95| =| to display 44.2992 ...

EXERCISE 4.7

In each question use sine or cosine or tangent to find the unknown, labelled, side.

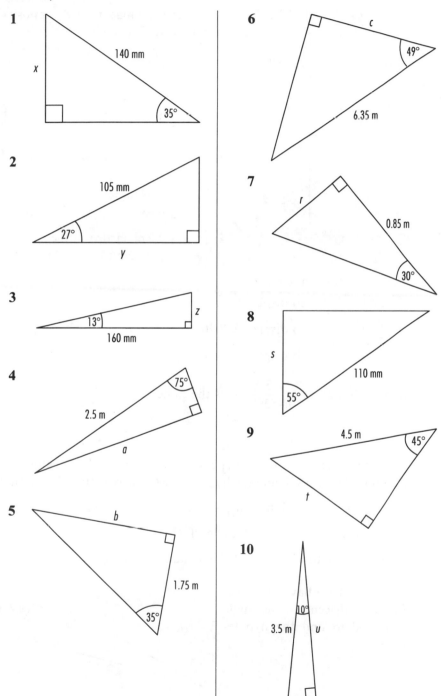

1   140 mm   x   35°

2   105 mm   27°   y

3   13°   z   160 mm

4   75°   2.5 m   a

5   b   1.75 m   35°

6   c   49°   6.35 m

7   r   0.85 m   30°

8   s   55°   110 mm

9   4.5 m   45°   t

10   10°   3.5 m   u

▬▬▬▬ **Example 4.17** ▬▬▬▬▬▬▬▬▬▬▬▬▬▬▬▬▬▬▬▬▬▬

In each case we find the side with the marked letter. Notice the method is similar based upon either sine or cosine or tangent.

i)

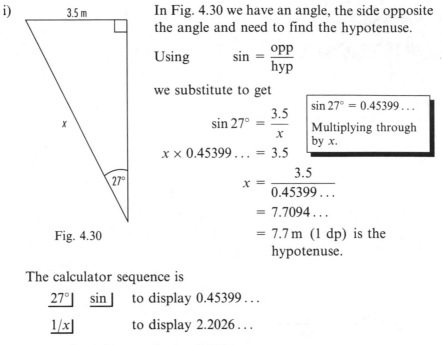

Fig. 4.30

In Fig. 4.30 we have an angle, the side opposite the angle and need to find the hypotenuse.

Using $\qquad \sin = \dfrac{\text{opp}}{\text{hyp}}$

we substitute to get

$$\sin 27° = \frac{3.5}{x}$$

> $\sin 27° = 0.45399 \ldots$
>
> Multiplying through by x.

$$x \times 0.45399 \ldots = 3.5$$

$$x = \frac{3.5}{0.45399 \ldots}$$

$$= 7.7094 \ldots$$

$$= 7.7 \text{ m} \ (1 \text{ dp}) \text{ is the hypotenuse.}$$

The calculator sequence is

27°| $\quad$ sin | $\quad$ to display 0.45399 ...

1/x| $\qquad\qquad$ to display 2.2026 ...

$\quad \times$| $\quad$ 3.5| $\quad$ to display 7.7094 ...

In our calculation $\dfrac{3.5}{0.45399 \ldots}$ is the same as $\dfrac{1 \times 3.5}{0.45399 \ldots}$

$$= \frac{1}{0.45399 \ldots} \times 3.5.$$

After this calculation we can check our answer. We have values for the opposite side (3.5) and the hypotenuse (approximately 7.7). Using them, $\dfrac{3.5}{7.7} = 0.\overline{45}$, and inv| sin| on the calculator give $27.03 \ldots °$. This is very close to our original angle of 27°.

We can use both ideas in the other example parts.

ii) In Fig. 4.31 we have an angle, the side adjacent to the angle and need to find the hypotenuse.

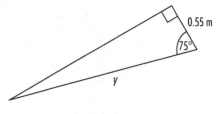

Fig. 4.31

Using $\qquad \cos = \dfrac{\text{adj}}{\text{hyp}}$

we substitute to get

$$\cos 75° = \frac{0.55}{y}$$

$$y \times 0.2588\ldots = 0.55$$

$$y = \frac{0.55}{0.2588\ldots}$$

$$= 2.125\ldots$$

$$= 2.1\,\text{m} \ (1\ \text{dp}) \ \text{is the hypotenuse.}$$

> $\cos 75° = 0.2588\ldots$
>
> Multiplying through by y.

The calculator sequence is

$\underline{75°	}$	$\underline{\cos	}$	to display $0.2588\ldots$
$\underline{1/x	}$		to display $3.8637\ldots$
$\underline{\times	}$	$\underline{0.55	}$	to display $2.125\ldots$

iii) In Fig. 4.32 we have an angle, the side opposite the angle and need to find the adjacent side.

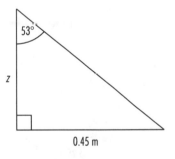

0.45 m

Fig. 4.32

Using $\qquad \tan = \dfrac{\text{opp}}{\text{adj}}$

we substitute to get

$$\tan 53° = \frac{0.45}{z}$$

$$z \times 1.327\ldots = 0.45$$

$$z = \frac{0.45}{1.327\ldots}$$

$$= 0.339\ldots$$

$$= 0.34\,\text{m} \ (2\ \text{dp}) \ \text{is the adjacent side.}$$

> $\tan 53° = 1.327\ldots$
>
> Multiplying through by z.

The calculator sequence is

$\underline{53°	}$	$\underline{\tan	}$	to display $1.327\ldots$
$\underline{1/x	}$		to display $0.75355\ldots$
$\underline{\times	}$	$\underline{0.45	}$	to display $0.339\ldots$

EXERCISE 4.8

In each question use sine or cosine or tangent to find the unknown, labelled, side.

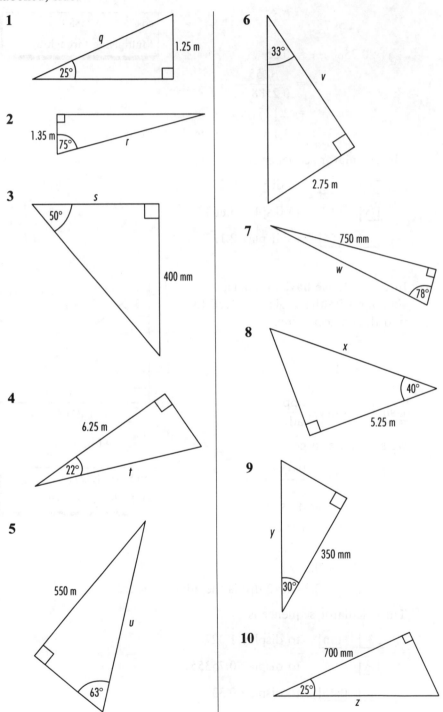

1

q

1.25 m

25°

2

1.35 m

75°

r

3

s

50°

400 mm

4

6.25 m

22°

t

5

550 m

u

63°

6

33°

v

2.75 m

7

750 mm

w

78°

8

x

40°

5.25 m

9

y

350 mm

30°

10

700 mm

25°

z

■■■■■■■ **Example 4.18** ■■■■■■■■■■■■■■■■■■■■■■■■■■■■■■■■■■■■

In this example we are going to **solve** a triangle, i.e. we are going to find all
the angles and sides. There are many method variations based upon
Examples 4.15, 4.16, 4.17 and Pythagoras' theorem.

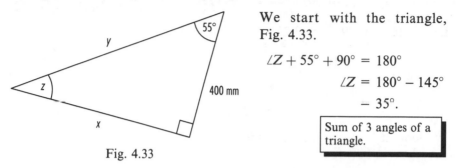

We start with the triangle,
Fig. 4.33.

$$\angle Z + 55° + 90° = 180°$$
$$\angle Z = 180° - 145°$$
$$- 35°.$$

Sum of 3 angles of a triangle.

Fig. 4.33

We can proceed in various ways. Our method is based on only the original
information, though sometimes this may *not* be possible. It avoids
compounding any errors we may make.

Now $\dfrac{x}{400} = \tan 55° = 1.428\ldots$

i.e. $x = 400 \times 1.428\ldots$
            $= 571.2592\ldots$
            $= 571\,\text{mm}$ (3 sf).

Also $\dfrac{400}{y} = \cos 55° = 0.573\ldots$

i.e. $y = \dfrac{400}{0.573\ldots}$
            $= 697.37872\ldots$
            $= 697\,\text{mm}$ (3 sf).

We can make a check of our answers before rounding using Pythagoras'
theorem, i.e.

$400^2 + 571.2592\ldots^2 = 697.3787\ldots^2$.

■■■■■■■ **Example 4.19** ■■■■■■■■■■■■■■■■■■■■■■■■■■■■■■■■■■■■

In Fig. 4.34 we have a horizontal
beam, *BC*, of length 3.25 m carrying
a load at *C*. It is supported by a rigid
member, *AC*, inclined at 70° to the
vertical wall *AB*. We are going to
find the length of this support.

We have an angle, the opposite side
and need to find the hypotenuse.

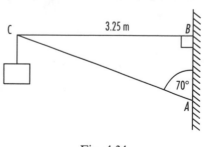

Fig. 4.34

Using $\sin = \dfrac{\text{opp}}{\text{hyp}}$

we substitute to get

$$\sin 70° = \frac{3.25}{AC}$$

i.e. $AC \times 0.939\ldots = 3.25$

i.e. $AC = \dfrac{3.25}{0.939\ldots}$

$$= 3.458\ldots$$

$$= 3.46\,\text{m} \ (2\ \text{dp}) \text{ is the length of the support.}$$

We look at further applications in Chapter 5.

EXERCISE 4.9

1
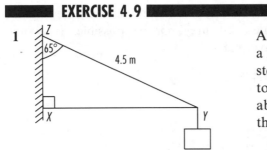
A horizontal beam, XY, carries a load at Y. It is supported by a steel cable, YZ, inclined at 65° to a vertical wall ZX. How high above the level of the beam is the cable attached to the wall?

2 The diagram shows a flat piece of glass. It is the front quarter light for a new battery operated town car. The two perpendicular sides are of lengths 120 mm and 300 mm. Find the unknown angles within the triangle.

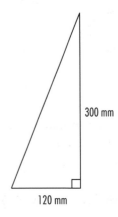

3
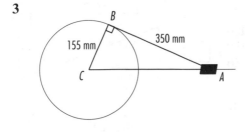
In this crank mechanism the length of the crank, BC, is 155 mm. The length of the connecting rod, AB, is 350 mm. When the rod is perpendicular to the crank, as shown, what is the inclination of AB to the horizontal?

4 The diagram shows a simplified roof truss. The supporting members are removed for clarity in this calculation.

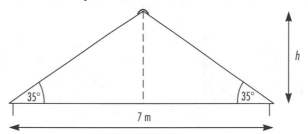

The span is 7 m and the inclination of the roof above the horizontal is 35°. Find the height of the ridge, *h*, above that horizontal.

5 *AB* is a horizontal beam of length 4.5 m carrying a load at *B*. It is bolted to a vertical wall at *A*. The support *CD* is inclined at 62.5° to the wall. Given that the ratio *AC* : *CB* is 2:1 find the length of *CD*.

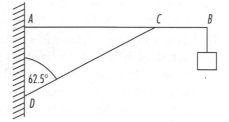

6 On a road a 10% gradient means the road rises for 1 m for every 10 m along that road. Draw a diagram showing this statement. What is the inclination of the road to the horizontal?
Suppose you double the angle of inclination. The percentage gradient is *not* doubled exactly. What is the new percentage?

7

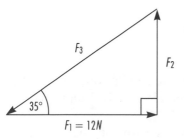

3 forces, F_1, F_2 and F_3, represented in sequence by the sides of the triangle are in equilibrium. The sizes of the forces are proportional to the lengths of the triangle sides. $F_1 = 12\,\text{N}$ and is perpendicular to F_2. F_3 is inclined at 35° to F_1. Find the values of F_2 and F_3.

8 A light inextensible string of length 1.35 m hangs vertically carrying a mass. The mass is pulled to one side a horizontal distance of 0.25 m. What is the height of the mass above its original position? What is the angle of inclination of the string to the vertical?

0.25 m

9

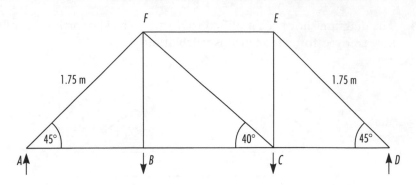

A light framework *ABCDEF* is supported at *A* and *D* while carrying loads at *B* and *C*. *AF* = *DE* = 1.75 m and both rods are inclined at 45° to the horizontal. *CF* is inclined at 40° to the horizontal. Find the lengths of *AB*, *BF* and *AD*.

10 The diagram shows a vertical aerial of height 11 m fixed at its base *E*. It is supported by 4 thin steel cables. The lengths of 2 cables have been calculated and removed from the diagram for clarity. You need to find the lengths of cables *AB* and *CD*. *B* is 1 m from the top and *D* is 3 m from the top of the aerial. *AB* is inclined at 65° to the horizontal. *CD* is inclined at 35° to the aerial. Write down the lengths of *BE* and *DE*. Find the length of each cable.

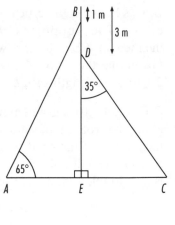

There are two interesting triangles that we should look at. The first one comes from bisecting an equilateral triangle of side 2 units. This creates the 30°, 60°, 90° triangle in Fig. 4.35

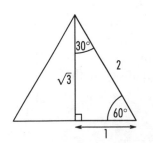

Fig. 4.35

We find the length of the bisector using Pythagoras' theorem. Notice we write it as $\sqrt{3}$ rather than as a decimal. The $\sqrt{}$ of $\sqrt{3}$ means this length is in surd form.

Then

$$\sin 30° = \frac{1}{2} = \cos 60°,$$

$$\sin 60° = \frac{\sqrt{3}}{2} = \cos 30°,$$

$$\tan 30° = \frac{1}{\sqrt{3}},$$

$$\tan 60° = \sqrt{3}.$$

The second one comes from bisecting a square of side 1 unit along a diagonal. This creates the 45°, 45°, 90° (isosceles) triangle in Fig. 4.36. We find the length of the diagonal using Pythagoras' theorem. Again we write it in surd form as $\sqrt{2}$.

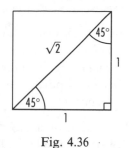

Then $\sin 45° = \dfrac{1}{\sqrt{2}} = \cos 45°,$

$\tan 45° = 1.$

Fig. 4.36

 ASSIGNMENT

We can return to our pulley problem and the belt's length around them. Fig. 4.37 shows both pulleys with 4 important points labelled. There is a line of symmetry joining their centres. Remember the benefits of a line of symmetry. In this case the top half is a mirror image, about the line of

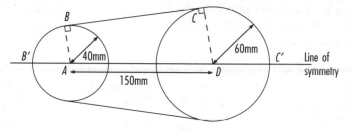

Fig. 4.37

symmetry, of the bottom half. It means we can look simply at the upper half and just double our answers. *ABCD* is an awkward shape. We can simplify it (Fig. 4.38) into a rectangle and a right-angled triangle.

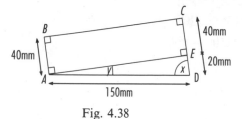

Fig. 4.38

We can now start to calculate some lengths and angles.

$AE = BC$, part of the belt.

In triangle *ADE*,

$$\cos x = \frac{20}{150} = 0.1333$$

$\therefore \qquad x = 82.34°$

and $\qquad y = 90° - 82.34° = 7.66°.$

Now we look at the curved portions of the belt, i.e. the arc lengths.

Fig. 4.39 shows the arc *PQ* of length *S*. This is the shorter arc *PQ*, called the **minor arc**. The longer one on the left of this particular circle is called the major arc.

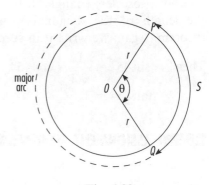

Fig. 4.39

The length of arc *S* depends on the size of the angle θ. The greater the angle θ then the longer the arc *PQ*. This will continue until one complete revolution with an angle of 2π radians. Then the arc has reached its maximum length in the form of the circumference 2π*r*. The formula for arc length, *S*, is $S = r\theta$ where *r* is the radius of the circle and θ is the angle in radians subtended at the centre of the circle.

For the larger pulley the upper belt section turns around *C* to *C'*. This is an angle of $180° - 82.34° = 97.66°$. It converts to approximately 1.70 radians. Arc length $= 60 \times 1.70 = 102.0$ mm.

For the smaller pulley the upper belt section turns around *B* to *B'*. This is an angle of $180° - (90° + 7.66°) = 82.34°$. It converts to approximately 1.44 radians. Arc length $= 40 \times 1.44 = 57.6$ mm.

Returning to triangle *ADE* we can calculate the length *AE* in one of many ways. By Pythagoras' theorem

$$AE^2 = 150^2 - 20^2$$
$$= 22500 - 400$$
$$AE = \sqrt{22100}$$

i.e. $AE = 148.7$ mm.

Now we have the 3 upper sections of the pulley belt which we can add together.

Upper section $= BB' + BC + CC'$.

The complete length is found by doubling this value.

$$\text{Belt length} = 2(57.6 + 148.7 + 102.0)$$
$$= 2 \times 308.3$$
$$= 616.6 \text{ mm}$$
$$= 617 \text{ mm} \qquad (3 \text{ sf}).$$

You might like to extend this Assignment. Calculate the area of metal required to make a guard for the belt and pulley system. You have to decide on the design, including the necessary overlap for safety.

The reciprocal ratios

These are 3 more **trigonometric ratios** (**trigonometric operations** or **functions**) that are true for all angles. They are **cosecant** (cosec), **secant** (sec) and **cotangent** (cot). We relate them to sine, cosine and tangent by the formulae

$$\text{cosec}\,\theta = \frac{1}{\sin\theta}, \quad \sec\theta = \frac{1}{\cos\theta}, \quad \cot\theta = \frac{1}{\tan\theta}.$$

We call them **reciprocal ratios** because they are defined using "**1 over**". Most calculators have only the keys for sine, cosine and tangent. We use these keys together with the reciprocal key, $1/x$.

━━━━━ **Examples 4.20** ━━━━━

Using a calculator find the values of

i) cosec 40°,   ii) sec 215°,   iii) cot x where $x = 2.5$ rads.

We can use either degrees or radians for any of our trigonometric functions.

i) Using our formula we know that cosec $40° = \dfrac{1}{\sin 40°}$.

   We use the calculator as

           40°| sin | to display 0.64278 . . .

   and   1/x| to display an answer of 1.556 (3 dp).

ii) Using our formula we know that sec $215° = \dfrac{1}{\cos 215°}$.

   We use the calculator as

          215°| cos| to display −0.81915 . . .

   and   1/x| to display an answer of −1.221 (3 dp).

ii) Using our formula we know that cot $2.5 = \dfrac{1}{\tan 2.5}$

   We use the calculator with radians as

          2.5| tan| to display −0.74702 . . .

   and   1/x| to display an answer of −1.339 (3 dp).

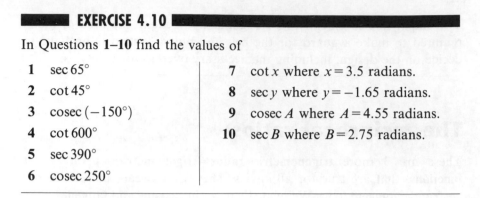

EXERCISE 4.10

In Questions **1–10** find the values of

1 $\sec 65°$

2 $\cot 45°$

3 $\operatorname{cosec}(-150°)$

4 $\cot 600°$

5 $\sec 390°$

6 $\operatorname{cosec} 250°$

7 $\cot x$ where $x = 3.5$ radians.

8 $\sec y$ where $y = -1.65$ radians.

9 $\operatorname{cosec} A$ where $A = 4.55$ radians.

10 $\sec B$ where $B = 2.75$ radians.

Some identities

An **identity** is a relationship that is **always true**. There are no exceptions.

For the first identity we will use a right-angled triangle and Pythagoras' theorem. You will remember that

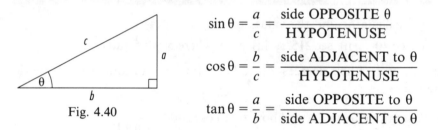

Fig. 4.40

$$\sin \theta = \frac{a}{c} = \frac{\text{side OPPOSITE } \theta}{\text{HYPOTENUSE}}$$

$$\cos \theta = \frac{b}{c} = \frac{\text{side ADJACENT to } \theta}{\text{HYPOTENUSE}}$$

$$\tan \theta = \frac{a}{b} = \frac{\text{side OPPOSITE to } \theta}{\text{side ADJACENT to } \theta}$$

Also the **hypotenuse** is the longest side of a right-angled triangle. It is opposite the right-angle, the largest angle.

Applying Pythagoras' theorem we have

$$a^2 + b^2 = c^2$$

i.e.

$$\frac{a^2}{c^2} + \frac{b^2}{c^2} = \frac{c^2}{c^2}$$

> Dividing by c^2.

i.e.

$$\left(\frac{a}{c}\right)^2 + \left(\frac{b}{c}\right)^2 = 1$$

then $(\sin \theta)^2 + (\cos \theta)^2 = 1$

i.e. $\sin^2 \theta + \cos^2 \theta = 1.$

Because this is an identity it is true for all angles and strictly we should use $\equiv$ instead of $=$.

You will see the slight re-arrangement from $(\sin\theta)^2$ to $\sin^2\theta$. These forms mean the same. $\sin^2\theta$ is the neater way. The same applies to $(\cos\theta)^2$ and $\cos^2\theta$. We can check out the order of operations using a calculator in the next set of examples. To find $\sin^2\theta$ we input θ, press the $\underline{\sin}$ key and then square the display to give the answer.

We have worked out (**derived**) the trigonometric identity using an acute angle in a right-angled triangle. Examples 4.21 show the identity working for a selection of other angles.

██████ **Examples 4.21** ████████████████████████████████

Check (**verify**) that the identity $\sin^2\theta + \cos^2\theta = 1$ is true for

i) $\theta = 150°$, ii) $\theta = 231°$, iii) $\theta = 327.2°$, iv) $\theta = 450°$, v) $\theta = 4.6$ radians.

We start with the left-hand side (LHS) of the identity and aim to finish with the value of 1 on the right.

i) $\theta = 150°$

$$\begin{aligned} \sin^2 150° + \cos^2 150° &= (0.5)^2 + (-0.866)^2 \\ &= 0.25 + 0.75 \\ &= 1. \end{aligned}$$

ii) $\theta = 231°$

$$\begin{aligned} \sin^2 231° + \cos^2 231° &= (-0.7771)^2 + (-0.6293)^2 \\ &= 0.6040 + 0.3960 \\ &= 1. \end{aligned}$$

iii) $\theta = 327.2°$

$$\begin{aligned} \sin^2 327.2° + \cos^2 327.2° &= (-0.5417)^2 + (0.8406)^2 \\ &= 0.2934 + 0.7066 \\ &= 1. \end{aligned}$$

iv) $\theta = 450°$

$$\begin{aligned} \sin^2 450° + \cos^2 450° &= 1^2 + 0^2 \\ &= 1. \end{aligned}$$

v) $\theta = 4.6$ radians,

$$\begin{aligned} \sin^2 4.6 + \cos^2 4.6 &= (-0.9937)^2 + (-0.1122)^2 \\ &= 0.9874 + 0.0126 \\ &= 1. \end{aligned}$$

For the second trigonometric identity we will combine all 3 of our trigonometric ratios.

Now $\dfrac{\sin\theta}{\cos\theta} = \dfrac{a/c}{b/c}$

$= \dfrac{a}{c} \times \dfrac{c}{b}$ | Rule for dividing fractions. |

$= \dfrac{a}{b}$ | Cancelling the *c*s. |

i.e. $\dfrac{\sin\theta}{\cos\theta} = \tan\theta.$

Again this is an identity true for all angles and strictly we should use $\equiv$ instead of $=$. The next set of examples shows a selection of angles combined with this identity.

■ Examples 4.22 ■

Verify that the identity $\dfrac{\sin\theta}{\cos\theta} = \tan\theta$ is true for

i) $\theta = 135°$, ii) $\theta = 262°$, ii) $\theta = 309.3°$, iv) $\theta = 540°$, v) $\theta = 1.25$ radians.

Separately we will work out the left- and right-hand sides of the identity. In each case we expect the same value for our answers.

i) $\theta = 135°$

$$\frac{\sin 135°}{\cos 135°} = \frac{0.7071}{-0.7071} = -1$$

and $\tan 135° = -1.$

ii) $\theta = 262°$

$$\frac{\sin 262°}{\cos 262°} = \frac{-0.9903}{-0.1392} = 7.115$$

and $\tan 262° = 7.115.$

iii) $\theta = 309.3°$

$$\frac{\sin 309.3°}{\cos 309.3°} = \frac{-0.7738}{0.6334} = -1.222$$

and $\tan 309.3° = -1.222.$

iv) $\theta = 540°$

$$\frac{\sin 540°}{\cos 540°} = \frac{0}{-1} = 0$$

and $\tan 540° = 0.$

v) $\theta = 1.25$ radians

$$\frac{\sin 1.25}{\cos 1.25} = \frac{0.9490}{0.3153} = 3.01$$

and $\tan 1.25° = 3.01$.

These reciprocal ratios are linked with our earlier ratios in some easy trigonometric identities. First we look at

$$\cot \theta = \frac{1}{\tan \theta} = \frac{1}{\dfrac{\sin \theta}{\cos \theta}} = \frac{\cos \theta}{\sin \theta}$$

> Dividing by a fraction we invert and multiply.

Earlier we deduced $\sin^2 \theta + \cos^2 \theta = 1$. Now we use this to deduce two more identities. In the first case we divide throughout by $\cos^2 \theta$ so that

$$\sin^2 \theta + \cos^2 \theta = 1$$

becomes

$$\frac{\sin^2 \theta}{\cos^2 \theta} + \frac{\cos^2 \theta}{\cos^2 \theta} = 1$$

i.e.

$$\tan^2 \theta + 1 = \sec^2 \theta.$$

This is usually quoted as

$$\mathbf{\sec^2 \theta = 1 + \tan^2 \theta}$$

In the second case we divide throughout by $\sin^2 \theta$ so that

$$\sin^2 \theta + \cos^2 \theta = 1$$

becomes

$$\frac{\sin^2 \theta}{\sin^2 \theta} + \frac{\cos^2 \theta}{\sin^2 \theta} = \frac{1}{\sin^2 \theta}$$

$$1 + \cot^2 \theta = \operatorname{cosec}^2 \theta.$$

This is usually quoted as

$$\mathbf{\operatorname{cosec}^2 \theta = 1 + \cot^2 \theta}$$

▀▀▀ Example 4.23 ▀▀▀

Using $\theta = 75°$ check that $\operatorname{cosec}^2 \theta = 1 + \cot^2 \theta$.

Now $\operatorname{cosec}^2 \theta = \operatorname{cosec}^2 75°$

$$= \frac{1}{\sin^2 75°}$$

$$= \frac{1}{(0.9659\ldots)^2}$$

$$= \frac{1}{0.9330\ldots}$$

$$= 1.072.$$

Also
$$1 + \cot^2 \theta = 1 + \cot^2 75°$$
$$= 1 + \frac{1}{\tan^2 75°}$$
$$= 1 + \frac{1}{(3.7320\ldots)^2}$$
$$= 1 + \frac{1}{13.9282\ldots}$$
$$= 1 + 0.072$$
$$= 1.072 \text{ as expected.}$$

EXERCISE 4.11

1   Using $\theta = 125°$ check that $\tan \theta = \dfrac{\sin \theta}{\cos \theta}$. Write down the value of 2θ.
    Now check that $\dfrac{1}{\tan 2\theta} = \dfrac{\cos 2\theta}{\sin 2\theta}$.

2   If $\theta = 90°$ show that $4\cos^2 \theta + 5\sin \theta = 5$. This relationship is *not* an identity because it is not true for all values of θ. Choose two more values for θ to test this statement.

3   Using $A = 30°$ check that $\sin^2 A + \cos^2(180° - A) = 1$.

4   For $A = 550°$ show that $\dfrac{\sin^2 A + \cos^2 A}{\cos^2 A}$ simplifies to $\sec^2 A$. Also show that it is the same as $1 + \tan^2 A$.

5   Let $\theta = 30°$. Write down the value of 2θ. Show that
    $\dfrac{1}{\tan \theta} - \dfrac{2}{\tan 2\theta} = \tan \theta$. Briefly explain why this is the same as $\cot \theta - 2 \cot 2\theta = \tan \theta$.

6   Using $\theta = 115°$ check that $\cot \theta = \dfrac{\cos \theta}{\sin \theta}$.

7   Given $\sec^2 \theta = 1 + \tan^2 \theta$ test that it is true for $\theta = 130°$.

8   Test the truth of $\operatorname{cosec}^2 \theta = 1 + \cot^2 \theta$ for $\theta = 235°$.

9   Why do we have to be careful with $\sec 90°$, $\operatorname{cosec} 0°$ and $\cot 0°$?. Which other angles in the range $0°$ to $360°$ need care?

10 Using $\theta = 5.45$ radians check that $\sec^2 \theta = 1 + \tan^2 \theta$. Is the identity true for $2\theta = 10.90$ radians?

Periodic properties

We complete this chapter by bringing together many of the features of the first 3 trigonometric functions.

The sine curve
 i) is continuous;
 ii) has a period of 360°;
 iii) has a value of 0 at 0°, 180°, 360°, ...;
 iv) has a maximum value of 1 at 90°, 450°, ...;
 v) has a minimum value of −1 at 270°, 630°, ...

The cosine curve
 i) is continuous;
 ii) has a period of 360°;
 iii) has a value of 0 at 90°, 270°, ...;
 iv) has a maximum value of 1 at 0°, 360°, ...;
 v) has a minimum value of −1 at 180°, 540°, ...

The tangent curve
 i) is discontinuous at 90°, 270°, 450°, ...;
 ii) has a period of 180°;
 iii) has a value of 0 at 0°, 180°, 360°, ...;
 iv) tends to infinity as it approaches 90°, 270°, 450°, ...

▆▆▆▆ MULTI-CHOICE TEST 4 ▆▆▆▆▆▆▆▆▆▆▆▆▆▆▆▆▆▆▆▆

Questions **1–5** share answer options A to D based upon the right-angled triangle XYZ.

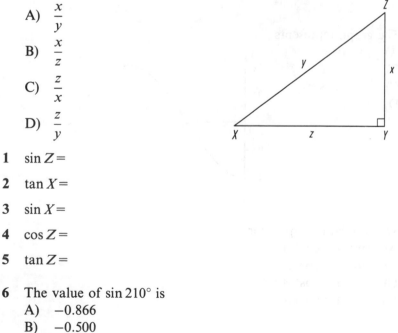

A) $\dfrac{x}{y}$

B) $\dfrac{x}{z}$

C) $\dfrac{z}{x}$

D) $\dfrac{z}{y}$

1 $\sin Z =$

2 $\tan X =$

3 $\sin X =$

4 $\cos Z =$

5 $\tan Z =$

6 The value of $\sin 210°$ is
  A) −0.866
  B) −0.500
  C) 0.500
  D) 0.866

7 In the range 0° to 360° the cosine is −0.5. The angles with this value are
A)   60° and 120°
B)   60° and 300°
C)   120° and 240°
D)   240° and 300°

8 The length of side *AB* is given by
A) $b \sin A$
B) $b \sec A$
C) $b \tan A$
D) $b \cos A$

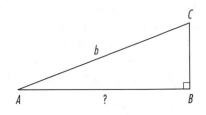

9 The cosine of an angle between 0 and 2π radians is positive in the range
A)   0 to π
B)   0 to $\dfrac{\pi}{2}$ and π to $\dfrac{3\pi}{2}$
C) π to 2π
D)   0 to $\dfrac{\pi}{2}$ and $\dfrac{3\pi}{2}$ to 2π

10 The graph represents
A) $y = \cos x$
B) $y = -\sin x$
C) $y = -\cos x$
D) $y = \sin x$

11 Which relationship is correct?
A) $\sin^2 A - \cos^2 A = 0$
B) $\sin^2 A + \cos^2 A = 0$
C) $\sin^2 A - \cos^2 A = 1$
D) $\sin^2 A = 1 - \cos^2 A$

12 In the range 0° to 360° the graph of $y = \tan \theta$ is

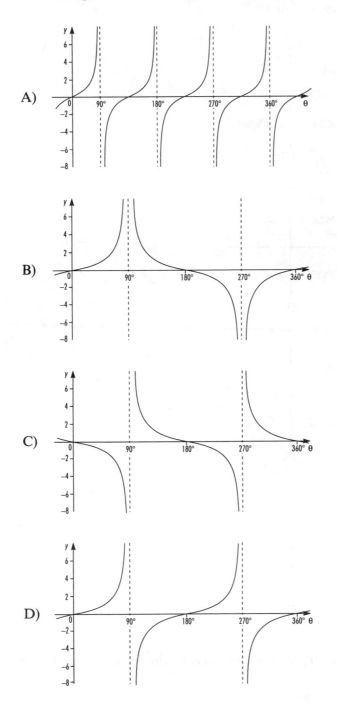

13 Which diagram correctly represents the positive trigonometric functions from 0° to 360°?

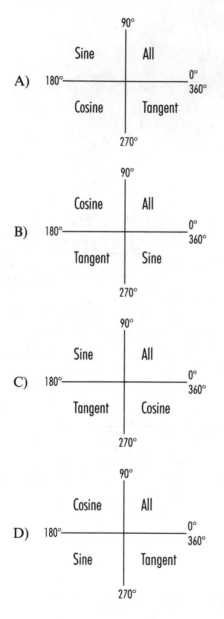

14 In the range 0° to 360° the identity $\sin^2\theta + \cos^2\theta = 1$ is true for angles from

A)   0° to 90° only
B)   0° to 180° only
C)   90° to 180° only
D)   0° to 360°

15 $\sin^2\theta + \cos^2\theta = 1$ is divided throughout by $\cos^2\theta$ to give
    A) $\tan^2\theta + 1 = \sec^2\theta$
    B) $\tan^2\theta + \sec^2\theta = 1$
    C) $\tan^2\theta + 1 = \mathrm{cosec}^2\theta$
    D) $\tan^2\theta + \sec^2\theta = \cot^2\theta$

Questions **16–18** share answer options A) to D)

    A) $\dfrac{1}{\sin\theta}$

    B) $\dfrac{1}{\cos\theta}$

    C) $\dfrac{1}{\tan\theta}$

    D) $\dfrac{\sin\theta}{\cos\theta}$

16 $\sec\theta =$

17 $\tan\theta =$

18 $\cot\theta =$

19   The value of $\sec 35°$ is
    A)   0.574
    B)   1.221
    C)   1.429
    D)   1.743

20   At 90° the cosine, sine and tangent in order are
    A)   0, 1, ∞
    B)   1, 0, ∞
    C) ∞, 1, 0
    D)   0, ∞, 1

5 Trigonometry and Areas

Element: Use trigonometry to solve engineering problems.

PERFORMANCE CRITERIA
- Trigonometrical ratios appropriate to the engineering problem are identified.
- Trigonometrical formulae are manipulated.
- Angles, sides and areas of any triangles are calculated.
- Numerical values are substituted and correct solutions to engineering problems are obtained.

RANGE
Engineering problems: mechanical.
Trigonometrical ratios: $\sin A$, $\cos A$; angles of any magnitude.
Trigonometrical formulae: sine rule, cosine rule, area of any triangle.

Introduction

We start this chapter with non-right-angled triangles. For such triangles we are *unable* to use our simple ratios for sine, cosine and tangent. Also we are *unable* to use Pythagoras' theorem.

Remember that the **sizes of angles correspond to the sizes of sides**. The smallest angle is opposite the smallest side. The largest angle is opposite the largest side. Also the 3 angles of a triangle add up to 180°.

ASSIGNMENT

A construction company has bought a plot of land adjacent to a road junction. The plan in Fig. 5.1 shows the plot as triangular together with its labelled dimensions. As you might expect, the site is not level. We will look at that aspect later in the chapter. The company has outline planning permission for new offices. Before drawing a site plan it wishes to calculate the angles of this triangle. Also it wishes to find the overall site area.

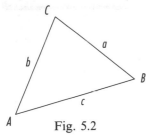

Fig. 5.1

The sine rule

The **sine rule** applies to any triangle. However, for right-angled triangles we use the simple sine ratio. In this section we look at non-right-angled triangles.

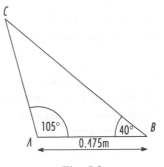

Fig. 5.2

The **sine rule** states $\dfrac{a}{\sin A} = \dfrac{b}{\sin B} = \dfrac{c}{\sin C}$. We do *not* always need all of these 3 sections. It is usual to use just 2 of them at a time, i.e.

$$\frac{a}{\sin A} = \frac{b}{\sin B} \text{ or } \frac{a}{\sin A} = \frac{c}{\sin C} \text{ or } \frac{b}{\sin B} = \frac{c}{\sin C}$$

We may use the sine rule in any triangle given

i) 2 angles and a side,

ii) 2 sides and a *non-included* angle (i.e. an angle that is *not* between the 2 sides).

For these cases, by using the sine rule repeatedly we can **solve a triangle**, i.e. we can find all its angles and sides.

███████ **Example 5.1** ███████

In triangle ABC (Fig. 5.3) if $\angle A = 105°$, $\angle B = 40°$ and $c = 0.475$ m solve the triangle.

This example applies the sine rule twice, given 2 angles and a side.

Fig. 5.3

Firstly $A + B + C = 180°,$

i.e. $105° + 40° + C = 180°$

$$\angle C = 180° - 145°$$

$$\angle C = 35°$$

> Sum of angles of a triangle.

This first simple calculation means we have values for all 3 angles of our triangle ABC. Now we know $\angle C$ and side c, meaning we must use $\dfrac{c}{\sin C}$ in our application of the sine rule. We have to find the sides a and b.

Using $\dfrac{a}{\sin A} = \dfrac{c}{\sin C}$ we can substitute our values

to get $\dfrac{a}{\sin 105°} = \dfrac{0.475}{\sin 35°}$

i.e. $\dfrac{a}{0.9659} = \dfrac{0.475}{0.5736}$

$$\dfrac{a}{0.9659} \times 0.9659 = \dfrac{0.475}{0.5736} \times 0.9659$$

> This multiplication leaves a alone on left.

i.e. $a = \dfrac{0.475}{0.5736} \times 0.9659$

i.e. $a = 0.800$ m.

Now using $\dfrac{b}{\sin B} = \dfrac{c}{\sin C}$ and substituting

we get $\dfrac{b}{\sin 40°} = \dfrac{0.475}{\sin 35°}$

i.e. $\dfrac{b}{0.6428} = \dfrac{0.475}{0.5736}$

i.e. $\dfrac{b}{0.6428} \times 0.6428 = \dfrac{0.475}{0.5736} \times 0.6428$

> This multiplication leaves b alone on left.

$$b = \dfrac{0.475}{0.5736} \times 0.6428$$

i.e. $b = 0.532$ m.

Knowing all angles and sides we can fully label our completely solved triangle ABC in Fig. 5.4.

We see the figure confirms that the smallest side is opposite the smallest angle. Also the largest side is òpposite the largest angle.

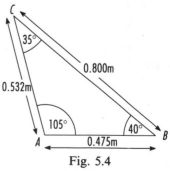

Fig. 5.4

Example 5.2

In triangle ABC (Fig. 5.5) if $\angle A = 55°$, $b = 0.85$ m and $a = 0.70$ m solve the triangle.

This example applies the sine rule given 2 sides and a non-included angle.

We have $\angle A$ and side a, meaning we must use $\dfrac{a}{\sin A}$ in the sine rule. We have no information about $\angle C$ or side c and so will look at these later.

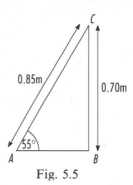

Fig. 5.5

Using $\qquad \dfrac{b}{\sin B} = \dfrac{a}{\sin A}$

we can substitute values

to get $\qquad \dfrac{0.85}{\sin B} = \dfrac{0.70}{\sin 55°}$

i.e. $\qquad \dfrac{0.85}{\sin B} = \dfrac{0.70}{0.8192}$

$$\sin B \times \dfrac{0.85}{\sin B} = \sin B \times \dfrac{0.70}{0.8192}$$

> This move puts sin B in the numerator.

$$0.85 \times 0.8192 = \times \dfrac{0.70}{0.8192} \times 0.8192$$

> The moves leave sin B alone. Left side is purely numbers.

$$\dfrac{0.85 \times 0.8192}{0.70} = \dfrac{\sin B \times 0.70}{0.70}$$

i.e. $\qquad 0.9947 = \sin B$

$\therefore \qquad \angle B = 84.1°, \ 180° - 84.1°$

> Sine is positive in the 1st and 2nd quadrants.

i.e. $\qquad \angle B = 84.1°, \ 95.9°$.

The mathematics has given us 2 possible values for $\angle B$. We need to consider them both.

Using $\qquad\qquad \angle B = 84.1°$

and $\qquad\quad A + B + C = 180°$

we have $\quad 55° + 84.1° + C = 180°$

i.e. $\qquad\qquad \angle C = 180° - 139.1°$

i.e. $\qquad\qquad \angle C = 40.9°$.

Using $\qquad\qquad \angle B = 95.9°$

and $\qquad\quad A + B + C = 180°$

we have $\quad 55° + 95.9° + C = 180°$

i.e. $\qquad\qquad \angle C = 180° - 150.9°$

i.e. $\qquad\qquad \angle C = 29.1°$.

In both cases we can apply the sine rule again:

$$\frac{c}{\sin C} = \frac{a}{\sin A}$$

i.e. $\dfrac{c}{\sin 40.9°} = \dfrac{0.70}{\sin 55°}$ and $\dfrac{c}{\sin 29.1°} = \dfrac{0.70}{\sin 55°}$

$\dfrac{c}{0.6547} = \dfrac{0.70}{0.8192}$ $\dfrac{c}{0.4863} = \dfrac{0.70}{0.8192}$

$c = \dfrac{0.70}{0.8192} \times 0.6547$ $c = \dfrac{0.70}{0.8192} \times 0.4863$

$c = 0.56\,\text{m.}$ $c = 0.42\,\text{m.}$

You can see there are 2 possible triangles for *ABC*. This is because the positive sine has solutions in the range 0° to 90° and 90° to 180°. It is called the **ambiguous case**. This possibility can arise when the first calculation finds an angle. Both answers are correct mathematically. In a practical situation you would have to choose which solution matched the physical constraints. Figs. 5.6 show both triangles.

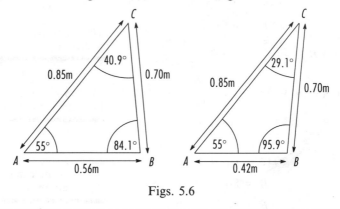

Figs. 5.6

So far our examples have correctly matched the diagrams and the mathematics. Not all triangle information is genuine. In this next example we see the solution quickly shows that a triangle cannot exist based on the original information.

Example 5.3

In triangle *ABC* if $\angle A = 75°$, $b = 0.85\,\text{m}$ and $a = 0.70\,\text{m}$ solve the triangle.

Notice that the format and numbers are similar to those of Example 5.2, with only a slight change. At this stage we will look at the mathematics and leave the diagram until later.

Let us use $\dfrac{b}{\sin B} = \dfrac{a}{\sin A}$

so that $\dfrac{0.85}{\sin B} = \dfrac{0.70}{\sin 75°}.$

Eventually this gives
$$\sin B = \frac{0.85 \times 0.9659}{0.70}$$
i.e. $\qquad \sin B = 1.1729.$

Remember that the maximum value of sine is 1. The value $\sin B = 1.1729$ shows that it is *not* possible to draw this particular triangle. You might like to check this out for yourself.

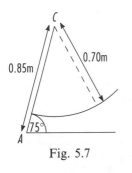

Fig. 5.7

Draw a horizontal line and label the left end A. With a protractor centred on A measure an angle of 75° above the line. From A along this inclined line measure 0.85 m (to your chosen scale) to C. You will find that C is greater than the necessary 0.70 m (to your chosen scale) from the original horizontal line. This is shown by an arc of radius 0.70 m with centre C. It does not reach the horizontal line. The calculation and Fig. 5.7 confirm that the original information will *not* create a triangle.

Let us return to real triangles that can be drawn. There is another section to the sine rule. This involves the circumcircle of the triangle.

The circumcircle of a triangle is the circle passing through each vertex of that triangle.

$$\frac{a}{\sin A} = \frac{b}{\sin B} = \frac{c}{\sin C} = 2R$$

where R is the radius of the circumcircle.

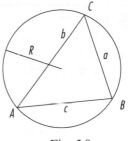

Fig. 5.8

▬▬▬ ASSIGNMENT ▬▬▬▬▬▬▬▬▬

The only triangular information we have for our building site is for the 3 sides. We know that the sine rule cannot be applied in this situation. This means we need some more theory if we are to solve our triangle successfully.

▬▬▬ EXERCISE 5.1 ▬▬▬▬▬▬▬▬▬

Use the sine rule to solve the triangles ABC in each question.

1 $\angle A = 40°$, $\angle B = 70°$, $a = 0.25$ m.

2 $\angle A = 55°$, $\angle C = 73°$, $c = 0.145$ m.

3 $\angle B = 34°$, $\angle C = 28°$, $a = 2.75$ m.

4 $c = 9.45$ m, $\angle B = 32°$, $\angle A = 68°$.

5   $a = 0.75$ m, $\angle C = 48°$,
    $\angle B = 46°$.

6   $\angle A = 50°$, $a = 0.29$ m,
    $b = 0.36$ m.

7   $\angle B = 25.5°$, $b = 0.42$ m,
    $c = 0.36$ m.

8   $\angle C = 120°$, $c = 1.7$ m,
    $a = 0.75$ m.

9   $\angle A = 41°$, $a = 0.24$ m,
    $c = 0.31$ m.

10 $\angle B = 48°$, $a = 0.49$ m,
    $b = 0.81$ m.

The cosine rule

The **cosine rule** applies to any triangle. However, for right-angled triangles we use the simple cosine ratio. In this section we look at non-right-angled triangles.

The **cosine rule** states

$$a^2 = b^2 + c^2 - 2bc \cos A.$$

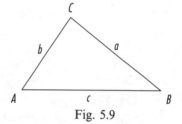

Fig. 5.9

Let us discuss some patterns of letters in our new formula. In $a^2 = b^2 + c^2 - 2bc \cos A$ notice how the side a is on the left of the formula as a^2. Its opposite angle, A, is on the opposite, right, side of the formula as $\cos A$. The other sides, b and c appear on the right. In separate appearances they are each squared and in the other, third appearance they are multiplied together.

We can re-arrange this formula to get an alternative version:

$$a^2 + 2bc \cos A = b^2 + c^2 - 2bc \cos A + 2bc \cos A$$

i.e. $a^2 + 2bc \cos A = b^2 + c^2$

$$a^2 + 2bc \cos A - a^2 = b^2 + c^2 - a^2$$

> Adding $2bc \cos A$ to both sides.
> Subtracting a^2.

i.e. $2bc \cos A = b^2 + c^2 - a^2$

$$\frac{2bc \cos A}{2bc} = \frac{b^2 + c^2 - a^2}{2bc}$$

> Dividing by $2bc$.

i.e. $\cos A = \dfrac{b^2 + c^2 - a^2}{2bc}.$

There are similar patterns in this formula too. Notice how $\angle A$ is separate as $\cos A$. Its opposite side, a, is squared and subtracted on the right. Again b and c appear twice. Each is squared. Also they are multiplied together in the denominator.

We may use the cosine rule in any triangle given

i)   2 sides and an included angle (i.e. angle between the 2 given sides) (version 1 of the formula),

ii) 3 sides (version 2 of the formula).

Already, in the sine rule section, we have seen information that is not genuine. Whenever we have 3 sides we can test to see if they will actually create a triangle. The test is that the sum of the 2 shorter lengths must exceed the third length.

Now let us return to the formulae. The first formula finds side a and the second formula finds $\angle A$. There are similar versions for the other sides and angles. The letters in each formula move around on a cyclical basis, i.e. b moves to a, c moves to b and a moves to c. This means you do *not* need to remember all the various versions. We have

$$b^2 = c^2 + a^2 - 2ca \cos B$$
and $$c^2 = a^2 + b^2 - 2ab \cos C.$$

Also $$\cos B = \frac{c^2 + a^2 - b^2}{2ca}$$

and $$\cos C = \frac{a^2 + b^2 - c^2}{2ab}.$$

The patterns we described for the early versions of our formulae apply to these versions. The same ideas are true with the letters moving around cyclically.

▆▆▆ Examples 5.4 ▆▆▆

In triangle *ABC* find side a if

i) $\angle A = 75°$, $b = 0.375\,\text{m}$ and $c = 0.415\,\text{m}$;
ii) $\angle A = 125°$, $b = 0.375\,\text{m}$ and $c = 0.415\,\text{m}$.

For both triangles we are finding the side a opposite the given $\angle A$. In each case 2 sides and an included angle are given, allowing us to use the cosine rule.

i) Fig. 5.10 shows the triangle labelled with our given information. We use the original version of our formula,

$$a^2 = b^2 + c^2 - 2bc \cos A.$$

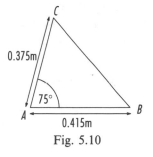

Fig. 5.10

We substitute our given values to get
$$a^2 = 0.375^2 + 0.415^2 - 2(0.375)(0.415)(\cos 75°)$$
$$= 0.375^2 + 0.415^2 - 2(0.375)(0.415)(0.2588)$$
$$= 0.1406 + 0.1722 - 0.0805$$
i.e. $$a^2 = 0.2323$$
∴ $$a = 0.48\,\text{m}.$$

ii) The given information in this case is very similar to that in the previous case. Fig. 5.11 shows the labelled triangle. Whilst b and c have the same values as before, $\angle A$ has been increased.

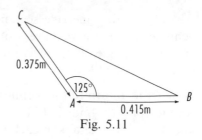

Fig. 5.11

Because of this we would expect the length of side a to increase as well. Also $\angle A$ is obtuse which means the cosine value will be negative. Using the same formula,

$$a^2 = b^2 + c^2 - 2bc\cos A$$

we substitute our given values to get

$$a^2 = 0.375^2 + 0.415^2 - 2\,(0.375)\,(0.415)\,(\cos 125°)$$
$$= 0.375^2 + 0.415^2 - 2\,(0.375)\,(0.415)\,(-0.5736)$$
$$= 0.1406 + 0.1722 + 0.1785$$

i.e. $a^2 = 0.491$

$\therefore$ $a = 0.70\,\text{m}.$

$(-)\,(-) = +.$

As we expected side a is indeed larger than in the previous case.

Examples 5.5

In triangle ABC we are given the lengths of the sides, $a = 1.15\,\text{m}$, $b = 1.30\,\text{m}$ and $c = 2.25\,\text{m}$.

Find   i)   the smallest angle,
        ii)   the largest angle

and   iii)   the third angle of the triangle.

Before using the cosine rule we ought to check that the 3 lengths do create a triangle. We need the sum of the 2 shorter lengths to be greater than the third length. Now $1.15 + 1.30 = 2.45$ which is greater than 2.25. The test works and so we do have a genuine triangle.

i) The smallest angle is opposite the smallest side, i.e. $\angle A$ is opposite a. We use version 2 of the cosine rule,

$$\cos A = \frac{b^2 + c^2 - a^2}{2bc}.$$

Substituting our given values we get

$$\cos A = \frac{1.30^2 + 2.25^2 - 1.15^2}{2(1.30)(2.25)}$$
$$= \frac{1.6900 + 5.0625 - 1.3225}{5.85} = \frac{5.43}{5.85}$$

i.e. $\cos A = 0.9282$

$\therefore$ $\angle A = 21.8°$ is the smallest angle

ii) The greatest angle is opposite the greatest side, i.e. $\angle C$ is opposite c. We use a similar version of the cosine rule, with the letters moved around cyclically,

$$\cos C = \frac{a^2 + b^2 - c^2}{2ab}.$$

Substituting our given values we get

$$\cos C = \frac{1.15^2 + 1.30^2 - 2.25^2}{2(1.15)(1.30)}$$

$$= \frac{1.3225 + 1.6900 - 5.0625}{2.99}$$

$$= \frac{-2.05}{2.99}$$

i.e. $\cos C = -0.6856$

∴ $\angle C = 133.3°$ is the largest angle.

iii) The third angle of the triangle is $\angle B$. We know that the sum of the angles of a triangle is 180°,

i.e. $A + B + C = 180°$.

Substituting our calculated angles we get

$$21.8° + B + 133.3° = 180°$$

i.e. $\angle B = 180° - 155.1°$

i.e. $\angle B = 24.9°$.

Alternatively, and longer, we could have found $\angle B$ using

$$\cos B = \frac{c^2 + a^2 - b^2}{2ca}.$$

Fig 5.12 shows our completely solved and fully labelled triangle ABC.

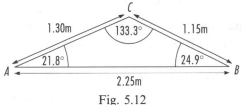

Fig. 5.12

Example 5.6

Given $a = 2.72\,\text{m}$, $b = 5.56\,\text{m}$ and $c = 5.35\,\text{m}$ solve the triangle ABC.

In the previous example we saw how to test whether the 3 lengths created a triangle. We checked that the sum of the 2 shorter lengths was greater than the third length.

We can apply the test again. Now $5.35 + 2.72 = 8.07$ which is greater than 5.56. The test works and so we do have a genuine triangle.

In the solution of our triangle we may start by finding any angle. Suppose we start with $\angle A$. Using

$$\cos A = \frac{b^2 + c^2 - a^2}{2bc}$$

we substitute our values to get

$$\cos A = \frac{5.56^2 + 5.35^2 - 2.72^2}{2(5.56)(5.35)}$$

$$= \frac{52.14}{59.49}$$

i.e. $\cos A = 0.8764$

$\therefore$ $\angle A = 28.8°$.

Let us draw our triangle with the information we have so far. This is shown in Fig. 5.13.

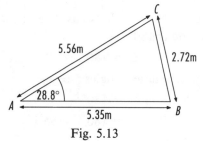

Fig. 5.13

Before we continue let us consider the angles in our triangle.

We know that $A + B + C = 180°$

i.e $28.8° + B + C = 180°$

i.e. $B + C = 180° - 28.8°$

i.e. $B + C = 151.2°$.

> Sum of angles of a triangle.

Now sides b and c are of very similar lengths, b being slightly longer than c. This means $\angle B$ and $\angle C$ will be very similar angles, $\angle B$ being slightly larger than $\angle C$, i.e. $\angle B$ being slightly more than half of $151.2°$ ($75.6°$) and $\angle C$ being slightly less than half of $151.2°$ ($75.6°$).

Next we can find either $\angle B$ or $\angle C$. Also we have enough triangle information to use either the sine rule or the cosine rule. Suppose we find $\angle B$.

Using the sine rule we have

$$\frac{b}{\sin B} = \frac{a}{\sin A}.$$

Substituting our values we get

$$\frac{5.56}{\sin B} = \frac{2.72}{\sin 28.8°}.$$

We can re-arrange this equation to get

$$\sin B = \frac{5.56 \times 0.4818}{2.72}$$

i.e. $\sin B = 0.9848$

∴ $\angle B = 80.0°.$

Alternatively using the cosine rule we have

$$\cos B = \frac{c^2 + a^2 - b^2}{2ca}$$

i.e. $\cos B = \dfrac{5.35^2 + 2.72^2 - 5.56^2}{2(5.35)(2.72)}$

$$= \frac{5.107}{29.10}$$

i.e. $\cos B = 0.1755$

∴ $\angle B = 79.9°.$

We have several points to mention. Notice the slight difference between the values for $\angle B$. Which is the more accurate? $\angle B = 79.9°$ is the more accurate value because it has been calculated directly from the original information. The other value of $\angle B$ was calculated from 28.8° and the sine rule. 28.8° was an approximation which was compounded further in the sine rule calculation. Also, as predicted $\angle B = 79.9°$ is slightly greater than 75.6°.

To find $\angle C$ we use

$$A + B + C = 180°$$

| Sum of angles of a triangle. |

and substitute our calculated angles to get

$$28.8° + 79.9° + C = 180°$$

i.e. $\angle C = 180° - 108.7°$

i.e. $\angle C = 71.3°.$

To complete our predictions $\angle C = 71.3°$ is slightly less than 75.6°.

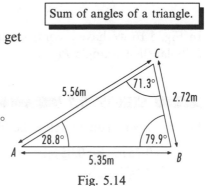

In Fig. 5.14 we have the complete and fully labelled triangle ABC.

Fig. 5.14

ASSIGNMENT

Our basic triangle PQR is drawn in Fig. 5.15 showing our 3 known sides.

We start our calculation by finding any angle. Suppose we find $\angle P$. We need to interpret our original formula and the patterns of letters so that

$$\cos P = \frac{q^2 + r^2 - p^2}{2qr}.$$

We substitute our values to get

$$\cos P = \frac{650^2 + 600^2 - 725^2}{2(650)(600)}$$

$$= \frac{256875}{780000}$$

i.e. $\cos P = 0.3293$

∴ $\angle P = 70.8°.$

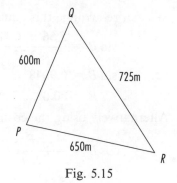

Fig. 5.15

Our experience in Example 5.6 suggests, for accuracy, that we apply the cosine rule again rather than the sine rule.

Using $\cos Q = \frac{600^2 + 725^2 - 650^2}{2(600)(725)}$

$$= \frac{463125}{870000}$$

i.e. $\cos Q = 0.5323$

∴ $\angle Q = 57.8°.$

The sum of the 3 angles of a triangle is 180°,

i.e. $P + Q + R = 180°$

so that $70.8° + 57.8° + R = 180°$

i.e. $\angle R = 180° - 128.6°$

∴ $\angle R = 51.4°.$

In Fig. 5.16 we have a complete and fully labelled triangle PQR.

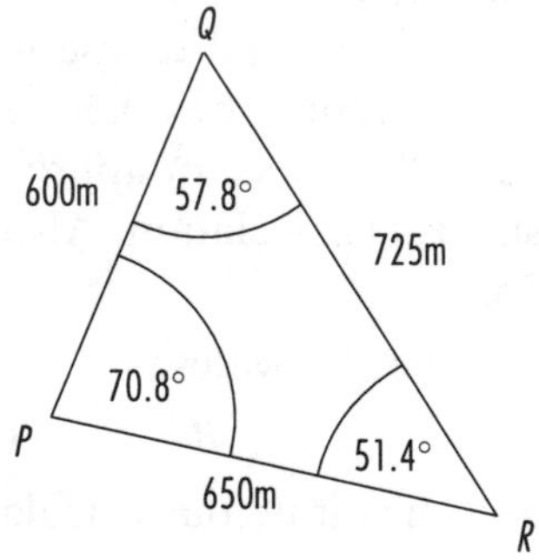

Fig. 5.16

![EXERCISE 5.2]

Use the cosine rule to solve the triangles ABC in each question.

1 $a = 0.75\,\text{m}$, $b = 0.85\,\text{m}$,
       $\angle C = 62°$

2 $b = 0.84\,\text{m}$, $c = 0.34\,\text{m}$,
       $\angle A = 43°$

3 $c = 0.56\,\text{m}$, $a = 0.14\,\text{m}$,
       $\angle B = 71°$

4 $b = 1.12\,\text{m}$, $a = 1.43\,\text{m}$,
       $\angle C = 25°$

5 $a = 0.32\,\text{m}$, $c = 0.53\,\text{m}$,
       $\angle B = 109°$

6 $a = 2\,\text{m}$, $b = 7\,\text{m}$, $c = 6\,\text{m}$

7 $a = 2.4\,\text{m}$, $b = 7.1\,\text{m}$,
       $c = 6.4\,\text{m}$

8 $a = 0.76\,\text{m}$, $b = 0.86\,\text{m}$,
       $c = 1.32\,\text{m}$

9 $a = 1.14\,\text{m}$, $b = 1.42\,\text{m}$,
       $c = 2.61\,\text{m}$

10 $a = 7.26\,\text{m}$, $b = 3.31\,\text{m}$,
        $c = 4.81\,\text{m}$

Area of a triangle

We are going to look at 3 different formulae for finding the area of a triangle. Which one you use depends on the original information of each particular problem.

Remember the units for area. If we have distances in metres (m) then the area is in metres squared (m^2).

1.

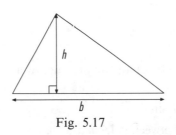

Fig. 5.17

The area of a triangle is half the product of the base and the altitude,

i.e.   Area $= \frac{1}{2}bh$.

The **altitude** is the perpendicular distance between the vertex and its opposite side.

This formula is the simplest of the triangle area formulae. We use it when the altitude (the perpendicular height from the base) is given or to calculate it is easy. The base does *not* have to be horizontal. Any side will do.

The triangles in Fig. 5.17 and Fig. 5.18 are identical. If we use the side labelled b_1 in Fig. 5.18 we would find

Area $= \frac{1}{2}b_1h_1$.

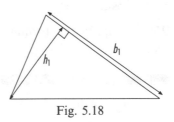

Fig. 5.18

Both area calculations would give exactly the same answers because the triangles are exactly the same. A variation from b to b_1 is compensated by a change from h to h_1 so the products $\frac{1}{2}bh$ and $\frac{1}{2}b_1h_1$ remain the same,

i.e. $\frac{1}{2}bh = \frac{1}{2}b_1h_1$

i.e. $bh = b_1h_1$

i.e. $\dfrac{b}{b_1} = \dfrac{h_1}{h}$. | Dividing by b_1 and h_1. |

This ratio of distances will occur in a triangle to keep the area calculations consistent.

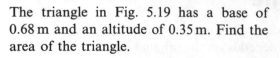

Example 5.7

The triangle in Fig. 5.19 has a base of 0.68 m and an altitude of 0.35 m. Find the area of the triangle.

We use $b = 0.68$ and $h = 0.35$ in our formula

$$\text{Area} = \tfrac{1}{2}bh$$

to get $\text{Area} = \tfrac{1}{2}(0.68)(0.35)$

$$= 0.12\,\text{m}^2$$

as the area of the triangle.

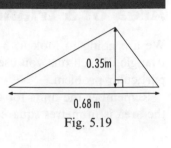

0.35m

0.68 m

Fig. 5.19

---

2.   The area of a triangle is half the product of any 2 sides and the sine of the included angle,

i.e. $\text{Area} = \tfrac{1}{2}ab\sin C$

where $\angle C$ is included between sides a and b.

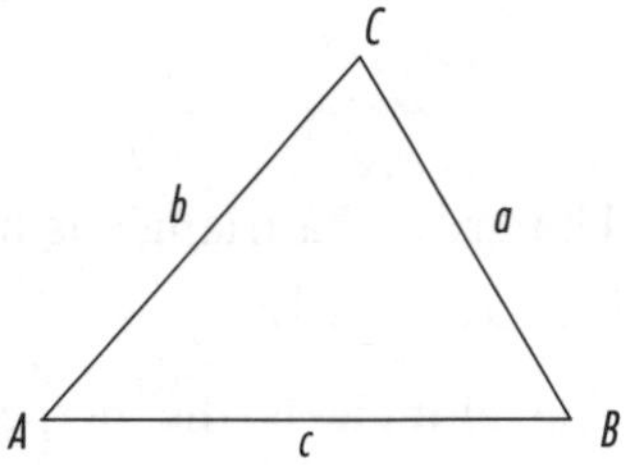

In every case we might not have these particular sides and angle. Other versions of the formula are

$$\text{Area} = \tfrac{1}{2}bc\sin A$$

and $\text{Area} = \tfrac{1}{2}ca\sin B.$

Fig. 5.20

When you compare all 3 versions of this area formula you can see the letters move around cyclically. Remember the cyclical movement of letters in the cosine rule?

We can apply the formula using the values from an earlier example, Example 5.4 i).

Example 5.8

In triangle $ABC = 75°$, $b = 0.375$ m and $c = 0.415$ m shown in Fig. 5.21. Find the area of the triangle.

Immediately you can see we have 2 sides and an included angle. Using the second version of our area formula we have

$$\text{Area} = \tfrac{1}{2}bc\sin A.$$

We substitute our given values to get

$$\text{Area} = \tfrac{1}{2}(0.375)(0.415)(\sin 75°)$$
$$= \tfrac{1}{2}(0.375)(0.415)(0.9659)$$
$$= 0.075\,\text{m}^2.$$

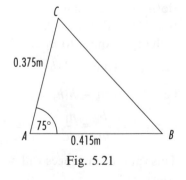

0.375m

75°

0.415m

Fig. 5.21

3. The final of our trio of area formulae is

$$\text{Area} = \sqrt{s(s-a)(s-b)(s-c)}$$

where $s = \frac{1}{2}(a+b+c)$

i.e. s is half the perimeter of the triangle.

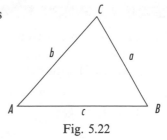

Fig. 5.22

As you can see this is the formula we choose when we know only the sides of the triangle.

████████ **Example 5.9** ████████████████████████████████

Calculate the area of triangle *ABC* given $a = 0.480$ m, $b = 0.375$ m and $c = 0.415$ m.

You might like to look again at Example 5.4i). These are the lengths of the sides of the triangle in that example.
    Before we can use our area formula we need to apply the half perimeter formula,

i.e. $s = \frac{1}{2}(a+b+c)$.

We substitute our given values to get

$$s = \frac{1}{2}(0.480 + 0.375 + 0.415)$$

$$= \frac{1}{2}(1.27)$$

i.e. $s = 0.635$ m.

Using s $= 0.635$

we have $s - a = 0.635 - 0.480 = 0.155,$

        $s - b = 0.635 - 0.375 = 0.260$

and $s - c = 0.635 - 0.415 = 0.220.$

Then $\text{Area} = \sqrt{s(s-a)(s-b)(s-c)}$

becomes $\text{Area} = \sqrt{0.635 \times 0.155 \times 0.260 \times 0.220}$

$$= \sqrt{5.63 \times 10^{-3}}$$

$$= 0.075 \, \text{m}^2.$$

You can compare this answer with the area answer in Examples 5.8. They are equal as both examples are based on a triangle from an earlier question. You would expect the answers to be the same, allowing for any calculation approximations.

We can return to our triangular plot of land and its area. Remember we have the lengths of the 3 sides; $p = 725$ m, $q = 650$ m and $r = 600$ m. This means we use the third of our formulae, first finding half the perimeter.

Using $\qquad s = \frac{1}{2}(p + q + r)$

we substitute to get

$$s = \frac{1}{2}(725 + 650 + 600)$$
$$= \frac{1}{2}(1975)$$
$$= 987.5 \text{ m.}$$

Applying $s \qquad\qquad\qquad\qquad = 987.5$

we have $\quad s - p = 987.5 - 725 \quad = 262.5,$

$\qquad\qquad s - q = 987.5 - 650 \quad = 337.5$

and $\qquad s - r = 987.5 - 600 \quad = 387.5.$

Then $\qquad$ Area $= \sqrt{s(s - p)(s - q)(s - r)}$

becomes $\quad$ Area $= \sqrt{987.5 \times 262.5 \times 337.5 \times 387.5}$

$$= \sqrt{3.39 \times 10^{10}}$$
$$= 1.84 \times 10^5 \text{ m}^2 \quad \text{is the area of the site.}$$

In each question find the area of the triangle ABC.

1   Base = 0.245 m, altitude = 0.780 m.

2   Base = 2.90 m, altitude = 1.35 m.

3   Base = 0.55 m, altitude = 0.55 m.

4   Base = 0.425 m, altitude = 0.150 m.

5   Base = 0.150 m, altitude = 0.425 m.

6   $a = 0.75$ m, $b = 0.85$ m, $\angle C = 62°$.

7   $b = 0.84$ m, $c = 0.34$ m, $\angle A = 43°$.

8   $c = 0.56$ m, $a = 0.14$ m, $\angle B = 71°$.

9   $b = 1.12$ m, $a = 1.43$ m, $\angle C = 25°$.

10   $a = 0.32$ m, $c = 0.53$ m, $\angle B = 109°$.

11   $a = 2$ m, $b = 7$ m, $c = 6$ m.

12   $a = 2.4$ m, $b = 7.1$ m, $c = 6.4$ m.

13   $a = 0.76$ m, $b = 0.86$ m, $c = 1.32$ m.

14   $a = 1.14$ m, $b = 1.42$ m, $c = 2.61$ m.

15   $a = 7.26$ m, $b = 3.31$ m, $c = 4.81$ m.

Area of a quadrilateral

A quadrilateral is a 4 sided plane figure. The square and rectangle are well known standard types with simple area formulae (Fig. 5.23 and Fig. 5.24).

1. The square

Perimeter $= 4a$

Area $= a^2$

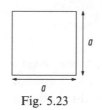

Fig. 5.23

2. The rectangle

Perimeter $= 2a + 2b = 2(a + b)$

Area $= ab$

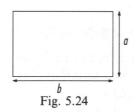

Fig. 5.24

3. The parallelogram

Perimeter $= 2a + 2b = 2(a + b)$

We can split the parallelogram into 2 triangles as in Figs. 5.25.

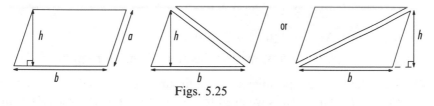

Figs. 5.25

Using these triangles and their area formula we have

Area of triangle $= \frac{1}{2}bh$.

Area of parallelogram $= 2 \times$ Area of triangle

$$= 2 \times \frac{1}{2}bh$$

$$= bh.$$

We can apply this same principle using another of the triangle area formulae, i.e.

Area of triangle $= \frac{1}{2}ab \sin C$

so that

Area of parallelogram $= ab \sin C$.

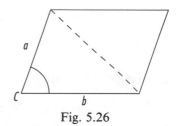

Fig. 5.26

The area of a triangle formula that uses the perimeter is less useful. For the area of a parallelogram we would not have a simple factor of 2 as for the other formulae.

4. The trapezium

Area = $\frac{1}{2}(a+b)h$.

To find the perimeter we need to include the lengths of the 2 unknown sides (Fig 5.27). We do this in the next example.

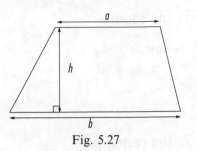

Fig. 5.27

━━━━━ **Example 5.10** ━━━━━

This trapezium has parallel sides of 0.55 m and 2.75 m separated by 0.80 m. Find its   i) area,
and ii) perimeter.

Firstly let us sketch and label the trapezium in Fig. 5.28.

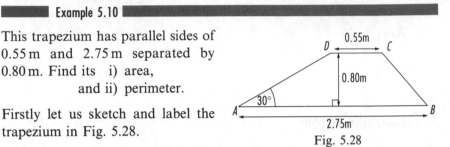

Fig. 5.28

i) In our area formula we substitute for $a = 0.55$ m, $b = 2.75$ m and $h = 0.80$ m.

Then   Area $= \frac{1}{2}(a+b)h$

becomes   Area $= \frac{1}{2}(0.55+2.75)0.80$

$= 1.32\,\text{m}^2$ is the area of the trapezium.

ii) To calculate the perimeter we split up the original trapezium in Figs. 5.29. This means we need to introduce two further labels E and F. The split allows us to find the unknown sides using convenient right-angled triangles. We apply simple trigonometric ratios for sine, cosine and tangent as necessary.

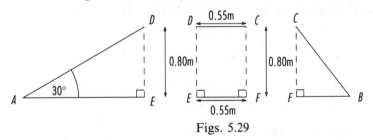

Figs. 5.29

In triangle AED using

$$\sin\theta = \frac{\text{opposite}}{\text{hypotenuse}}$$

we get $\sin 30° = \dfrac{0.80}{AD}$

i.e. $0.50 = \dfrac{0.80}{AD}.$

Then $\qquad 0.50 \times AD = \dfrac{0.80}{AD} \times AD$ | Multiplying by AD. |

i.e. $\qquad 0.50 \times AD = 0.80.$

Also $\qquad \dfrac{0.50 \times AD}{0.50} = \dfrac{0.80}{0.50}$ | Dividing by 0.50. |

i.e. $\qquad AD = \dfrac{0.80}{0.50} = 1.60\,\text{m}.$

Also $\qquad \tan\theta = \dfrac{\text{opposite}}{\text{adjacent}}$

becomes $\qquad \tan 30° = \dfrac{0.80}{AE}$

i.e. $\qquad 0.5774 = \dfrac{0.80}{AE}.$

Using the same steps as before, eventually we get

$$AE = \dfrac{0.80}{0.5774} = 1.39\,\text{m}.$$

In the memory we retain the calculator value to avoid too much inaccuracy later in triangle BCF (i.e. $AE = 1.3856\ldots$).

In the trapezium we have

$$AE + EF + FB = AB$$

i.e. $\qquad 1.39 + 0.55 + FB = 2.75$

i.e. $\qquad FB = 2.75 - 1.94$

$$FB = 0.81.$$ | Calculator value is $0.814\ldots$ |

In triangle BCF, by Pythagoras' theorem

$$BC^2 = FB^2 + CF^2$$
$$= 0.81^2 + 0.80^2$$
$$= 0.66 + 0.64$$

i.e. $\qquad BC^2 = 1.30\ldots$ | Square root of both sides. |

$\therefore \qquad BC = 1.14\,\text{m}$

Finally, $\qquad$ Perimeter $= AB + BC + CD + DA$
$$= 2.75 + 1.14 + 0.55 + 1.60$$
$$= 6.04\,\text{m}$$

This technique of splitting up a figure can be useful when finding the area of a quadrilateral. We apply it to an area in the next example.

Example 5.11

The diagram, Fig. 5.30, shows a rectangular piece of metal measuring 0.75 m by 0.25 m. We are interested in the shaded area once the 2 triangles and quadrant have been removed. Find the shaded area.

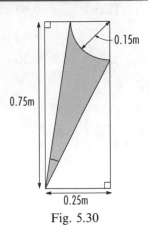

Fig. 5.30

We think of the shaded area as a rectangle minus the triangles and quadrant. For ease we re-draw these components in Figs. 5.31.

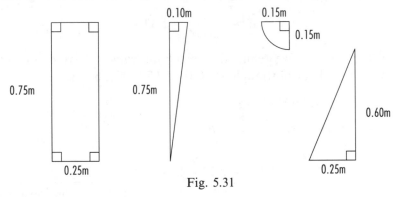

Fig. 5.31

For accuracy in our final answers we quote part answers but retain more decimal places in the calculator's memory.

Rectangle: Area $= 0.25 \times 0.75$ $= 0.1875 \, \text{m}^2$

Top triangle: Area $= \dfrac{1}{2} \times 0.75 \times 0.10$ $= 0.0375 \, \text{m}^2$

Quadrant: Area $= \dfrac{1}{4} \pi 0.15^2$ $= 0.0177 \, \text{m}^2$

Lower triangle: Area $= \dfrac{1}{2} \times 0.25 \times 0.60$ $= 0.075 \, \text{m}^2$

Shaded area $= 0.1875 - 0.0375 - 0.177 - 0.075$

                        $= 0.057 \, \text{m}^2$.

The next set of exercises is based on areas, using formulae from this and the circular measure chapters.

■ EXERCISE 5.4 ■

1 The diagram shows the cross-section of a V-block. What cross-sectional area has been removed to create the V? What area of metal remains?

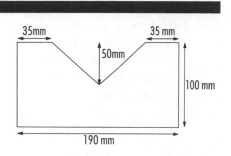

2 A small swimming pool is 1.25 m deep at the shallow end and 2.15 m deep at the deep end. The slope of the bottom is uniform. The length and width of the pool are 10.00 m and 6.50 m respectively. Make a sketch of the pool. What area of tiling is required to line the pool completely? The new owner decides to have a 0.50 m width of tiling to surround the pool. What is this extra area of tiling? Express the extra tiling as a percentage of the original tiling correct to 1 decimal place.

3 In the diagram we have a circle of radius 125.0 mm and a sector of angle 75°. Find the area of the shaded segment. (The area of a sector, A, is $A = \frac{1}{2}r^2\theta$ where θ is in radians.)

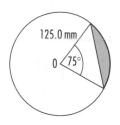

4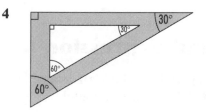

The diagram shows a large 30°, 60° and 90° triangular set square used for demonstrations. The length of the larger hypotenuse is 1.00 m and of the shorter is 0.50 m. Calculate the area of material in this set square.

5 From a piece of high density polymer in the form of a trapezium we have removed a circle of radius 0.20 m. Some other dimensions are shown on the diagram. What is the area remaining?

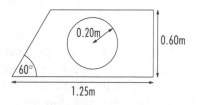

6 A parallelogram has sides of 0.725 m and 0.895 m. One diagonal measures 0.820 m. Draw this parallelogram to aid your calculation of its area.

7 The diagram shows a trapezium shaped piece of sheet metal. A semi-circle of diameter 0.60 m has been removed from the trapezium as shown. What is the area of remaining metal?

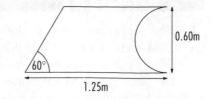

8 Find the area of triangle *ABC*. You are given the coordinates to be *A*(1, 1), *B*(7, 4) and *C*(10, 10).

9 Find the area of the field labelled *ABCD*.

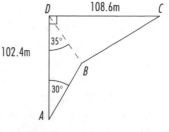

10 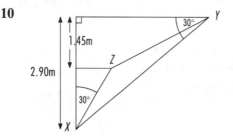 Find the area of triangle *XYZ* using the information on the accompanying diagram.

Angles of elevation and depression

For an **angle of elevation** you are looking up from the horizontal. For an **angle of depression** you are looking down from the horizontal.

Suppose we have a scaffolding tower and a tool lying on the ground some distance away (Fig. 5.32). From the tool, the angle of elevation of the top of the tower is $x°$ above the horizontal. From the top of the tower, the angle of depression of the tool is $y°$. This is $y°$ below the horizontal. Simple geometry tells us that these are the same angles,

i.e. $x = y$.

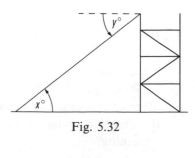

Fig. 5.32

Alternate angles.

We can put some numbers to this general idea in the next example.

▬▬▬▬▬ **Example 5.12** ▬▬▬▬▬▬▬▬▬▬▬▬▬▬▬▬▬▬▬

Our scaffolding tower is 7.4 m high and an angle grinder is 12 m away from its base on the horizontal ground.

i) What is the angle of elevation of the top of the tower from the grinder?

A man is three-quarters of the way up the tower when he sees the tool he has forgotten.

ii) What is the angle of depression of the grinder?

In each case we can draw a diagram for our information and apply some trigonometry (Fig. 5.33 and Fig. 5.34).

i)

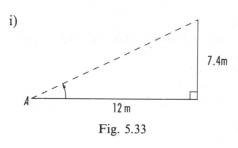

Fig. 5.33

$$\tan A = \frac{7.4}{12} = 0.61\bar{6}$$

i.e. $\angle A = 31.7°$ is the angle of elevation of the top of the tower from the grinder.

ii)

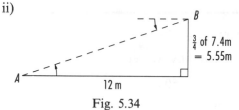

¾ of 7.4 m
= 5.55 m

Fig. 5.34

$$\tan B = \frac{12}{5.55} = 2.16$$

i.e. $\angle B = 65.2°$.

∴ $90° - 65.2° = 24.8°$ is the angle of depression of the forgotten tool.

▬▬▬▬▬ **ASSIGNMENT** ▬▬▬▬▬▬▬▬▬▬▬▬▬▬▬▬▬▬▬▬

We can take another look at our triangular building plot. Suppose the plot, as expected, is *not* level. Hence our dimensions follow the lie of the land rather than the horizontal.

From P the angle of elevation of Q is found to be 2.25°. From P the angle of depression of R is found to be 0.80°. Using our theory we can find the height of Q above the level of P, and the height of R below P. We can follow this with simple arithmetic to find the height difference between Q and R.

First we deal with P and Q, drawing a supplementary diagram (Fig. 5.35) from our extra information. Q' is vertically below Q along the level of P.

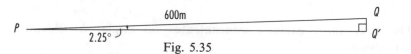

Fig. 5.35

In triangle $PQ'Q$, $\sin 2.25° = \dfrac{QQ'}{600}$

i.e. $0.0393 = \dfrac{QQ'}{600}$

i.e. $0.0393 \times 600 = \dfrac{QQ'}{600} \times 600$ | Multiplying by 600. |

so that $0.0393 \times 600 = QQ'$.

We may complete the multiplication to get

$$QQ' = 23.6\,\text{m}$$

i.e. Q is 23.6 m above the level of P.

Now we look at P and R, again drawing another separate diagram, Fig. 5.36, for our information.

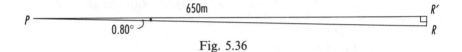

Fig. 5.36

In triangle PRR', $\tan 0.80° = \dfrac{RR'}{650}$.

Using the same method as before eventually we get

$$RR' = 0.0140 \times 650$$
$$= 9.1\,\text{m},$$

i.e. R is 9.1 m below the level of P.

Finally, linking together our results (Fig. 5.37) we can find the difference in heights of Q and R.

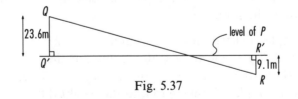

Fig. 5.37

Difference in heights $= 23.6 + 9.1$
$$= 32.7\,\text{m}.$$

The next short exercise concentrates on angles of elevation and depression.

▬▬▬ EXERCISE 5.5 ▬▬▬

1 A ladder of length 4 m has its foot on horizontal ground. It leans against a vertical wall. If its angle of elevation is 76° how far is its foot from the wall? What is the acute angle between the top of the ladder and the wall?

2 A railway line has a gradient of 1 in 125 measured along the line. What is its angle of elevation?

3 The eyes of a man 1.82 m tall are 1.725 m above horizontal ground. Unknowingly he drops his pay slip and then something catches his attention. If he sees the pay slip 10 m away what is its angle of depression? Before he can run towards it the wind blows it a further 2 m away. What is the new angle of depression?

4 On a small hillock stands a tree. Along the line of sight the base of the tree is 150 m away at an angle of elevation of 4.05°. From the same point the angle of elevation of the top of the tree is 7.84°. Estimate the height of the tree.

5 A church tower needs a new weather vane. The height of the vane is 1.35 m. The blacksmith places the vane on top of the tower at the front. From 100 m away the angle of elevation of the top of the vane is 14.4°. How high is the tower? It is placed at the back on top of the tower. From the same place the angle of elevation is 13.6°. What is the depth of the tower, front to back?

Angle between a line and a plane

In Fig. 5.38 $WXYZ$ is any plane and AC is a line. A lies on the plane and B, vertically below C, also lies on the plane. This means we have a right-angled triangle ABC. $\angle CAB$ is the angle between the line and the plane.

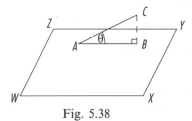

Fig. 5.38

The **projection** of AC onto the plane is AB, $\qquad \dfrac{AB}{AC} = \cos\theta$

$$\text{i.e.} \qquad AB = AC \times \cos\theta$$

Suppose A does *not* lie on the plane $WXYZ$ as in Fig. 5.39.

Again B is vertically below C. Also A and B are at the same height above the plane. The projections of A and B onto the plane are A' and B', i.e. A' and B' lie on the plane, vertically below A and B respectively. Hence AB and $A'B'$ are equal in length and parallel. Also we have parallel lines AA' and CBB'. Notice also there are quite a few right-angles marked in the diagram.

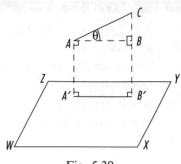

Fig. 5.39

Again $\angle CAB$ is the angle between the line and the plane.

Also AB or $A'B' = AC \times \cos\theta$.

Example 5.13

Fig. 5.40 shows a rectangular metal block measuring 0.25 m by 0.15 m by 0.10 m. We have labelled it for reference.
What is the angle between the line SY and the plane $WXYZ$?

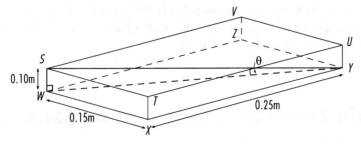

Fig. 5.40

We start with the projection of SY onto the plane $WXYZ$, WY. To show this more clearly we draw a supplementary diagram, Fig. 5.41.

In triangle WXY, by Pythagoras' theorem

$$WY^2 = WX^2 + XY^2$$
$$= 0.15^2 + 0.25^2$$
$$= 0.0225 + 0.0625$$

i.e. $WY^2 = 0.085$

$\therefore$ $WY = \sqrt{0.085}$

$$= 0.29\,\text{m}.$$

Fig. 5.41

Calculator value is 0.2915...

Picking out $\angle\theta$ from the block we have

$$\tan\theta = \frac{0.10}{0.29\ldots} = 0.3430$$

$$\therefore \quad \theta = 18.9°$$

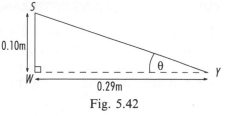

Fig. 5.42

i.e. the angle between the line SY and the plane $WXYZ$ is 18.9°.

There are 4 lines joining opposite corners like this. They are SY, TZ, UW and VX. Each of them is inclined at 18.9° to the planes $WXYZ$ and $STUV$ (Fig. 5.42).

Alternate angles.

Angle between two intersecting planes

In Fig. 5.43 XY is the line of intersection of the 2 planes and A is some point along this common line. AC is a line lying in one plane whilst AB is lying in the other plane. Both AC and AB are perpendicular to XY. Then $\angle CAB$ is the angle between the 2 planes.

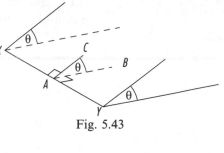

Fig. 5.43

If C is vertically above B (Fig. 5.44) then the projection of AC onto the other plane is AB.

Again $\quad \dfrac{AB}{AC} = \cos\theta$

i.e. $\quad AB = AC \times \cos\theta.$

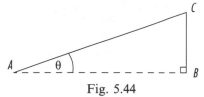

Fig. 5.44

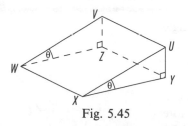

Fig. 5.45

Similarly we can project areas. In Fig. 5.45 U is vertically above Y and V is vertically above Z. WX is common to both planes as their line of intersection. The projection of area $WXUV$ onto the other plane is $WXYZ$, i.e.

Area $WXYZ$ = Area $WXUV \times \cos\theta$.

Suppose $WXYZ$ is a horizontal plane and plane $WXUV$ is inclined at the angle θ. On $WXUV$ there are many lines inclined at many angles to various other lines of reference. An important line is a **line of greatest slope**. A line of greatest slope is inclined at θ to the horizontal, just like its plane.

Example 5.14

The diagram in Fig. 5.46 shows a pyramid with a horizontal rectangular base $WXYZ$ measuring 0.42 m by 0.30 m. The apex of the pyramid, A, is vertically above the centre of the base. The length of each sloping edge is 0.54 m. What are the angles of inclination of the sloping sides to the base?

The centre of the rectangular base $WXYZ$ is B, vertically below A. The opposite sides of a rectangle are equal. Hence for our pyramid the opposite sloping sides are equally inclined to the base.

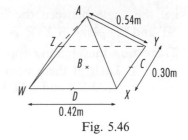

Fig. 5.46

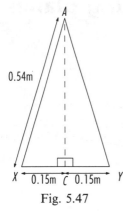

Fig. 5.47

Let us look at the sloping side AXY. XY is the common line of intersection between the plane AXY and the base plane. The symmetry of the pyramid means that triangle AXY is isosceles with $AX = AY$. We can draw just this triangle with AC perpendicular to XY, where C is the mid-point of XY (Fig. 5.47).

In triangle XCA, by Pythagoras' theorem

$$AC^2 + CX^2 = AX^2$$

i.e. $\quad AC^2 + 0.15^2 = 0.54^2$

i.e. $\quad\quad AC^2 = 0.2916 - 0.0225$

$$AC^2 = 0.2691$$

$\therefore \quad\quad\quad AC = \sqrt{0.2691}$

$$= 0.52\,\text{m}.$$

> Calculator value is 0.5187...

Remember B is at the centre of the rectangle and C is the mid-point of side, XY. Hence BC is perpendicular to XY. This means we have lines in each plane both perpendicular to the planes' line of intersection. Again we use a supplementary diagram, Fig. 5.48. We note the length of BC is 0.21 m ($\frac{1}{2} \times 0.42$ m), half the side of the rectangle.

In triangle ABC,

$$\cos\theta = \frac{BC}{AC}$$

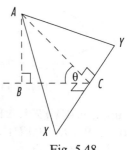

Fig. 5.48

i.e. $\quad \cos \theta = \dfrac{0.21}{0.51 \ldots} = 0.4048.$

$\therefore \qquad \angle \theta = 66.1°$

is the angle of inclination of both sloping sides AXY and AWZ to the rectangular base.

We apply the same technique to the other 2 opposite sides (Fig. 5.49). For the sloping side AWX, WX is the common line of intersection with the base plane. The symmetry of the pyramid means triangle AWX is isosceles with $AW = AX$. We can draw just this triangle with AD perpendicular to WX, where D is the mid-point of WX.

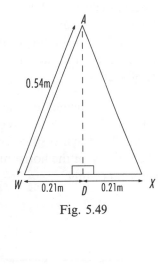

Fig. 5.49

In triangle WDA, by Pythagoras' theorem

$$AD^2 + DW^2 = AW^2$$

i.e. $\quad AD^2 + 0.21^2 = 0.54^2$

i.e. $\qquad AD^2 = 0.2916 - 0.0441$

$\qquad\qquad AD^2 = 0.2475$

$\therefore \qquad\quad AD = \sqrt{0.2475}$

$\qquad\qquad\quad = 0.50\,\text{m}.$

Calculator value is 0.497...

Remember B is the centre of the rectangle and D is the mid-point of side WX. Hence BD is perpendicular to WX. This means we have lines in each plane perpendicular to the planes' line of intersection. Again we use a supplementary diagram, Fig. 5.50. We note the length of BD is 0.15 m ($\frac{1}{2} \times 0.30$ m), half the other side of the rectangle.

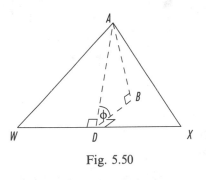

Fig. 5.50

In triangle ABD,

$$\cos \phi = \frac{BD}{AD}$$

i.e. $\quad \cos \phi = \dfrac{0.15}{0.497\ldots} = 0.3015$

$\therefore \qquad \angle \phi = 72.5°$

is the angle of inclination of both sloping sides AWX and AYZ to the rectangular base.

Our final example in this chapter looks at both these angle techniques. We distinguish between the 2 angles by looking at the inclinations of an edge and a sloping face to the horizontal. Much of the numerical work is omitted and left as an exercise for you to do.

Example 5.15

We have a porch to the front of a house with rectangular floor dimensions of 1.6 m and 2.0 m, drawn and labelled in Fig. 5.51. The height of the porch to the internal ceiling is 2.20 m and the overall height is 3.00 m. Calculate the angle of inclination to the horizontal of AX (and hence AY) and of the sloping hipped roof sections.

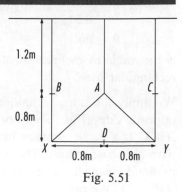

Fig. 5.51

Firstly the height of the roof alone is $3.00 - 2.20 = 0.80$ m. We can project vertically down 0.80 m from A to A' so that A' lies on the horizontal plane of the internal ceiling.

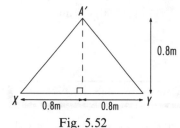

Fig. 5.52

Fig. 5.52 shows $A'XY$ to be an isosceles triangle.

Using Pythagoras' theorem

$$A'X = 1.13\,\text{m}.$$

The projection of the sloping edge AX onto the horizontal is $A'X$. With the aid of Fig. 5.53 we have used the tangent ratio to get $\theta = 35.3°$, i.e. to the horizontal the angle of inclination of AX, and AY, is $35.3°$.

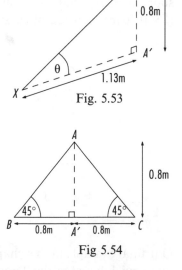

Fig. 5.53

Fig. 5.54 shows a cross-sectional triangle through $A'A$ in the form of triangle ABC. To the horizontal the angle of inclination of the similar sloping sides, using tangent, is $45°$.

Fig 5.54

Similarly we use tangent in right-angled triangle $AA'D$. To the horizontal the angle of inclination of side AXY is $45°$.

■ EXERCISE 5.6 ■

1 A pyramid has a horizontal square base *PQRS* of side 0.50 m. The apex of the pyramid, *A*, is vertically above the centre of the square. Each sloping edge is of length 1.00 m. Calculate the angle between
   i) a sloping edge and the base
and ii) a sloping face and the base.

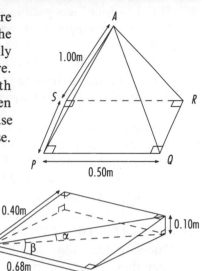

2 In the fully labelled diagram we have a wedge. Find the 2 marked angles α and β.

3 The pyramid in the diagram has a square base of side 0.60 m. Its apex is at an altitude of 0.80 m above the centre of the base. Calculate the length of a sloping edge. What are the angles of inclination of each sloping edge and of each sloping face to the horizontal base?

4 In this diagram of a triangular wedge *A* is vertically above *D* and *B* is vertically above *E*. Both vertical heights are 0.44 m. In triangle *ABC* *AC* = 1.24 m, *BC* = 1.50 m and *AB* = 0.68 m. What is the area of this triangle? What is the area of triangle *CDE*? Find ∠*DCE*.

5 A weight is supported by 4 chains of equal length, 1.4 m. All the chains meet at a point to carry the weight. The other ends of the chains form a horizontal square bolted to a network of steel beams. The weight hangs 0.8 m below the square. What is the angle of inclination between a chain and the vertical? What is the length of a side of the square?

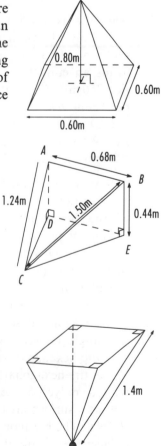

6 The diagram shows a wedge *ABCD* of four faces each in the shape of a triangle. *A* is vertically above *D*. *AB*=0.98 m, *AC*=1.06 m, *BC*=0.56 m and *AD*=0.30 m. Calculate the total surface area of the wedge.

7 The figure shows a hipped roof with a horizontal rectangular base *WXYZ* measuring 8 m by 14 m. Each roof section is inclined at 39° to the horizontal. Estimate the area of roof to be tiled. Ridge tiles will be used along *AB*, *AW*, *AX*, *BY* and *BZ*. Calculate the total length requiring ridge tiles.

8 *ABCDEF* is a 1.50 m length of moulding of triangular cross-section. That cross-section forms an equilateral triangle of side 250 mm. Calculate the angle between *BE* and the plane *ABCD*. *G* is the mid-point of *BC*. What is the angle between *EG* and the plane *ABCD*?

9 A hexagon is formed from a horizontal circle of radius 0.60 m. From each vertex of the hexagon stiff wires create a hexagonal pyramid, joining together at an apex *O*. *O* is 0.80 m above the centre of the circle (and hence the hexagon). Calculate the
  i) angle between a wire and the hexagonal base,
  ii) angle between a sloping face and the hexagonal base.
The entire hexagonal pyramid is covered with a polyester membrane. Calculate the total surface area of the pyramid.
(HINT: The hexagon may be split into 6 adjacent equilateral triangles of side 0.60 m.)

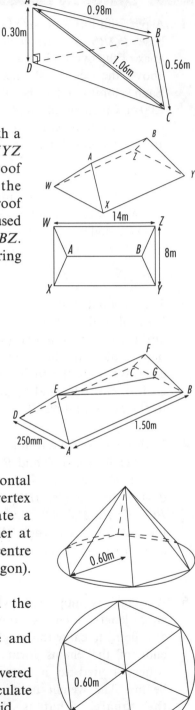

10 The diagram shows a plan of a conservatory *ABCDEF* with marked dimensions. *GH* is a horizontal roofing beam 1.00 m above the tops of the side panels.

$$\angle ABC = 150°$$
$$\angle BCD = 120°$$
$$\angle CDE = 120°$$
$$\angle DEF = 150°$$

What is the angle between the horizontal and
  i) *BG*?
  ii) *CG*?
  iii) plane *ABGH*?
  iv) plane *BCG*?
  v) plane *CDG*?

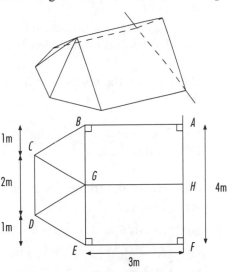

MULTI-CHOICE TEST 5

Questions **1–5** share answer options A) to D) based upon triangle *XYZ*.
  A) sine rule
  B) cosine rule
  C) neither the cosine nor sine rules
  D) both the cosine and sine rules

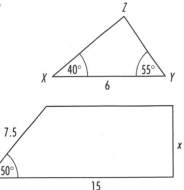

1 Given *x*, *y*, *z* which option would you use attempting to find *Z*?

2 Given *X*, *Y*, *Z* which option would you use attempting to find *x*?

3 Given *x*, *X* and *z* which option would you use attempting to find *Z*?

4 Given *z*, *X* and *Y* which option would you use attempting to find *Z*?

5 Given *z*, *X* and *Y* which option would you use attempting to find *x*?

6 The length of the side *YZ* is given by
  A) $6\cos 55°$
  B) $6\sin 40°$
  C) $6\sin 40° \cos 55°$
  D) $6\sin 40°\operatorname{cosec} 85°$

7 From the diagram $x =$
  A) $7.5\cos 50°$
  B) $7.5\sin 50°$
  C) $15\cos 50°$
  D) $15\sin 50°$

8   We may use the sine rule in any triangle given
   A)   3 angles
   B)   3 sides
   C)   2 sides and a non-included angle
   D)   2 sides and an included angle

9   We may use the cosine rule in any triangle given
   A)   3 angles
   B)   3 sides
   C)   2 sides and a non-included angle
   D)   2 angles and a side

10 The area of the triangle is given by
   A) $\frac{1}{2}ab\sin C$
   B) $\frac{1}{2}ab\cos C$
   C) $\frac{1}{2}(a+b)c$
   D) $\frac{1}{2}(a+b)\sin C$

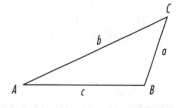

11 A triangle has sides a, b and c with $s=\frac{1}{2}(a+b+c)$. The formula for its area is
   A) $\sqrt{(s-a)(s-b)(s-c)}$
   B) $\sqrt{(s+a)(s+b)(s+c)}$
   C) $\sqrt{s(s-a)(s-b)(s-c)}$
   D) $s(s-a)(s-b)(s-c)$

12 The area of the parallelogram is given by
   A) $\frac{1}{2}bc\sin A$
   B) $ad\sin C$
   C) $bd\sin A$
   D) $ac\sin C$

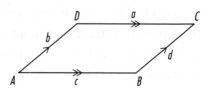

Questions **13** and **14** refer to the trapezium.

13 The area, A, is given by the formula
   A) $\frac{1}{2}ab+h$
   B) $\frac{1}{2}(a+b)h$
   C) $\frac{1}{2}(ab+h)$
   D) $\frac{1}{2}a+bh$

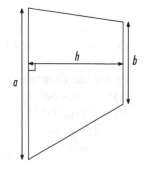

14 Now the lengths in this trapezium are given as $a = 6\,\text{m}$, $b = 3\,\text{m}$ and $h = 2\,\text{m}$. The area, A, is

A) $9\ \text{m}^2$
B) $12\ \text{m}^2$
C) $18\ \text{m}^2$
D) $36\ \text{m}^2$

15 Decide whether each of the statements is true.

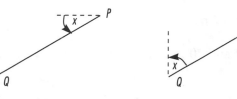

i)   A person stands at P and looks down at Q. x is an angle of depression.

ii)   A person stands at Q and looks up to P. y is an angle of elevation.

Which option best describes the two statements?

A)   i)   T   ii)   T
B)   i)   T   ii)   F
C)   i)   F   ii)   T
D)   i)   F   ii)   F

16 The angle between the line PQ and the plane $WXYZ$ is

A) α
B) β
C) γ
D) δ

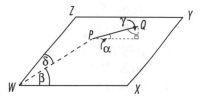

17 The roof $KLMN$ makes an angle of $30°$ with the horizontal ceiling $IJKL$. The area of the ceiling is

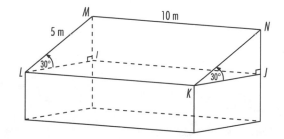

A) $8.66\ \text{m}^2$
B) $25\ \text{m}^2$
C) $43.3\ \text{m}^2$
D) $50\ \text{m}^2$

18   The area of the plane *WXYZ* is 50 m². It
is inclined at 60° to the horizontal plane
UVWX. Hence the area of *UVWX* is
A)   approximately 25 m²
B)   exactly 25 m²
C)   approximately 43 m²
D)   less than 43 m²

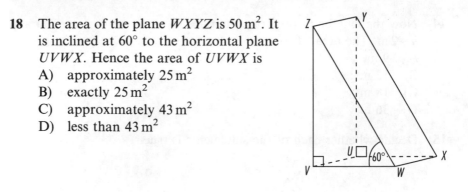

Questions **19** and **20** refer to the diagram of a gable ended roof.

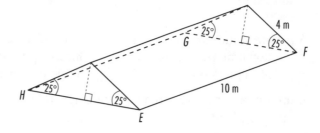

19   The length of *FG* (m) is
A) $4\cos 25°$
B) $8\sin 25°$
C) $8\cos 25°$
D) $10\sin 25°$

20   The area (m²) of the loft floor *EFGH* is
A) $40\cos 25°$
B) $80\sin 25°$
C) $80\cos 25°$
D)   80

6 Using Graphs: I – Trigonometrically

Element: Use graphs to solve engineering problems.

PERFORMANCE CRITERIA
- Coordinates and scales appropriate to the engineering problem are selected.
- Data are accurately plotted.
- The relationship between variables is identified.
- Values are accurately determined from graphs, and engineering problems are solved.

RANGE
Engineering problems: electrical, electronic.
Coordinates: Cartesian.
Scales: linear.
Relationship: sinusoidal.
Values: two variables.

Element: Use phasors to solve engineering problems.

PERFORMANCE CRITERIA
- Phasors appropriate to the engineering problem are identified.
- Phasors are correctly represented.
- Phasors are correctly manipulated.
- Numerical values are substituted and solutions relevant to the engineering problem are obtained.

RANGE
Engineering problems: electrical, electronic.
Representation: magnitude, direction; modulus, argument.
Manipulation: addition, subtraction; graphical.

Element: Use trigonometry to solve engineering problems.

PERFORMANCE CRITERIA
- Trigonometrical ratios appropriate to the engineering problem are identified.
- Trignometrical formulae are manipulated.
- Angles and sides of any triangles are calculated.
- Numerical values are substituted and correct solutions to engineering problems are obtained.

RANGE
Engineering problems: electrical, electronic.
Trigonometrical ratios: $\sin A$, $\cos A$, $\tan A$; angles of any magnitude.
Trigonometrical formulae: sine rule, cosine rule.

Introduction

The majority of this chapter looks at sine and cosine **graphs (waveforms)**. We start with the standard waves for $\sin x$ and $\cos x$ and develop them with different multipliers. We can use either degrees or radians and so remind you of the conversion techniques.

■■■■ ASSIGNMENT ■■■■

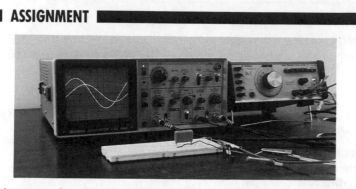

Our Assignment for this chapter looks at a pair of parallel a.c. voltages, $v_1 = 15 \sin 5t$ and $v_2 = 8 \cos 5t$. We will look at the waveforms of these voltages separately. Later we will add them to find a total source voltage.

Graphs of sin *x* and cos *x*

In Fig. 6.1 we refresh our memories with the graphs of $y = \sin x$ and $y = \cos x$ on a pair of axes. You need to look at them carefully and understand their shapes. Look for any similarities and any differences between them. Pay attention to where they cross each axis. They are important because much of our later work is based on these waves. We are going to discuss their features in the next few sections.

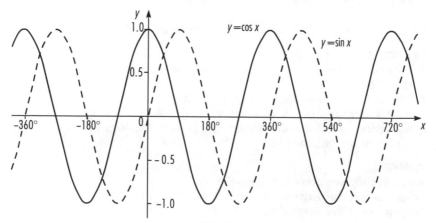

Fig. 6.1

Amplitude

The **amplitude** (**peak value**) is the maximum value from some mean (or equilibrium) position. For many, but by no means all, trigonometric **graphs** (**curves** or **waveforms**) the mean position is the horizontal axis. We understand the amplitude to be a positive value.

The amplitudes of the sine and cosine waves are both 1. On the vertical axis the waveforms lie between -1 and 1. Their mean positions are at the horizontal axis. We may think of

$$y = \sin x \qquad \text{and} \qquad y = \cos x$$

as $\quad y = 1 \sin x \qquad$ and $\qquad y = 1 \cos x.$

> 1 is the amplitude.

More generally $y = A \sin x$ and $y = A \cos x$, where each have an amplitude of A.

Example 6.1

In Fig. 6.2 we sketch the graphs of
 i) $\quad y = \sin x,$
 ii) $\quad y = 2 \sin x,$
 iii) $\quad y = \frac{1}{2} \sin x.$

> Amplitude $= 1.$
> Amplitude $= 2.$
> Amplitude $= \dfrac{1}{2}.$

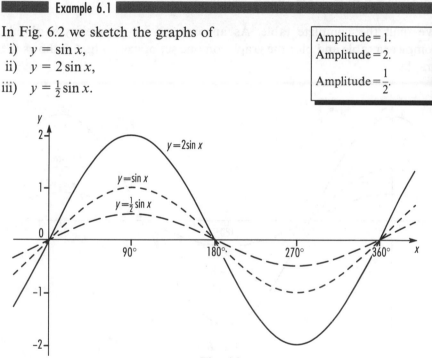

Fig. 6.2

Sometimes you may see $\frac{1}{2} \sin x$ written as $\dfrac{\sin x}{2}$

We omit the complete table but include two specimen calculations.

For $\quad y = 2 \sin x$ when $x = 60°$ we use the calculator as

$$60° \mid \quad \sin \mid \quad \text{to display } 0.8660\ldots$$

and $\qquad \qquad \times \mid \quad 2 \mid \quad$ to display an answer of 1.732 (4 sf).

For $y = \frac{1}{2}\sin x$ when $x = 135°$ we use the calculator as

$\boxed{135°}$ $\boxed{\sin}$ to display 0.7071...

and $\boxed{\times}$ $\boxed{0.5}$ to display an answer of 0.354 (3 dp).

As an exercise for yourself you should complete a table and plot the graphs on one set of axes.

■■■■■■ **Examples 6.2** ■■■■■■

In Fig. 6.3 we sketch the graphs of

i) $y = \cos x$, | Amplitude $= 1$.
ii) $y = 2\cos x$, | Amplitude $= 2$.
iii) $y = \frac{1}{2}\cos x$. | Amplitude $= \frac{1}{2}$.

Sometimes you may see $\frac{1}{2}\cos x$ written as $\dfrac{\cos x}{2}$.

We omit the complete table. As an exercise for yourself you should complete a table and plot the graphs on one set of axes. Fig. 6.3 shows the graphs.

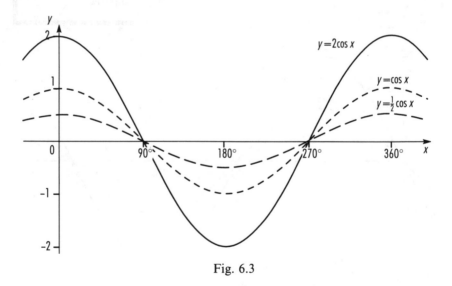

Fig. 6.3

Period

The sine and cosine graphs are continuous waveforms that repeat themselves. This means they are periodic. $y = \sin x$ and $y = \cos x$ repeat themselves every 360° (2π radians). We may think of

$$y = \sin x \qquad \text{and} \qquad y = \cos x$$

as $\quad y = \sin 1x \qquad \text{and} \qquad y = \cos 1x$

$$\boxed{\frac{360°}{1} = 360°.}$$

More generally $y = \sin ax$ and $y = \cos ax$ repeat themselves every $\dfrac{360°}{a}$.

This means the graphs of $y = \sin x$ and $y = \cos x$ are periodic, with period $360°$ (2π radians).

Examples 6.3

In Fig. 6.4 we sketch the graphs of

i) $\quad y = \sin x, \qquad a = 1, \qquad\qquad \text{period} = \dfrac{360°}{1} = 360°.$

ii) $\quad y = \sin 2x, \qquad a = 2, \qquad\qquad \text{period} = \dfrac{360°}{2} = 180°.$

iii) $\quad y = \sin \tfrac{1}{2}x, \qquad a = \tfrac{1}{2}, \qquad\qquad \text{period} = \dfrac{360°}{\tfrac{1}{2}} = 720°.$

The cycle for $y = \sin 2x$ is compressed into $180°$. In contrast, the cycle for $y = \sin \tfrac{1}{2}x$ is extended over $720°$. We omit the complete table but include two specimen calculations.

For $\quad y = \sin 2x$ when $x = 45°$ we use the calculator with

$\boxed{45°} \quad \boxed{\times} \quad \boxed{2} \quad \boxed{=}$ to display $90°$

and $\boxed{\sin}$ to display an answer of 1.

For $\quad y = \sin \tfrac{1}{2}x$ when $x = 150°$ we use the calculator as

$\boxed{150°} \quad \boxed{\times} \quad \boxed{0.5} \quad \boxed{=}$ to display $75°$

and $\boxed{\sin}$ to display an answer of 0.966.

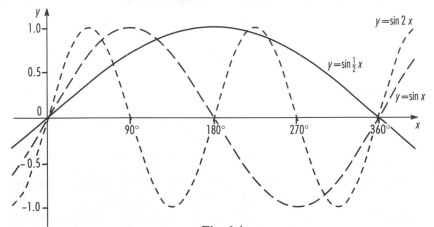

Fig. 6.4

━━━━━━ **Examples 6.4** ━━━━━━

In Fig. 6.5 we sketch the graphs of

i) $y = \cos x$, $a = 1$, period $= \dfrac{360°}{1} = 360°$.

ii) $y = \cos 2x$, $a = 2$, period $= \dfrac{360°}{2} = 180°$.

iii) $y = \cos \tfrac{1}{2} x$, $a = \tfrac{1}{2}$, period $= \dfrac{360°}{\frac{1}{2}} = 720°$.

The cycle for $y = \cos 2x$ is compressed into $180°$. In contrast, the cycle for $y = \cos \tfrac{1}{2} x$ is extended over $720°$.

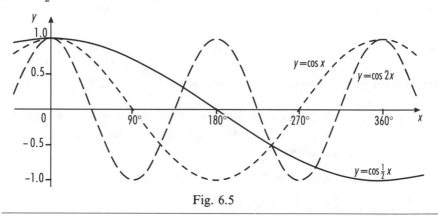

Fig. 6.5

We often refer to **period** as **time period**. The units for time are *not* degrees. Instead of $y = \sin ax$ and $y = \cos ax$ we use time, t, $y = \sin \omega t$ and $y = \cos \omega t$ with radians.

Time period, $T = \dfrac{2\pi}{\omega}$ seconds.

━━━━━━ **Examples 6.5** ━━━━━━

In terms of time, t, we look at time period using radians.

i) $y = \cos t$, $\omega = 1$, time period $= \dfrac{2\pi}{1} = 2\pi$.

ii) $y = \cos 2t$, $\omega = 2$, time period $= \dfrac{2\pi}{2} = \pi$.

iii) $y = \cos \tfrac{1}{2} t$ $\omega = \tfrac{1}{2}$, time period $= \dfrac{2\pi}{\frac{1}{2}} = 4\pi$.

We have looked at the basic waves of $y = A \sin ax$ and $y = A \cos ax$ separately, altering A and a. Just as easily, we can alter them both together. In Examples 6.6 we look at the necessary order of calculator operations. A complete table and plot is left for you to do in Exercise 6.1.

Examples 6.6

i) For $y = 3\sin 4x$, let $x = 75°$. We find y by using

| 4 | × | 75° | = | to display 300°,

sin| to display $-0.8660\ldots$

and   × | 3 | = | to display an answer of -2.598.

ii) We can think of $y = 2.5\cos\dfrac{1}{3}x$ as $y = 2.5\cos\dfrac{x}{3}$. Let $x = 150°$ and find y by using

| 150° | ÷ | 3 | = | to display 50°,

cos| to display $0.642\ldots$

and   × | 2.5 | = | to display an answer of 1.607.

For our next variation we see what happens when we add a number rather than simply use multiplication.

Example 6.7

For $y = 2 + \sin x$ our original sine wave is shifted vertically upwards by 2. This vertical shift has no effect on the size of the amplitude, angular velocity, period or frequency. For this example involving $\sin x$ these values remain as 1, 1, 2π and $\dfrac{1}{2\pi}$ respectively. Again we omit the complete table but include a specimen calculation. We let $x = 35°$ and find y by using

| 35° | sin| to display $0.5735\ldots$

and   + | 2 | = | to display 2.574.

We show the wave in Fig. 6.6.

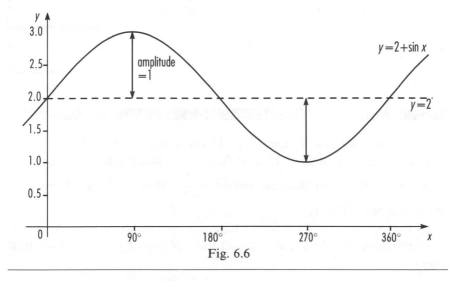

Fig. 6.6

Frequency

We define **frequency**, **f**, to be the number of cycles every second. The units of frequency are Hertz (Hz).

$$\textbf{Frequency} = \frac{\textbf{Number of cycles}}{\textbf{1 second}}$$

$$= \frac{1 \text{ cycle}}{\text{Time period}}$$

$$= \frac{1}{T} = \frac{1}{\dfrac{2\pi}{\omega}}$$

i.e. $f = \dfrac{\omega}{2\pi}$ **Hz**.

$\boxed{\text{Also } f = \dfrac{1}{T}.}$

$\omega = 2\pi f$ is the **angular velocity** with units of radians per second (rads^{-1}).

Example 6.8

$$y = 4\sin\frac{3}{5}t$$

$\boxed{y = A\sin\omega t.}$

has $\omega = \dfrac{3}{5}$ or 0.6.

This means we have an angular velocity of 0.6 rads^{-1}.

We have a time period of $\dfrac{2\pi}{0.6} = 10.5$ seconds.

$\boxed{T = \dfrac{2\pi}{\omega}.}$

Also we have a frequency of $\dfrac{0.6}{2\pi} = 0.095$ Hz.

$\boxed{f = \dfrac{\omega}{2\pi}.}$

ASSIGNMENT

Let us look at our a.c. voltages, $v_1 = 15\sin 5t$ and $v_2 = 8\cos 5t$.

For v_1 the amplitude is 15 and for v_2 the amplitude is 8. For both voltages $\omega = 5$ and so the time period is $\dfrac{2\pi}{5} = 0.4\pi = 1.256$ s. Using $f = \dfrac{1}{T}$ the frequency is 0.80 Hz.

In Fig. 6.7 we plot these waves together on one set of axes. Again we omit the table of values, but you will have plenty of practice in the next exercise.

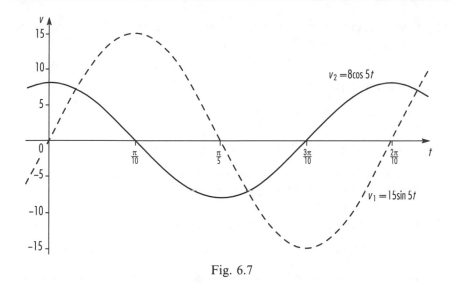

Fig. 6.7

■■■■ EXERCISE 6.1 ■■■■

For the trigonometrical waveforms in Question **1–5** write down the
i)   amplitude,
ii) angular velocity,
iii) period (degrees and radians), and
iv) frequency.

1 $y = \sin 3x$.

2 $y = 5 \cos 2t$.

3 $y = \cos 3x$.

4 $y = 2 \sin 3x$.

5 $y = -1 + \cos t$.

6 Over 1 cycle sketch the waveforms in Questions **1–5**.

7 Write down the equation of the sinusoidal wave with an amplitude of 4 and a frequency of 2.5.

8 You are given the equation $y = 2 \cos 3t$. Write down the sine wave with the same amplitude and twice the frequency.

9 On one pair of labelled axes over 1 cycle sketch the graphs of
i) $y_1 = \sin 3x$, ii) $y_2 = 2 \sin 3x$, iii) $y_3 = 4 + 2 \sin 3x$. Clearly label where each curve cuts the axes.

10 For the wave $y = 3 \sin 4x$ construct a table of values and plot the graph on a fully labelled pair of axes. Use values of x from $0°$ to $100°$ at intervals of $5°$.

11 In the range 0 to 2π radians on one set of labelled axes sketch
i) $y_1 = \cos x$, ii) $y_2 = \cos 2x$, iii) $y_3 = 1 + \cos 2x$. Clearly label where each curve cuts the axes.

12 Over 1 cycle on the same axes sketch the graphs of $y = 2 \sin x$ and $y = -2 \sin x$. Clearly label each curve.

13 For the wave $y = 4 + 2 \sin 3t$ construct a table of values and plot the graph on a fully labelled pair of axes. Use values of t from 0 to $\frac{2\pi}{3}$ radians at intervals of $\frac{\pi}{18}$.

14 What is the period of the wave $y = \frac{1}{2}(1 - \cos 2x)$? Construct a table of values from $0°$ over this period at intervals of $15°$. Hence plot a fully labelled graph of y against x.

15 Plot the graph of $y = \frac{1}{2}(1 + \cos 2x)$ over 1 cycle. Your table of values should be from $x = 0°$ at intervals of $15°$. Fully label your graph.

Graphs of $\sin^2 x$ and $\cos^2 x$

One of our original waves was $y = \sin x$. We base the graph of $y = \sin^2 x$ on this wave, understanding $\sin^2 x$ to be $(\sin x)^2$, i.e. we square the values of our original sine wave. This makes all the negative values positive and so the wave does *not* lie below the horizontal axis. We omit the complete table but include a specimen calculation. Let $x = 240°$ so our order of calulator operations is

$$240° \underline{| \quad \sin |} \quad \text{to display } -0.8660 \ldots$$

and $\underline{\quad x^2 |} \quad$ to display 0.75.

Fig. 6.8 shows both waves on one set of axes.

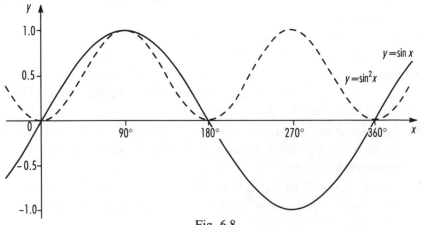

Fig. 6.8

Exactly the same ideas apply to $y = \cos^2 x$, being the squared values of our original cosine wave. Again, we include a specimen calculation with $x = 135°$ so that

$$\boxed{135°} \quad \boxed{\cos} \quad \text{to display } -0.7071 \dots$$

and $\boxed{x^2}$ to display 0.5.

Fig. 6.9 shows both waves on one set of axes

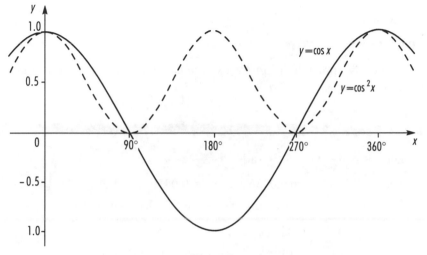

Fig. 6.9

In Fig. 6.10 we bring both $y = \sin^2 x$ and $y = \cos^2 x$ together on one set of axes. Notice how they oscillate about the line $y = 0.5$, i.e. this is their mean position. Then the amplitude (maximum displacement from this mean position) is 0.5. Also, the waves repeat themselves every 180° (π radians).

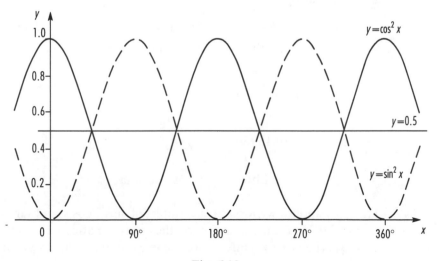

Fig. 6.10

Approximations

We have **approximations in radians** for $\sin x$, $\cos x$ and $\tan x$. These are, provided x is small, generally less than $0.10\,\text{rad}$ ($5°$ or $6°$). Remember that $1° = \dfrac{\pi}{180}$ radians.

$\sin x \approx x$.

$\cos x \approx 1 - \dfrac{x^2}{2}$.

$\tan x \approx x$.

To avoid early approximation we use the calculator memory and quite a few decimal places.

Examples 6.9

i) Let $x = 2.5° = 0.0436\ldots$ rad.

Then $\quad \sin x = \sin 0.0436\ldots \qquad = 0.0436\ldots$

only differs in the 5th decimal place.

$\tan x = \tan 0.0436\ldots \qquad = 0.0436\ldots$

only differs in the 5th decimal place.

$\cos x = \cos 0.0436\ldots \qquad = 0.999048\ldots$

and $\quad 1 - \dfrac{x^2}{2} = 1 - \dfrac{(0.0436\ldots)^2}{2} = 0.999048\ldots$

is correct to the first 6 decimal places.

ii) Let $x = 8° = 0.1396$ rad.

Then $\quad \sin x = \sin 0.1396 \qquad = 0.13917\ldots$

only differs in the 4th decimal place.

$\tan x = \tan 0.1396 \qquad = 0.1405\ldots$

differs in the 2nd decimal place.

$\cos x = \cos 0.1396\ldots \qquad = 0.990268\ldots$

and $\quad 1 - \dfrac{x^2}{2} = 1 - \dfrac{(0.13966\ldots)^2}{2} = 0.990252\ldots$

only differs in the 5th decimal place.

In Figs. 6.11(a), (b) and (c) we look at the graphs of $\sin x$, $\cos x$ and $\tan x$ for small values of x. You can see how good the approximations are close to the vertical axis (i.e. $x = 0$). Only as we move away with the size of x increasing do the graphs and their approximations diverge.

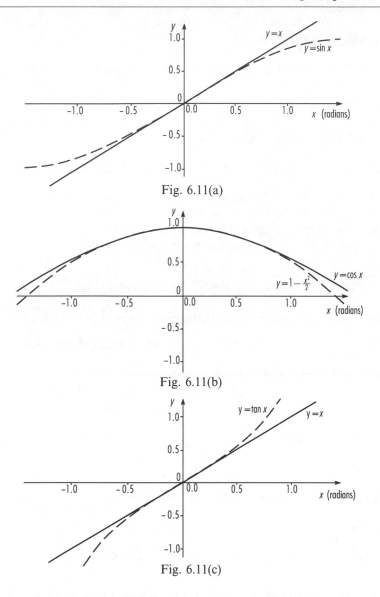

Fig. 6.11(a)

Fig. 6.11(b)

Fig. 6.11(c)

EXERCISE 6.2

For $\sin x$, $\cos x$ and $\tan x$ test the accuracy of the approximations. In each case state at which decimal place the trigonometric function and its approximation start to differ.

1	0.100 rad	6	8.5°
2	0.125 rad	7	$-10°$
3	-0.150 rad	8	$-7.5°$
4	0.200 rad	9	12.5°
5	-0.185 rad	10	11°

Phase angle

So far we have seen sine and cosine waves crossing the horizontal axis at
90°, 180°, 270° and 360° etc. When we alter the frequency we get multiples
and sub-multiples of these values. Also we have seen a vertical shift of our
graphs, e.g. $y = 2 + \sin x$. In this section we look at a horizontal shift. A
horizontal shift creates a **phase difference**.

▮▮▮▮▮▮ Example 6.10 ▮▮▮▮▮▮

In Fig. 6.12 we sketch the graphs of $y = \sin x$ and $y = \sin(x + 30°)$. You
can see we shift the first wave horizontally by 30° ($\pi/6$ radians) to get the
second wave. Again, we omit the complete table but include two specimen
calculations. When $x = 20°$ we use the calculator with

$\underline{20°|}$ $\underline{+|}$ $\underline{30°|}$ $\underline{=|}$ to display 50°

and $\underline{\sin|}$ to display an answer of 0.766...

In radians we can rewrite 30° and our equation as either $y = \sin\left(x + \dfrac{\pi}{6}\right)$ or
$y = \sin(x + 0.523\ldots)$. When $x = 1.375$ we use the calculator with

$\underline{1.375|}$ $\underline{+|}$ $\underline{0.523\ldots|}$ $\underline{=|}$ to display 1.898...

and $\underline{\sin|}$ to display an answer of 0.946...

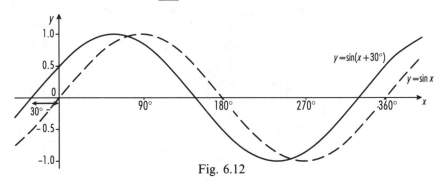

Fig. 6.12

We have marked the **phase angle** of 30° on our diagram. Notice that we
have shifted the graph of $y = \sin x$ horizontally to the left to get
$y = \sin(x + 30°)$. $y = \sin(x + 30°)$ **leads** $y = \sin x$. We can think of this as
reaching the peak value first. Alternatively, $y = \sin x$ **lags** $y = \sin(x + 30°)$.

▮▮▮▮▮▮ Example 6.11 ▮▮▮▮▮▮

In Fig. 6.13 we sketch the graphs of $y = \sin t$ and $y = \sin\left(t - \dfrac{\pi}{4}\right)$. You can
see we shift the first wave horizontally $\pi/4$ radians to get the second wave.
Again, we omit the complete table. The calculations follow like those in
Example 6.10 using subtraction rather than an addition. We have marked
the **phase angle** of $\pi/4$ on our diagram. Notice that we have shifted the

graph of $y = \sin t$ horizontally to the right to get $y = \sin\left(t - \dfrac{\pi}{4}\right)$. $y = \sin\left(t - \dfrac{\pi}{4}\right)$ **lags** $y = \sin t$. We can think of this as reaching the peak value later. Alternatively, $y = \sin t$ **leads** $y = \sin\left(t - \dfrac{\pi}{4}\right)$.

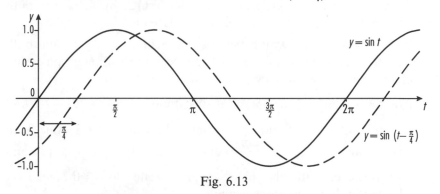

Fig. 6.13

Generally we can shift our waves horizontally and compare them against $y = R \sin \omega t$.

Then $y_1 = R \sin(\omega t + \alpha)$ **leads** $y = R \sin \omega t$

and $y_2 = R \sin(\omega t - \alpha)$ **lags** $y = R \sin \omega t$,

each with a phase angle of α.

Exactly the same ideas apply to cosine graphs:

$y_1 = R \cos(\omega t + \alpha)$ **leads** $y = R \cos \omega t$

and $y_2 = R \cos(\omega t - \alpha)$ **lags** $y = R \cos \omega t$,

each with a phase angle of α.

Let us return to our basic sine and cosine curves, shown in Fig. 6.14. You can see that the cosine curve leads the sine curve by 90° ($\pi/2$ radians). We may write that $y = \cos x$ is equivalent to $y = \sin(x + 90°)$. Alternatively, the sine curve lags the cosine curve. We may write that $y = \sin x$ is equivalent to $y = \cos(x - 90°)$.

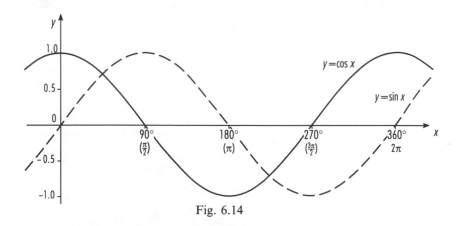

Fig. 6.14

▰▰▰▰ EXERCISE 6.3 ▰▰▰▰▰▰▰▰▰▰▰▰▰▰▰▰▰▰

1   Write down the cosine graph that leads $y = 4\cos t$ by $\dfrac{\pi}{3}$. Sketch both waves on one set of axes with clear labels.

2   Write down the sine graph that lags $y = 3\sin 2t$ by $\dfrac{\pi}{6}$. Sketch both waves on one set of axes with clear labels.

3   Write down the sine graph that leads $y = \sin x$ by $90°$. Sketch both these waves on one set of axes. We have seen the leading graph earlier in this chapter. Write down its alternative equation.

4   Plot the graphs of current, i, against time, t, $i_1 = \sin\left(2t + \dfrac{\pi}{6}\right)$ and $i_2 = \sin 2t$. Construct your table using values of t over 1 cycle from $t = 0$ radians at intervals of $\dfrac{\pi}{12}$ radians. Clearly label your graphs and axes together with the phase angle. Write down the lead/lag relationship between these currents.

5   Plot the graphs of voltage, v, against time, t, $v_1 = \sin\left(t + \dfrac{\pi}{3}\right)$ and $v_2 = \sin\left(t - \dfrac{\pi}{4}\right)$. Construct your table using values of t from $t = -\dfrac{\pi}{2}$ to $t = \dfrac{5\pi}{2}$. Clearly label your graphs and axes together with the phase angle. Write down the lead/lag relationship between the voltages.

Graphical wave addition – same frequencies

This is simply done. We construct a table of values and add those values where necessary. In Example 6.12 we do this in stages, noting the result is sinusoidal.

▰▰▰▰▰ Example 6.12 ▰▰▰▰▰▰▰▰▰▰▰▰▰▰▰▰▰

Plot the graph of $y = 3\sin x + 4\cos x$ over 1 cycle.

We break this down in stages, looking separately at $3\sin x$ and $4\cos x$. Our table shows values of x in degrees but radians will work just as easily. Create your own table from $x = 0°$ to $x = 360°$ at intervals of either $10°$ or $5°$.

x	$0°$	$30°$	$45°$	...	
$\sin x$	0	0.500	0.707	...	
$3\sin x$	0	1.500	2.121	...	y_1
$\cos x$	1	0.866	0.707	...	
$4\cos x$	4	3.464	2.828	...	y_2
$3\sin x + 4\cos x$	4	4.96	4.95	...	$y_1 + y_2$

We have plotted all three graphs in Fig. 6.15. We are particularly interested in the combined wave, $y = 3\sin x + 4\cos x$. It retains the sinusoidal shape with a phase shift. Notice the amplitude is 5 and the phase angle is 53.13° compared with $y = 5\sin x$. In fact, our graph may be alternatively labelled as $y = 5\sin(x + 53.13°)$. $y = 3\sin x + 4\cos x$ is equivalent to $y = 5\sin(x + 53.13°)$.

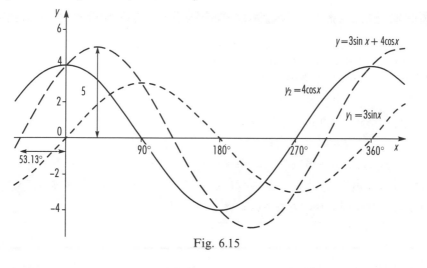

Fig. 6.15

ASSIGNMENT

We can add our a.c. voltages $v_1 = 15\sin 5t$ and $v_2 = 8\cos 5t$ graphically using the method from Example 6.12. This time we ought to use radians over 1 cycle from $t = 0$ to $t = 1.26$ s. As an exercise for yourself, construct a table and plot the combined wave of $v = 15\sin 5t + 8\cos 5t$ at intervals of 0.05 or 0.10 seconds. You can check your own attempt against our graph in Fig. 6.16. Unlike Example 6.12, we have highlighted the combined wave by plotting it alone.

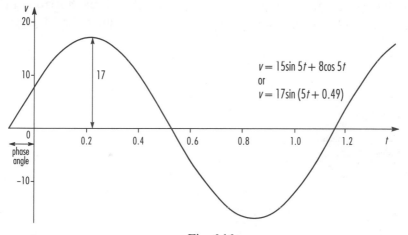

Fig. 6.16

Notice the amplitude is 17. Notice the size of our phase shift is shown as $\frac{0.49}{5} = 0.10$ (2dp) because we have plotted t rather than $5t$ horizontally. In fact, our graph may be alternatively labelled as $v = 17\sin(5t + 0.49)$. $v = 15\sin 5t + 8\cos 5t$ is equivalent to $v = 17\sin(5t + 0.49)$.

■ EXERCISE 6.4 ■

In the following questions, graphically add the given waves. Your table should cover 1 cycle of values. Clearly label the phase angle. State the lead/lag relationship between your equivalent sine wave and a sine wave of the same frequency.

1   $y_1 = 5\sin x$ and $y_2 = 12\cos x$.

2   $y_1 = \sin 2t$ and $y_2 = 2\cos 2t$.

3   $y_1 = 12\sin x$ and $y_2 = 12\cos x$.

4   $y_1 = 3\sin 2t$ and $y_2 = 6\cos 2t$.

5   $y_1 = 9\sin 3x$ and $y_2 = 6\cos 3x$.

Phasor addition

We only apply this method to waves of the **same frequency**. We know from earlier graphs that the sine and cosine waves of the same frequency are 90° out of phase, i.e. the cosine leads the sine by 90°. This allows us to use them at right-angles to each other. We use the amplitudes in each case. Our method involves drawing the amplitude of the sine wave horizontally and of the cosine wave vertically. Then we complete the rectangle. The diagonal from the origin is the amplitude of the combined waves. Its inclination to the horizontal is the phase angle. Above the horizontal the phase angle is positive, and negative below. We can use either degrees or radians.

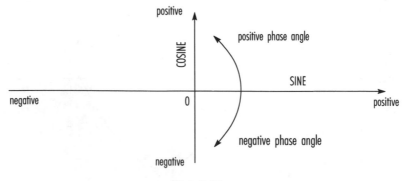

Fig. 6.17

Example 6.13

Use phasor addition to find the sine wave representing $y = 3 \sin x + 4 \cos x$.

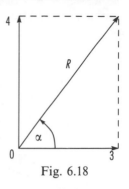

We saw this combination of sine and cosine in Example 6.12. This time we draw a horizontal line of length 3 (amplitude of $3 \sin x$) and a vertical line of length 4 (amplitude of $4 \cos x$), as in Fig. 6.18.

We complete the rectangle and can measure the diagonal, R. Alternatively we can apply Pythagoras' theorem,

$$R^2 = 3^2 + 4^2$$

i.e. $R = \sqrt{9 + 16} = 5,$ using the positive root.

Fig. 6.18

For the phase angle, α, we can measure the inclination of the diagonal to the horizontal. Alternatively we can use some simple trigonometry

$$\tan \alpha = \frac{4}{3} = 1.\overline{3}$$

i.e. $\alpha = 53.13°.$

Obviously we *cannot* expect such accuracy with measurement. As an alternative we could have used radians just as easily as degrees. Finally we write our combined wave as $y = 5 \sin(x + 53.13°)$, as in Example 6.12.

Example 6.14

Use phasor addition to find the sine wave representing $y = 12 \sin x - 5 \cos x$.

We draw a horizontal line of length 12 (amplitude of $12 \sin x$) and a vertical line of length 5 (amplitude of $5 \cos x$) in Fig. 6.19. Notice the vertical line is downwards, to be consistent with a negative cosine.

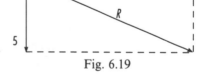

Fig. 6.19

We complete the rectangle and can measure the diagonal, R, or apply Pythagoras' theorem,

$$R^2 = 12^2 + 5^2$$

i.e. $R = \sqrt{144 + 25} = 13,$ using the positive root.

For the phase angle, α, we can measure the inclination of the diagonal to the horizontal or use some simple trigonometry,

$$\tan \alpha = -\frac{5}{12}$$

i.e. $\alpha = -22.62°.$

This means our combined wave is $y = 13 \sin(x - 22.62°)$.

Use phasor addition to find the sine wave representing
$y = 7.5 \sin x + 4 \sin(x + 60°)$.

   We draw a horizontal line of length 7.5 and a line of length 4 inclined at
60° above the horizontal (sine wave with leading phase angle 60°) in
Fig. 6.20. This time we complete the parallelogram. Again we can measure
for our approximate values and calculate for greater accuracy. Without
right-angled triangles we need to use the sine and cosine rules.

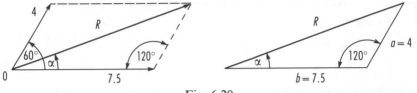

Fig. 6.20

Using the cosine rule,
$$a^2 = b^2 + c^2 - 2bc \cos A$$
we substitute to get
$$R^2 = 4^2 + 7.5^2 - 2(4)(7.5) \cos 120°$$
$$= 16 + 56.25 - 2(4)(7.5)(-0.5)$$
i.e. $\qquad R^2 = 102.25$
$\therefore \qquad R = \sqrt{102.25} = 10.1 \qquad$ is the amplitude.

Using the sine rule,
$$\frac{a}{\sin A} = \frac{b}{\sin B}$$
we substitute to get
$$\frac{R}{\sin 120°} = \frac{4}{\sin \alpha}$$
i.e. $\qquad \sin \alpha = \dfrac{4 \times 0.866}{10.1\ldots} = 0.3426$
$\therefore \qquad \alpha = 20.0° \qquad$ is the phase angle.

This means our combined wave is $y = 10.1 \sin(x + 20.0°)$.

ASSIGNMENT

We can add our a.c. voltages
$v_1 = 15 \sin 5t$ and $v_2 = 8 \cos 5t$ using
phasor addition. We draw a horizontal line of length 15 and a vertical
line of length 8 in Fig. 6.21.

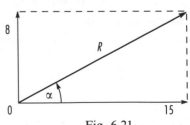

Fig. 6.21

We complete the rectangle and apply Pythagoras' theorem,

$$R^2 = 15^2 + 8^2$$

i.e. $R = \sqrt{225 + 64} = 17,$ using the positive root.

For the phase angle, α, we use simple trigonometry,

$$\tan \alpha = \frac{8}{15} = 0.5\bar{3}$$

i.e. $\alpha = 0.49$ rads.

Finally we write our combined wave as $y = 17 \sin(5t + 0.49)$.

■■■■ EXERCISE 6.5 ■■■■■■■■■■■■■■■■■■■■■■■■■■■■■■■

Use phasor addition to add the given waves. In each case calculate the amplitude and phase angle. Write down the combined wave in the form of a sine.

1   $y_1 = 12 \sin x$ and $y_2 = 12 \cos x.$

2   $y_1 = 6 \sin 2t$ and $y_2 = 4 \cos 2t.$

3   $y_1 = 7 \sin 3x$ and $y_2 = 3.5 \cos 3x.$

4   $i_1 = \sin\left(2t + \dfrac{\pi}{6}\right)$ and $i_2 = \sin 2t.$

5   $v_1 = \sin\left(t + \dfrac{\pi}{3}\right)$ and $v_2 = \sin\left(t - \dfrac{\pi}{4}\right).$

Graphical wave addition – different frequencies

We need to add waves of different frequencies graphically. This is the only useful method. The different frequencies mean the resultant wave is no longer sinusoidal. However, we still have continuous cycles (repetition).

▨▨▨▨ Example 6.16 ▨▨▨▨▨▨▨▨▨▨▨▨▨▨▨▨▨▨▨▨▨▨▨▨

Plot the graph of y against x where $y = 3 \sin x + 4 \sin 2x$.

As usual, we start with a table of values. Complete your own table from $x = 0°$ to $x = 360°$ at intervals of 5° or 10°. Use some of our specimen table values as a check.

x	0°	15°	30°	45°	60°	...
$\sin x$	0	0.259	0.500	0.707	0.866	...
$3 \sin x$	0	0.78	1.50	2.12	2.60	...
$2x$	0°	30°	60°	90°	120°	...
$\sin 2x$	0	0.500	0.866	1.000	0.866	...
$4 \sin 2x$	0	2.00	3.46	4.00	3.46	...
$3 \sin x + 4 \sin 2x$	0	2.78	4.96	6.12	6.06	...

We have drawn the graph in Fig. 6.22. You can see it has a period of 360°
but is *not* sinusoidal. We *cannot* find a simpler equation to represent it.

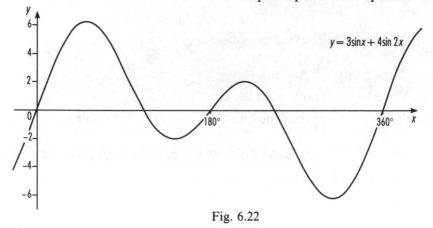

Fig. 6.22

Our final selection of questions looks at graphical addition for waves of
different frequencies.

■■■■■ EXERCISE 6.6 ■■■■■

1   $y = 5\sin 2x + 10\sin x$. Construct a table of values for y from $x = 0°$ to
$x = 360°$ at intervals of 30°. Plot a graph of y against x on a clearly
labelled pair of axes.

2   Plot the graph of y against t where $y = \sin t + \sin 3t$. Your table of
values should start with $t = 0$ and finish with $t = 2\pi$ radians at
intervals of $\dfrac{\pi}{6}$ radians.

3   Using values of x from $x = -180°$ to $x = 180°$ at intervals of 30° plot
the graph of $y = 12(\sin 2x + \cos x)$. From your graph read off the
value of y when $x = 100°$. Check the accuracy of your answer by
substituting for $x = 100°$ in the original equation.

4   Plot the graph of y against time, t, where $y = \sin\left(t + \dfrac{\pi}{6}\right) + \sin 2t$.
Construct your table using values of t from $t = 0$ to $t = 2\pi$ radians at
intervals of $\dfrac{\pi}{9}$ radians. Clearly label your graph and axes. When
$y = 0.5$ what are the values of t?

5   Plot the graph of y against time, t, where
$y = \sin\left(t + \dfrac{\pi}{3}\right) - \sin\left(t - \dfrac{\pi}{4}\right)$.
Construct your table from $t = -\pi$ to $t = \pi$ radians. Clearly label your
graph and axes.

■ MULTI-CHOICE TEST 6 ■

1 Where $x = 215°$ and $y = 2 \sin 3x$ we get
   A) -1.9319
   B) -1.1472
   C) -0.9659
   D) 22.5102

2 If you plot the wave $y = \sin \dfrac{\theta}{4}$ what will be the largest value of y?
   A) -4
   B) 1
   C) $\dfrac{1}{4}$
   D) 4

3 The graph shows
   A) $y = \sin x$
   B) $y = 3 \cos x$
   C) $y = \cos x$
   D) $y = 3 \sin x$

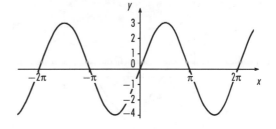

4 Where ω is the angular velocity the formula for time period is given by
   A) $\dfrac{\pi}{2\omega}$

   B) $\dfrac{\omega}{2\pi}$

   C) $\dfrac{2\pi}{\omega}$

   D) $\dfrac{2\omega}{\pi}$

5 The graph shows a particular sinusoidal waveform. Its period is
   A) 4
   B) 5
   C) 9
   D) 10

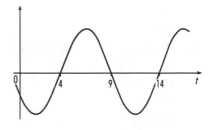

6   The graph shows $y=$
   A) $\frac{1}{2}\cos x$
   B) $\cos\frac{1}{2}x$
   C) $\frac{1}{2}\cos 2x$
   D) $2\cos\frac{1}{2}x$

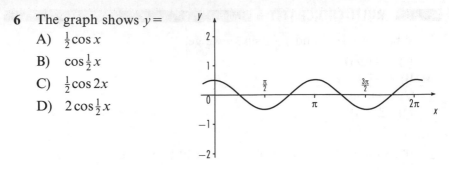

7   The waveform $y=3\cos\dfrac{x}{3}$ is represented by

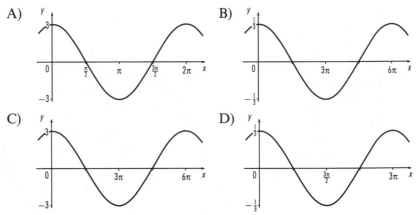

   A)

   B)

   C)

   D)

8   If you plot the wave $y=\cos\dfrac{\theta}{3}$ after how many degrees will it repeat itself?
   A)   120°
   B)   360°
   C)   540°
   D)   1080°

9   The distance, d (m), a mass oscillates is related to time, t (s), by $d=\sin\frac{1}{2}t$. Its frequency is

   A) $\dfrac{1}{4\pi}$

   B) $\dfrac{1}{2\pi}$

   C) 4π
   D) 8π

10   The waveform $y=\sin 2.5x$ has a period of
   A)   0.4 rad
   B)   2.5 rad
   C)   144°
   D)   360°

11 The graph shows the curve $y =$

A) $2\cos\dfrac{\theta}{3}$

B) $2\cos 3\theta$

C) $3\cos\dfrac{\theta}{2}$

D) $3\cos 2\theta$

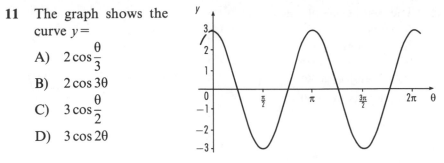

12 An alternating current, i (A), in terms of time, t (s) is given by $i = \sin 50\pi t$.

Decide whether each of the statements is True (T) or False (F).

i) The angular velocity is 50π

ii) The time period is $\dfrac{1}{25}$

Which option best describes the two statements?

A) i) T   ii) T
B) i) T   ii) F
C) i) F   ii) T
D) i) F   ii) F

13 The graph shows $y =$

A) $\cos 2t$
B) $\sin 2t$
C) $\sin^2 t$
D) $\cos^2 t$

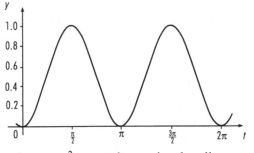

14 The graphs of $y = \sin^2 t$ and $y = \cos^2 t$ are shown in the diagram. Adding these curves gives

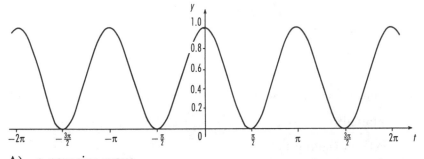

A) a new sine wave
B) a straight line $y = 1$
C) a straight line $y = 0.5$
D) a straight line $y = 2$

15 The equation of the curve is

A) $y = \sin(x + 60°)$
B) $y = \sin(x - 60°)$
C) $y = \cos(x + 60°)$
D) $y = \cos(x - 60°)$

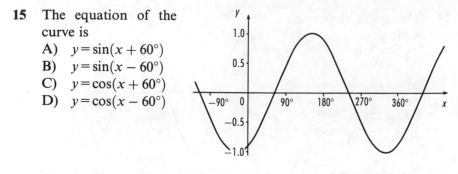

16 The angular velocity of the sine wave $y = 3\sin(2\theta - 45°)$ is

A) 3
B) 2
C) $\frac{1}{2}$
D) 45°

17 A cosine wave has an amplitude of 5 and a phase angle of 0.2 radians lagging a standard cosine waveform. Its equation is

A) $y = 5\cos(\theta - 0.2)$
B) $y = 5\cos(\theta + 0.2)$
C) $y = 0.2\cos(\theta + 5)$
D) $y = 0.2\cos(\theta - 5)$

18 The phasor for $\sin\theta + \cos\theta$ is shown by

A)   B)

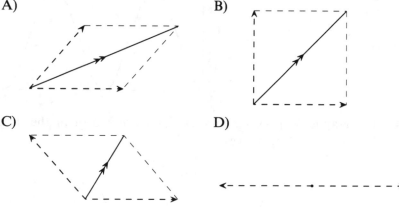

C)   D)

19 On one set of axes from 0 to 2π radians the following graphs are plotted: $y_1 = \cos t$, $y_2 = \cos\left(t + \dfrac{\pi}{3}\right)$ and $y_3 = \cos\left(t - \dfrac{\pi}{4}\right)$

A) y_3 leading y_2
B) y_1 lagging y_3
C) y_2 leading y_3
D) y_2 lagging y_1

20 $y_1 = 2 \sin 3t$, $y_2 = 3 \cos 2t$ and $y = y_1 + y_2$. Plotting y against t gives a

A) sine graph

B) cosine graph

C) graph that is neither a sine nor a cosine

D) straight line graph

7 Introducing Vectors

Element: Use phasors and vectors to solve engineering problems.

PERFORMANCE CRITERIA
- Phasor and vector quantities appropriate to the engineering problem are identified.
- Phasor and vector quantities are correctly represented.
- Phasor and vector quantities are correctly manipulated.
- Numerical values are substituted and solutions relevant to the engineering problem are obtained.

RANGE
Engineering problems: electrical, electronic; mechanical.
Representation: magnitude, direction; modulus, argument.
Manipulation: addition, subtraction; resolution, graphical.

Introduction

In this chapter we will concentrate on vectors. However, to begin with we do distinguish between scalars and vectors. There are many examples of vectors in engineering and science, and many different ways of representing them. As a start we will use a 3 dimensional system. All the axes will be mutually perpendicular.

███ ASSIGNMENT ███

Two problems form the Assignments for this chapter.
1. A construction firm is currently engaged in building a bungalow on a remote site. It is accessible only across a river, 120 m wide, with a downstream current of $4\,\mathrm{ms}^{-1}$. A boat with a top speed of $5\,\mathrm{ms}^{-1}$ is available for transporting materials. For operational efficiency should the boat cross the river by the shortest route or in the fastest possible time?

202

2. An aircraft is capable of flying at $600\,\text{kmh}^{-1}$ in still air. It is attempting to fly due East but is being blown off course by a wind from the North East at $75\,\text{kmh}^{-1}$. What is the resultant velocity?

Vectors and scalars

We may divide physical quantities into two groups: **scalar quantities** and **vector quantities**. Temperature is one type of **scalar quantity**, e.g. 32°F, 100°C.

A scalar is defined as a quantity that has magnitude only. It is the numerical value (or **size** or **magnitude** or **modulus**) of 32 or 100 that is important on the particular temperature scale. 32 and 100 are said to be **scalars**.

A vector is defined in terms of its magnitude and direction. Hence a vector includes a scalar in part of its definition. A person may have a mass of 80 kg and hence a weight of $80g\,\text{N}$ (approximately 800 N taking the acceleration due to gravity, g, as $10\,\text{ms}^{-2}$). The mass is a scalar because it acts in no particular direction. This contrasts with the weight. It is a vector because it has direction, acting vertically downwards.

There are some subtle differences between scalars and vectors. We look at these in the following examples.

Examples 7.1

i) A speed of $15\,\mathrm{ms}^{-1}$ is a scalar, whereas a velocity of $15\,\mathrm{ms}^{-1}$ due East is a vector. In fact speed is the magnitude of the velocity vector.
ii) A distance of 7.2 m is a scalar, whereas a displacement of 7.2 m horizontally to the left is a vector. Similarly distance is the magnitude of the displacement vector.

We may represent a vector in a diagram by a directed line in various ways. The length of the line is the magnitude of the vector and its direction is the direction of the vector (see Fig. 7.1).

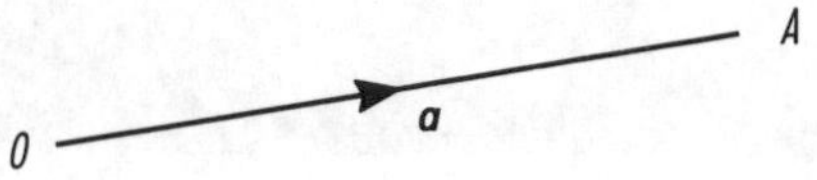

Fig. 7.1

We start at O and finish at A, shown by the directional arrow. Textbooks denote this vector by $\boldsymbol{OA}$ or $\boldsymbol{a}$ in bold type. In hand-written form they would be $\overrightarrow{OA}$ or $\underset{\sim}{a}$. $\boldsymbol{OA}$ (or $\overrightarrow{OA}$) means that the vector starts at O and ends at A. Really $\boldsymbol{a}$ is a ***free vector*** because its start and finish points are not given.

Equal vectors

Two vectors are equal if they have the same magnitude and the same direction, as in Figs. 7.2.

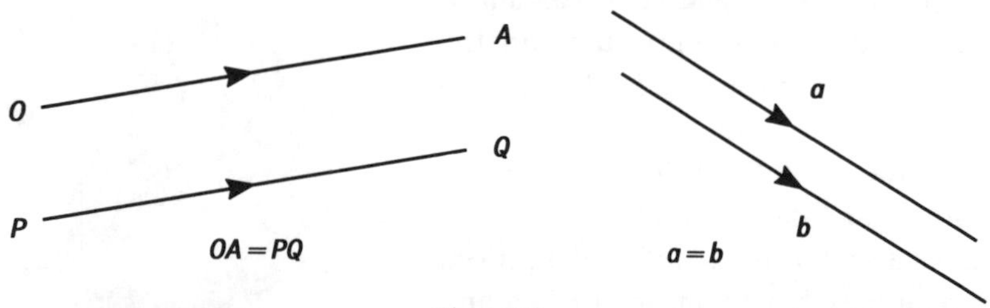

Figs. 7.2

We know already that both $\boldsymbol{a}$ and $\boldsymbol{b}$ are **free vectors** because the starting and finishing points are not specified. A free vector may be shifted provided its magnitude and direction remain unaltered. In some cases we re-label fixed vectors as free vectors and shift them too.

The direction is important. $\boldsymbol{OA}$ is different from $\boldsymbol{QP}$ because their directions are *not* the same.

Negative vectors

Let us look at two vectors *x* and *y* that are parallel and have equal magnitude. Figs. 7.3 show the two possibilities.

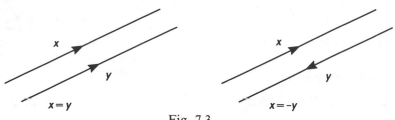

Fig. 7.3

The second option, of opposite direction, shows the negation of a vector, i.e. a minus sign means 'in the opposite direction' to the original vector.

Subtraction is a simplified form of "addition of a negative quantity", i.e. $a - b = a + (-b)$. We interpret this as the addition of *a* and a vector in the opposite direction to *b*.

Scale drawing

We can use an accurate scale drawing to represent a vector, e.g. A velocity of $15\,\text{ms}^{-1}$ due East can be represented by

where $1\,\text{cm} = 5\,\text{ms}^{-1}$.

Using the same scale the next line represents $15\,\text{ms}^{-1}$ due West.

$$15\,\text{ms}^{-1}$$

Figs. 7.4

If $\quad v_1 = 15\,\text{ms}^{-1}$ due East

and $\quad v_2 = 15\,\text{ms}^{-1}$ due West

then $\quad v_1 \neq v_2$ because the directions are different.

In fact these opposite directions are interpreted with a minus sign as $v_1 = -v_2$. Both examples have the same length of line, i.e. $|v_1| = |v_2|$. The horizontal lines represent the size of each vector. A simpler way is to ignore the vector symbols and write this as $v_1 = v_2$. (Notice the lack of bold type.)

Similarly $OA = -AO$ and so $OA = AO$.

███████ **ASSIGNMENT** ███████

We can return to our first problem. For the
bungalow problem we have 2 different choices.
Should we steer the boat directly across the river
and allow the current to take it downstream as
in Fig. 7.5a? Alternatively, should we steer it

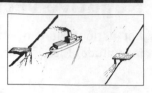

partially upstream? Then the downstream current can bring it to the other
bank. This ought to be directly opposite its starting point, Fig. 7.5b. The
calculations support the reasoning behind the scale drawing.

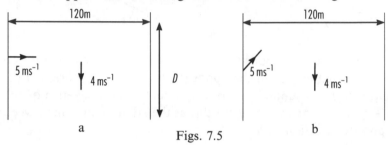

a   Figs. 7.5   b

Let us look at the first option. We attempt to steer the boat directly across
the river and let the current take it downstream. This means the vertical
and horizontal velocities and displacements (from Fig. 7.5a) will be in the
same ratio. In scale terms, for every $5 \, \text{ms}^{-1}$ across the river the current will
take the boat at $4 \, \text{ms}^{-1}$ downstream, i.e. for each second the distances are
$5 \, \text{m}$ across the river and $4 \, \text{m}$ downstream. We compare these as a ratio of

$$\frac{\text{downstream}}{\text{across the river}}$$

so that $\qquad \dfrac{D}{120} = \dfrac{4}{5}$

i.e. $\qquad D = 120 \times \dfrac{4}{5}$

$$D = 96 \, \text{m}.$$

This means the boat reaches the other side of the river at a distance of
$96 \, \text{m}$ downstream.

Also, using $\quad \text{Speed} = \dfrac{\text{Distance}}{\text{Time}}$

horizontally $\qquad 5 = \dfrac{120}{t} \qquad$ or vertically $\quad 4 = \dfrac{96}{t}$

i.e. $\qquad 5t = 120 \qquad\qquad\qquad\qquad 4t = 96$

$$t = \dfrac{120}{5} \qquad\qquad\qquad\qquad t = \dfrac{96}{4}$$

i.e. $\qquad t = 24 \, \text{s}$ in both cases.

This means the journey time is consistent at $24 \, \text{s}$ across the river and
downstream.

Now for the second option. We attempt to steer the boat partially upstream. Part of its velocity cancels out the effect of the current. The remaining part moves the boat directly across the river as though the motion was in still water. This produces a right-angled triangle with $5\,\text{ms}^{-1}$ as the hypotenuse (see Fig. 7.6). According to Pythagoras' theorem, the unknown side is $3\,\text{ms}^{-1}$, i.e. the boat moves directly across the river with a reduced velocity of $3\,\text{ms}^{-1}$.

Again using $\text{Speed} = \dfrac{\text{Distance}}{\text{Time}}$

i.e. $3 = \dfrac{120}{t}$

$t = 40 \text{ s}$

i.e. the journey time is 40 s taking the shortest distance directly across the river.

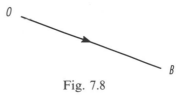

Fig. 7.6

Equivalent vectors

Let us look at a person walking 2 km due South and then 4 km due East. This may be drawn as in Fig. 7.7.

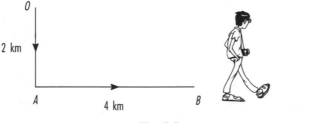

Fig. 7.7

Alternatively the person might have walked directly along *OB* to reach the same point, *B*. This is the same resultant vector or net displacement (see Fig. 7.8).

Fig. 7.8

i.e. ***OA***   +   ***AB***   =   ***OB***

i.e. ***OA*** 'together with' ***AB*** 'is the same as' ***OB***.

The common letter either side of + allows this addition.

When two sides of a triangle are taken in order their resultant is represented by the third side. This is called the **triangle law of vectors**.

Suppose the person continues walking from *B* for another 2 km in a North Westerly (i.e. N 45° W) direction. The extended diagram becomes that in Fig. 7.9.

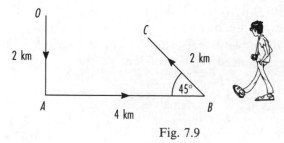

Fig. 7.9

The person started at *O* and finished at *C*. This gives the resultant vector as *OC*. If we take all stages of the walk in turn, *O* to *A*, *A* to *B* and *B* to *C* we get

$$OA + AB + BC = OC.$$

For vectors in pairs

$$OA + AB = OB \quad \text{or} \quad AB + BC = AC$$

and $OB + BC = OC$ and $OA + AC = OC$.

> Using the common letter either side of $+$.

Thus we may add vectors in pairs, either the first pair or the last pair. In the case of some long expression we may add any intermediate pair. Figs. 7.10 look at the combinations of pairs of vectors.

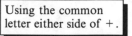

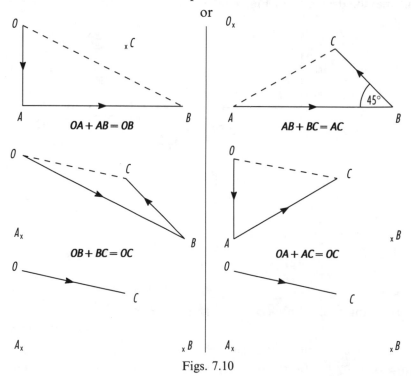

Figs. 7.10

OC is the third side of both triangles OBC and OAC. We see how the triangle law of vectors is obeyed.

▰▰▰▰ Example 7.2 ▰▰▰▰

i) Simplify $AB + BC - DC$,
   i.e. find the resultant of $AB + BC - DC$.

ii) Simplify $AB + BC + DC$.

i) The first pair of vectors are separated by a + sign with a common letter on either side. We deal with this pair first.

Now $\qquad AB + BC - DC$

$\qquad\qquad = AC - DC$

$\qquad\qquad = AC + CD$

$\qquad\qquad = AD.$

> Using the common B with side of +.

> $CD = -DC.$

ii) For this second example once more we can look at the first pair of vectors.

This time $\qquad AB + BC + DC$

$\qquad\qquad = AC + DC.$

We cannot simplify this result any further because there is no common letter either side of +. Immediately to the left of the + is C and to the right is D.

▰▰▰▰ Example 7.3 ▰▰▰▰

Given $OA = a$, $OB = b$ and $OC = c$ find expressions for AB and CA using a, b and c.

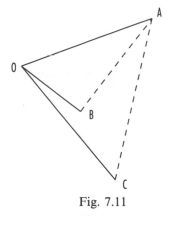

Fig. 7.11

We can write the direct vector AB in terms of other vectors involving O,

i.e. $\qquad AB = AO + OB$

$\qquad\qquad = -OA + OB$

$\qquad\qquad = -a + b$

$\qquad\qquad\qquad$ or $b - a$.

Similarly $\quad CA = CO + OA$

$\qquad\qquad = -OC + OA$

$\qquad\qquad = -c + a$

$\qquad\qquad\qquad$ or $a - c$.

███████ **Example 7.4** ███████████████████████

Simplify $KL + MN - ON + OK - ML$.

There are many ways to simplify this expression of vectors. To avoid confusion we find it easier to make sure they are all positive.

$$KL + MN - ON + OK - ML$$
$$= KL + MN + (NO + OK) + LM$$
$$= KL + (MN + NK) + LM$$
$$= KL + MK + LM$$
$$= KL + (LM + MK)$$
$$= KM + MK$$
$$= KM - KM$$
$$= 0.$$

> $NO = -ON$ and
> $LM = -ML$.

> Common letters either side of +.

> $MK = -KM$.

0 represents the **zero vector** (or **null vector**) which has neither magnitude nor direction.

The brackets highlight the order of addition, but are not essential to the mathematics.

███████ **ASSIGNMENT** ███████████████████████

We can apply the principles of scale drawing and equivalent vectors to our aircraft problem. The resultant velocity is the added effect of the wind velocity on to the aircraft's velocity. In Fig. 7.12 we use accurate drawing and measurement. The third side of the triangle is this resultant velocity. It is approximately $550 \, \text{ms}^{-1}$ inclined at an angle of $5.5°$ to the original easterly direction.

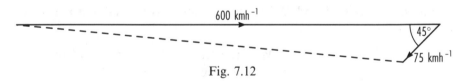

Fig. 7.12

███████ **EXERCISE 7.1** ███████████████████████

1 Decide whether the following are scalars or vectors:
   i) $3.7 \, \text{m}$
   ii) $24 \, \text{ms}^{-1}$ vertically up
   iii) a weight of $19 \, \text{N}$
   iv) $0.36 \, \text{m}$ due North
   v)   a crowd of $17\,362$ people.

2 By drawing find the resultant of a person walking $5 \, \text{m}$ due North, $2 \, \text{m}$ due West, $3 \, \text{m}$ due South and $2 \, \text{m}$ due East.

3 Simplify the following vectors:
   i)   *CA + AD + DB*
   ii) *DB + CA + AD*
   iii) *AB − CB*
   iv) *AD − DC + CE − FE*
   v) *BA − EA + FE*

4 Draw diagrams to represent the following relationships:
   i)   *AB + BC = AC*
   ii) *CA + AB + BC = 0*

5 PQRS is a square. Decide, stating your reason, whether the following statements are true or false:
   i)   *PQ = SR*
   ii) *PQ = SR*
   iii) *PQ = RS*
   iv) *SR = RQ*
   v) *PQ + QR + RS = PS*

Components and resultants

A vector may exist in any direction. Sometimes it is easier to consider it in parts, i.e. in **components**. Together these components have exactly the same effect as the original vector. The most useful way is to **resolve** a vector into 2 components that are at 90° (at right-angles) to each other. These components are perpendicular components. Then we can use some basic trigonometry to evaluate them. Frequently we use horizontal and vertical directions for the components. However, there are other helpful alternatives. They can involve directions that are parallel and perpendicular to an inclined plane.

Figs. 7.13 show a vector *d* in two different, but equivalent, ways. In one of them it is inclined at an angle θ above the horizontal. In the other it is shown in its equivalent perpendicular components.

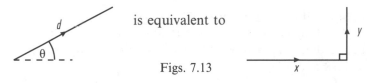

Figs. 7.13

We may superimpose these two figures to form a right-angled triangle. Using some simple trigonometry from Chapter 4 we have

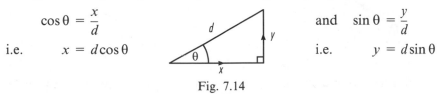

$$\cos\theta = \frac{x}{d} \qquad\qquad \text{and} \quad \sin\theta = \frac{y}{d}$$

i.e. $x = d\cos\theta$ i.e. $y = d\sin\theta$

Fig. 7.14

Fig 7.14 shows $d\cos\theta$ acting horizontally together with $d\sin\theta$ acting vertically. They have the same effect as the original d.

In Fig. 7.13 we started with a vector and split it into 2 perpendicular components. Instead Fig. 7.15 shows us starting with the same perpendicular x and y. We may apply Pythagoras' theorem to find the third side, say d,

i.e. $\qquad d^2 = x^2 + y^2$

i.e. $\qquad d = \sqrt{x^2 + y^2}.$

Also $\quad \tan\theta = \dfrac{y}{x}.$

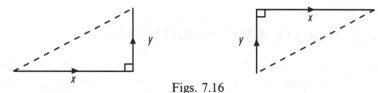

Fig. 7.15

We can look at pairs of perpendicular vectors in several ways. Because the start and finish points for y are not specified it is a free vector. Similarly x is a free vector. We know already that free vectors may be shifted. In Figs. 7.16 we have 2 possibilities. In each case we find the resultant of x and y by completing the triangle.

Figs. 7.16

Figs. 7.17 show the vectors x and y acting at a point. In one case they are coming from a point. In the other case they are going to a point. We find their resultant by completing the rectangle. The diagonal is the resultant vector. You can see there are similarities between Figs. 7.16 and 7.17.

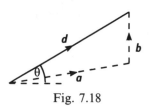

Figs.7.17

Equally valid, though generally not as useful, are non-perpendicular components. Right-angled trigonometry and Pythagoras' theorem cannot be applied if the components are not perpendicular. By way of example are Figs. 7.18 and 7.19:

or

Fig. 7.18   Fig. 7.19

vector d has components a and b vector d has components e and f.

Example 7.5

Evaluate the horizontal and vertical components of a velocity vector of 5ms^{-1} inclined at 30° above the horizontal.

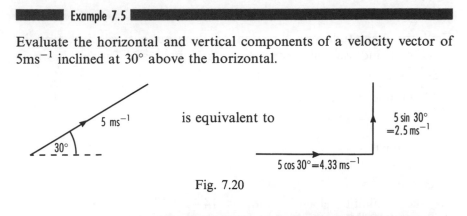

Fig. 7.20

Example 7.6

A force of 14 N acts horizontally as shown below. Find its components parallel and perpendicular to the plane inclined at 60° to the horizontal.

12.12 N and 7 N are the sides of a rectangle where 14 N is the diagonal.

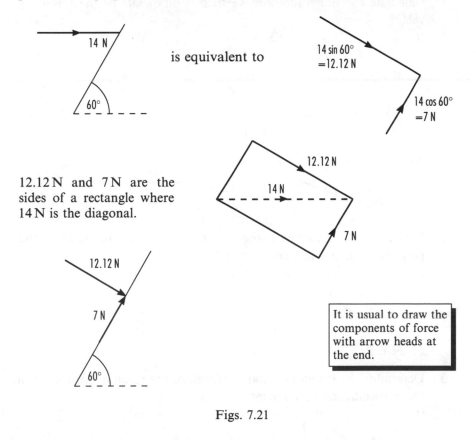

It is usual to draw the components of force with arrow heads at the end.

Figs. 7.21

Now we look at an example with the two perpendicular components given. We find the magnitude and direction of the original vector.

▬▬▬ **Example 7.7** ▬▬▬▬▬▬▬▬▬▬▬▬▬▬▬▬▬▬

Find the magnitude and direction of the resultant velocity given its horizontal and vertical components of $12\,\text{ms}^{-1}$ and $8\,\text{ms}^{-1}$ shown in Fig. 7.22.

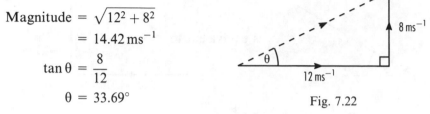

$$\text{Magnitude} = \sqrt{12^2 + 8^2}$$
$$= 14.42\,\text{ms}^{-1}$$
$$\tan\theta = \frac{8}{12}$$
$$\theta = 33.69°$$

Fig. 7.22

i.e. the resultant velocity vector is $14.42\,\text{ms}^{-1}$ at $33.69°$ above the horizontal.

▬▬▬ **EXERCISE 7.2** ▬▬▬▬▬▬▬▬▬▬▬▬▬▬▬▬

1   Calculate the horizontal and vertical components of the following vectors:

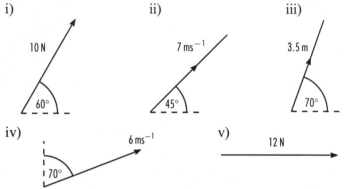

i)   ii)   iii)

10 N   7 ms⁻¹   3.5 m
60°   45°   70°

iv)   v)
          6 ms⁻¹   12 N
70°

2   For each force calculate the components that are parallel and perpendicular to the inclined plane:

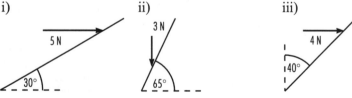

i)   ii)   iii)

5 N   3 N   4 N
30°   65°   40°

3   Determine the magnitude and direction of the resultant vector given the perpendicular components:

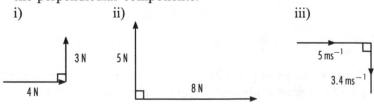

i)   ii)   iii)

3 N   5 N   5 ms⁻¹
4 N   3.4 ms⁻¹
                   8 N

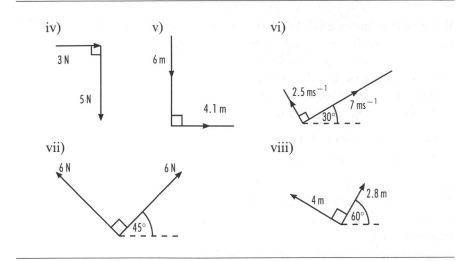

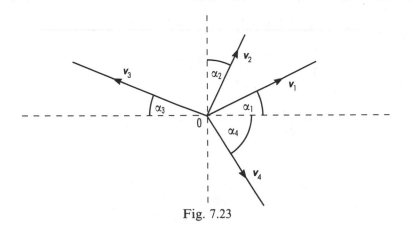

Fig. 7.23

These two techniques of finding components and finding a resultant may be extended to find the resultant of many vectors.

In Fig. 7.23 we look at a system of vectors all applied at some point.

Each of them, v_1, v_2, v_3, v_4, may be resolved into horizontal and vertical components, as in Fig. 7.24.

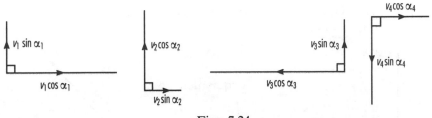

Figs. 7.24

We see that not all the horizontal directions are to the right. Similarly not all the vertical directions are upwards. We need to decide which are to be the positive directions. We are going to be consistent with axes and graphical work. Horizontally right and vertically up will be taken as positives. Let the resultant of all these vectors be represented by a single vector with components (Fig. 7.25) X and Y.

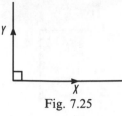

Fig. 7.25

Then $X = v_1 \cos \alpha_1 + v_2 \sin \alpha_2 - v_3 \cos \alpha_3 + v_4 \cos \alpha_4$

and $Y = v_1 \sin \alpha_1 + v_2 \cos \alpha_2 + v_3 \sin \alpha_3 - v_4 \sin \alpha_4$

In a numerical example X and Y can be simplified into one value each. Then we find the magnitude of the resultant of X and Y using Pythagoras' theorem. We find the direction of the resultant using the tangent from trigonometry.

Sometimes we find the value of X or Y may turn out to be negative. The magnitude remains the same. The negative sign means the true direction is opposite to the one we have chosen, i.e. horizontally to the left instead of the right, or vertically down instead of up.

Example 7.8

Find the magnitude and direction of the resultant force for the system in Fig. 7.26.

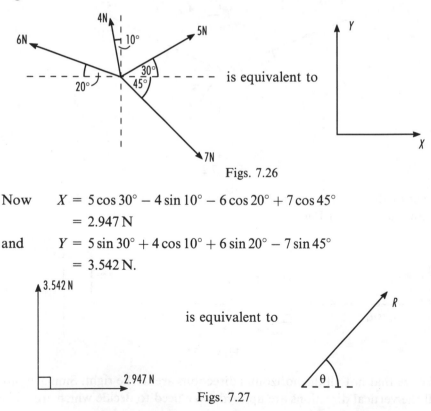

Figs. 7.26

Now $X = 5 \cos 30° - 4 \sin 10° - 6 \cos 20° + 7 \cos 45°$

$= 2.947 \, \text{N}$

and $Y = 5 \sin 30° + 4 \cos 10° + 6 \sin 20° - 7 \sin 45°$

$= 3.542 \, \text{N}.$

Figs. 7.27

By Pythagoras' theorem

$$R = \sqrt{2.947^2 + 3.542^2} \qquad \tan \theta = \frac{3.542}{2.947}$$

$$= 4.61 \, \text{N} \qquad\qquad \theta = 50.24°$$

i.e. the resultant force is 4.61 N inclined at 50.24° above the horizontal, shown in Figs. 7.27.

████ Example 7.9 ████

For the velocity vectors in Figs. 7.28 find the magnitude and direction of their resultant.

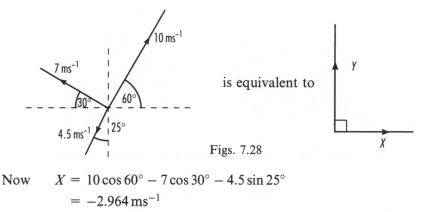

is equivalent to

Figs. 7.28

Now $X = 10 \cos 60° - 7 \cos 30° - 4.5 \sin 25°$

$$= -2.964 \, \text{ms}^{-1}$$

and $Y = 10 \sin 60° + 7 \sin 30° - 4.5 \cos 25°$

$$= 8.082 \, \text{ms}^{-1}.$$

The negative value for X means that it should be directed towards the left whilst retaining the same magnitude.

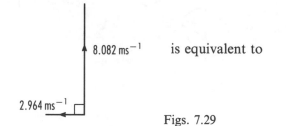

is equivalent to

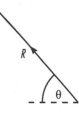

Figs. 7.29

By Pythagoras' theorem

$$R = \sqrt{2.964^2 + 8.082^2}$$

$$= 8.61 \, \text{ms}^{-1}$$

$$\tan \theta = \frac{8.082}{2.964}$$

$$\theta = 69.86°$$

i.e. the resultant velocity is 8.61 ms^{-1} inclined at 69.86° above the horizontal as shown in Figs. 7.29.

ASSIGNMENT

We have developed the theory a little further. Now it is possible to re-consider the Assignment problems from the beginning of the chapter.

In the construction problem we have two options. The first one is to cross the river as quickly as possible. This means letting the current take the boat downstream. Fig. 7.30 shows the resultant velocity, v_1, along the hypotenuse of the right-angled triangle.

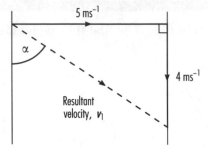

Fig. 7.30

By Pythagoras' theorem

$$v_1 = \sqrt{5^2 + 4^2}$$
$$= 6.40 \, \text{ms}^{-1}$$

$$\tan \alpha = \frac{5}{4} = 1.25$$
$$\alpha = 51.34°$$

i.e. the boat's resultant velocity is $6.40 \, \text{ms}^{-1}$ at $51.34°$ to the river bank.

For the second option we attempt to steer the boat partially upstream. The downstream effect of the current means the resultant velocity, v_2, is directly across the river. In this case Fig. 7.31 shows it is the $5 \, \text{ms}^{-1}$ that forms the hypotenuse of the right-angled triangle.

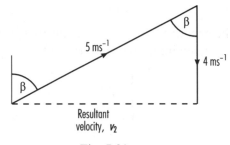

Fig. 7.31

The velocities of $5 \, \text{ms}^{-1}$ and $4 \, \text{ms}^{-1}$ in Figs. 7.32 are not perpendicular. This means a little more care is necessary with the components and the final resultant, i.e.

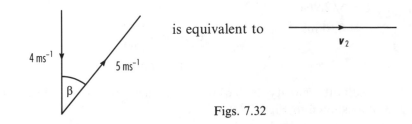

is equivalent to

Figs. 7.32

Now across the river $v_2 = 5 \sin \beta$

and downstream $0 = 5 \cos \beta - 4$

The second equation gives a solution of $\beta = 36.87°$. By substitution the first equation gives $v_2 = 3\,\text{ms}^{-1}$, i.e. the helmsman must attempt to steer the boat at 36.87° to the bank. This is because the effect of the current produces a resultant velocity of $3\,\text{ms}^{-1}$ directly across the river.

We can apply this last technique to the aircraft problem. The velocities of $600\,\text{kmh}^{-1}$ due East and $75\,\text{kmh}^{-1}$ from the North East are shown in Fig. 7.33.

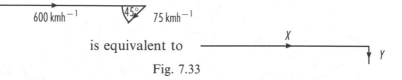

is equivalent to

Fig. 7.33

Our choice of Y vertically downwards is consistent with our problem. Alternatively we could have chosen vertically upwards as positive. We would interpret the resulting negative sign as the 'opposite direction'.

Now horizontally $X = 600 - 75 \cos 45°$

$\qquad\qquad\qquad = 546.97\,\text{kmh}^{-1}$

and vertically $Y = 75 \sin 45°$

$\qquad\qquad\qquad = 53.03\,\text{kmh}^{-1}$

Now we can combine our 2 perpendicular components to find their resultant. Thus

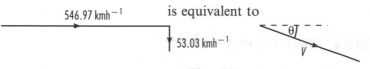

is equivalent to

Figs. 7.34

By Pythagoras' theorem

$$V = \sqrt{546.97^2 + 53.03^2}$$

$$= 549.5\,\text{kmh}^{-1}$$

$$\tan \theta = \frac{53.03}{546.97}$$

$$\theta = 5.54°$$

The bearing is $90° + \theta$

$$= 095.54°$$

i.e the aircraft's resultant velocity is $549.5\,\text{kmh}^{-1}$ on a bearing of 095.54°.

EXERCISE 7.3

1 In each case find the magnitude and direction of the resultant vector:

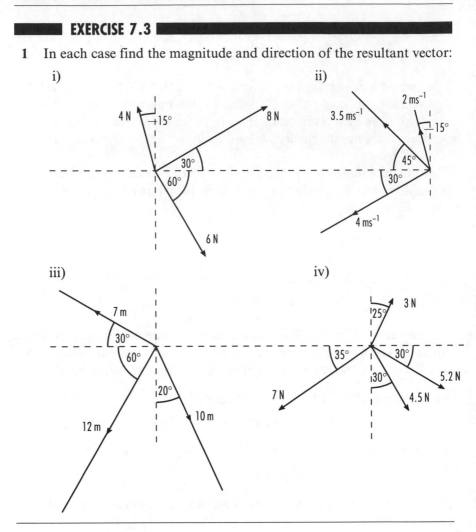

i)

ii)

iii)

iv)

Scalar multiplication

Scalar multiplication of a vector is simple arithmetical multiplication. By way of example in Figs. 7.35 $3a$ is a vector in the same direction as a but three times the length. Also $-3a$ is the same length as $3a$ but in the opposite direction.

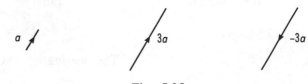

Figs. 7.35

These are free vectors so we have not given any start or finish points.

In contrast in Figs. 7.36 we have OA, $3OA$ and $-3OA$ where those points are important.

Figs. 7.36

The parallelogram law

In earlier examples we formed a triangle by vectors when one followed another. The third side represented their resultant. Alternatively we have looked at an original vector in terms of its components. These have been equivalent systems. More recently our examples have concentrated on horizontal and vertical components for easier calculation.

Now let us look at vectors at a point with any angle between them. **The parallelogram law, for two vectors acting at a point**, uses vector simplification similar to the triangle law where one vector follows another. It is based upon the opposite sides of a parallelogram being equal and parallel, and that free vectors may be shifted.

We may think of the parallelogram $OABC$ in Fig. 7.37 as pairs of triangles:

Fig. 7.37

either   triangles OAB and OCB with a common side OB

or   triangles OAC and BAC with a common side AC

Let us consider two vectors, a and b, acting at one point. Because they are free vectors acting at a point either of them may be shifted. Figs. 7.38 shows the 2 possible variations.

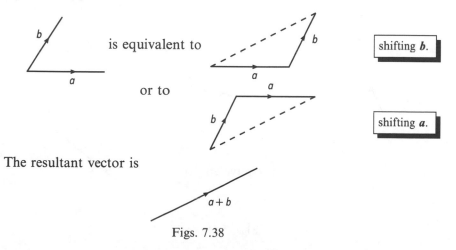

is equivalent to

shifting b.

or to

shifting a.

The resultant vector is

$a + b$

Figs. 7.38

We may apply the parallelogram law to vector subtraction. We represent
$a - b$ as $a + (-b)$ for vectors acting at one point.

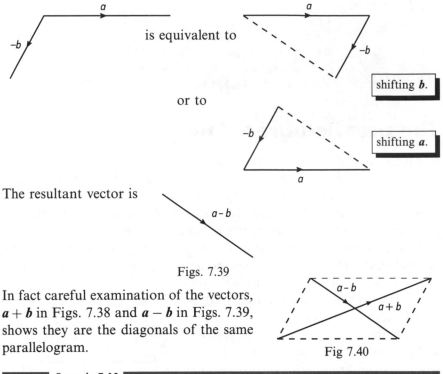

Figs. 7.39

In fact careful examination of the vectors,
$a + b$ in Figs. 7.38 and $a - b$ in Figs. 7.39,
shows they are the diagonals of the same
parallelogram.

Fig 7.40

Example 7.10

Apply the parallelogram law to represent $a + 3b$ where a and $3b$ act at a
point O.

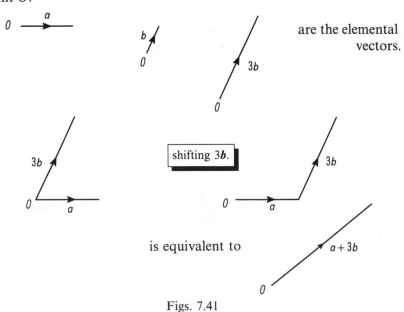

Figs. 7.41

Example 7.11

Represent $2a - 3b$, where the elemental vectors act at O, using a parallelogram.

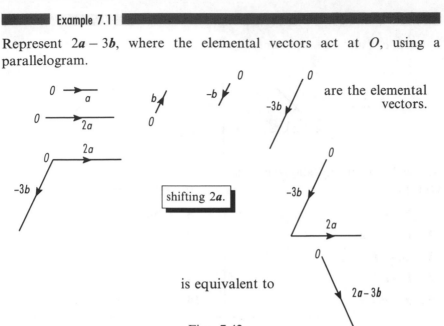

Figs. 7.42

When we deal with more than two vectors the parallelogram law may be used repeatedly on pairs of vectors. We apply this technique in the next example.

Example 7.12

Use the parallelogram law to represent $a + b + 2d$ where

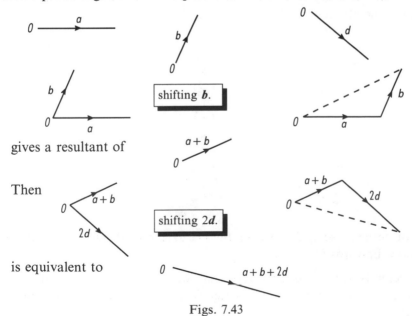

Figs. 7.43

EXERCISE 7.4

Let *a*, *b*, *c*, *d* and *e* be represented by lines of 4 cm in the directions shown:

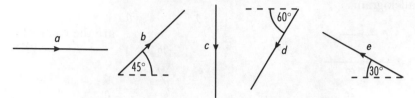

In the following questions use the parallelogram law to represent the vectors.

1 $b + c$	6 $0.5b - a$
2 $b + c + d$	7 $3e + 2d$
3 $2a + d$	8 $2c - 3d + 0.5a$
4 $b - e$	9 $a + b + c + d + e$
5 $a + b + c$	10 $a - b + c - d + e$

Rectangular form

Rectangular (or **Cartesian**) form expresses a vector in terms of components that are mutually perpendicular. A sheet of paper has only two dimensions. The components usually act along axes that are horizontal and vertical. It is difficult to draw 3-dimensional axes on paper. In the reality of the 3-dimensional world the axes are defined by the 'right-handed screw rule' shown in Fig. 7.44.

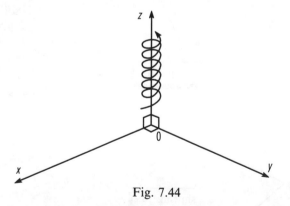

Fig. 7.44

If you turned a screwdriver from the Ox axis towards Oy you could drive a screw upwards along Oz.

A vector is such a versatile tool that we can extend it to many more dimensions. For more than 3 dimensions it is not possible to create a diagram.

To identify the direction of the components we need some standard notation. We use a unit vector in the direction of each axis. As the name suggests a **unit vector** has a size of 1.

i is the unit vector in the Ox direction
j is the unit vector in the Oy direction
k is the unit vector in the Oz direction

$a = 2i + 4j + 5k$ is an example of a vector. It has component lengths 2 along Ox, 4 along Oy and 5 along Oz. It is the vector line joining the origin, $O(0,0,0)$ to $A(2,4,5)$, shown in Fig. 7.45.

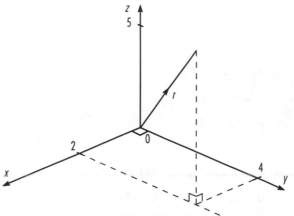

Fig. 7.45

Our example has positive values. As might be expected some components may be zero and/or some may be negative.

Rectangular form is just one form of vector representation. Other useful ones are cylindrical polars and spherical polars, though the mathematics becomes too involved for this level. All the general theories developed throughout the chapter apply to any of these vector forms.

We will concentrate on the rectangular system. The next 2 sets of examples look at some simple calculations. In Examples 7.13 we look at scalar multipliers, both positive and negative ones. In Examples 7.14 we look at addition and subtraction. Just using the usual rules of algebra we add/subtract **like terms**, i.e. we add/subtract the **same type of components**.

 Example 7.13

Given $a = 2i + 4j + 5k$ represent in rectangular form
  i) $-a$,   ii) $3a$,   iii) $-3a$,   iv) $0.2a$.

  i) $-a = -(2i + 4j + 5k) = -2i - 4j - 5k$.
  ii) $3a = 3(2i + 4j + 5k) = 6i + 12j + 15k$.
 iii) $-3a = -3(2i + 4j + 5k) = -6i - 12j - 15k$.
 iv) $0.2a = 0.2(2i + 4j + 5k) = 0.4i + 0.8j + k$.

▬▬▬ **Example 7.14** ▬▬▬▬▬▬▬▬▬▬▬▬▬▬▬▬▬▬▬▬▬▬▬

Given $a = 2i + 4j + 5k$ and $b = 6i - 3k$ in rectangular form represent as simply as possible

i) $a + b$,   ii) $b - a$,   iii) $2a + 3b$,   iv) $a - 5b$.

i) $a + b$ $= (2i + 4j + 5k) + (6i - 3k)$
   $= (2 + 6)i + (4 + 0)j + (5 - 3)k$
   $= 8i + 4j + 2k$

> An equally valid solution is $2(4i + 2j + k)$ where a common factor of 2 is extracted from all the components.

ii) $b - a$ $= (6i - 3k) - (2i + 4j + 5k)$
   $= (6 - 2)i + (0 - 4)j + (-3 - 5)k$
   $= 4i - 4j - 8k$

   or $-4(i - j - 2k)$.

iii) $2a + 3b$ $= 2(2i + 4j + 5k) + 3(6i - 3k)$
   $= 4i + 8j + 10k + 18i - 9k$
   $= (4 + 18)i + (8 + 0)j + (10 - 9)k$
   $= 22i + 8j + k$

iv) $a - 5b$ $= (2i + 4j + 5k) - 5(6i - 3k)$
   $= 2i + 4j + 5k - 30i + 15k$
   $= (2 - 30)i + (4 + 0)j + (5 + 15)k$
   $= -28i + 4j + 20k$

   or $4(-7i + j + 5k)$.

▬▬▬ **EXERCISE 7.5** ▬▬▬▬▬▬▬▬▬▬▬▬▬▬▬▬▬▬▬▬▬▬

Given $a = 2i + 0.5j - 3k$, $b = -i + j + 4k$ and $c = 6j - 5k$ evaluate and simplify the following vectors in i, j, k form.

1	$-b$	6	$b - 2a$
2	$2a$	7	$c + b + a$
3	$b + c$	8	$a + 2b + 3c$
4	$-b + 2a$	9	$2a + 0.4b + 0.5c$
5	$3b + 4c$	10	$3c - 2b + 4c$

■ ASSIGNMENT ■

And so once again the theory allows us to re-consider the two problems posed at the beginning of the chapter.

Firstly we will consider our construction problem. In option one for the fastest crossing we ended 96 m downstream.

By considering a unit vector i across the river and a unit vector j upstream we may interpret the velocities as $5i$ and $-4j$ (see Fig. 7.46). Using simple vector addition their resultant is

$$v_1 = 5i - 4j \text{ ms}^{-1}.$$

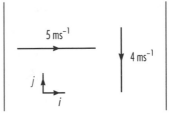

Fig. 7.46

We can find the magnitude and direction of v_1 using Pythagoras' theorem and trigonometry as before.

For the direct crossing

$$v_2 = (5\sin\beta)i + (5\cos\beta - 4)j \text{ ms}^{-1}$$

We have $5\cos\beta - 4 = 0$ because v_2 must act directly across the river. Again the values of v_2 and β are as before.

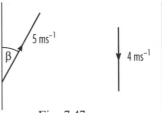

Fig. 7.47

Secondly it is just as easy to consider the aircraft with a unit vector i due East and a unit vector j due North.

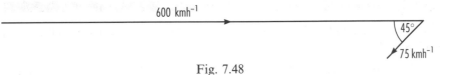

Fig. 7.48

Then we can represent the overall velocity by

$$V = (600 - 75\cos 45°)i - (75\sin 45°)j \text{ kmh}^{-1}$$

Once again the magnitude and direction of V are the same as before.

The following examples are to consolidate the theory we have developed during this chapter. They concentrate on rectangular form before calculating any magnitudes or directions.

███████ **Example 7.15** ███████████████████████████████

Triangle OAB has vertices $(0, 0)$, $(2, -3)$ and $(4, 5)$ respectively. Calculate the lengths of sides OA and AB.

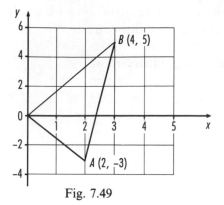

Fig. 7.49

In vector form $OA = 2i - 3j$

The magnitude of OA, or $|OA|$, or OA

$$= \sqrt{2^2 + (-3)^2}$$

> Pythagoras' theorem.

$$= 3.61 \text{ units.}$$

Also $AB = AO + OB$

$$= -OA + OB$$

$$= -(2i - 3j) + (4i + 5j)$$

$$= 2i + 8j.$$

Then $|AB| = \sqrt{2^2 + 8^2}$

$$= 8.25 \text{ units.}$$

███████ **Example 7.16** ███████████████████████████████

Calculate the length of side AB in triangle OAB with vertices $A(2, -3, 1)$ and $B(4, 5, 7)$.

To draw an accurate 3-dimensional diagram in 2 dimensions is difficult. Fig. 7.50 is a representation rather than an accurate drawing.
In vector form

$$AB = AO + OB$$

$$= -OA + OB$$

$$= -(2i - 3j + k) + (4i + 5j + 7k)$$

$$= 2i + 8j + 6k.$$

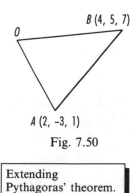

Fig. 7.50

Then $|AB| = \sqrt{2^2 + 8^2 + 6^2}$

$$= 10.20.$$

> Extending
> Pythagoras' theorem.

■■■■■ EXERCISE 7.6 ■■■■■■

1 Triangle OXY has vertices given by the origin, $X(-1, 4)$ and $Y(3, 10)$. Calculate the lengths of the triangle's sides.

2 In triangle MNO calculate the length of MN given that $OM = 2i - 3j + 4k$ and $ON = i - 5k$.

3 In rectangular form find the resultant force of $-i + 2j$, $3i - 4k$, $6j + 2k$ and $i - 5j + 7k$ N. What is its magnitude?

4 Four forces are in equilibrium, i.e. the sum of the forces is zero, $\Sigma F = 0$. They are $2i + j$, $6i - 3j$, $4j$ and $ai + bj$ N. Find the values of a and b.

5 $ABCD$ is a quadrilateral with $A(1, 0, 3)$, $B(-2, 4, 9)$, $C(2, 5, -1)$ and $D(-3, 8, 6)$. Calculate the lengths of AB and BC. Also calculate the perimeter of $ABCD$.

6 A ring of weight 4 N is pulled horizontally by a force of 7 N. What is the magnitude and direction of the resultant force?

7 It is mid-February and snow is falling vertically at a gentle $2\,\text{ms}^{-1}$. The weather changes with a horizontal breeze now blowing at $1.8\,\text{ms}^{-1}$. What is the magnitude and direction of a snowflake's velocity now?

8 The figure shows a rowing boat travelling directly across a river at $10\,\text{ms}^{-1}$. There is a northerly current of $8\,\text{ms}^{-1}$. Calculate the magnitude and direction of the resultant velocity. The river width is 200 m. How far downstream will the boat be when it reaches the other bank?

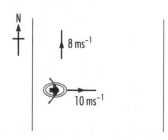

9 A failed engine has left a boat at sea at the mercy of the wind, blowing South East at $12\,\text{ms}^{-1}$, and the current running South West at $18\,\text{ms}^{-1}$. In which direction is it drifting? Dangerous rocks are close and the engineer manages a temporary repair just in time. Headed due South at $10\,\text{ms}^{-1}$ calculate the actual resultant velocity in these weather conditions.

10 A surveyor starts from his mobile office, O, and walks 100 m due West. He turns and walks 50 m South West followed by 120 m due South, reaching a tower crane, T. Take due North as j and due East as i. Represent each of these displacements in rectangular vector form. Write down their resultant OT in the same form. Calculate the direct distance of the crane's base from the office. If the cab on the crane is 35 m above ground level find its elevation from the surveyor's office.

■ MULTI-CHOICE TEST 7 ■

1   Decide whether each of the statements is True (T) or False (F).
    i)   A vector is defined in terms of its magnitude and direction.
    ii) A scalar is defined in terms of its magnitude only.
    Which option best describes the two statements?
    A) i) T   ii)   T
    B) i) T   ii)   F
    C) i) F   ii)   T
    D) i) F   ii)   F

2   Decide whether the following quantities are vectors or scalars.
    i)   A mass of 25 kg.
    ii) A weight of approximately 250 N.
    iii) A speed of $3.5 \, \text{ms}^{-1}$
    iv) A distance of 75 m.
    These four statements may be classified by:
    A) 3 vectors and 1 scalar
    B) 2 vectors and 2 scalars
    C) 1 vector and 3 scalars
    D) 0 vectors, only 4 scalars

3   Decide whether each of the statements is True (T) or False (F).
    i)   A temperature of 23° is a scalar.
    ii) A force of 33 N is a scalar.
    Which option best describes the two statements?
    A) i) T   ii)   T
    B) i) T   ii)   F
    C) i) F   ii)   T
    D) i) F   ii)   F

Questions **4–7** refer to the following information. *PQRS* is a parallelogram

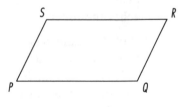

4   A correct scalar statement is:
    A) $PR = QS$
    B) $PQ + QR = PR$
    C) $PQ = RS$
    D) $PQ = QR$

5   A correct vector statement is:
    A) $\mathbf{PS} + \mathbf{SR} = \mathbf{RP}$
    B) $\mathbf{PQ} + \mathbf{SR} = \mathbf{0}$
    C) $\mathbf{PQ} + \mathbf{QR} = \mathbf{PR}$
    D) $\mathbf{PR} = \mathbf{QS}$

6 A correct vector statements is:
A) $PQ + QR = RS + SP$
B) $SP = QR$
C) $PS + QR = 0$
D) $SQ + QR + RS = 0$

7 $PQRS$ is changed from a parallelogram to a square. A correct vector statement is:
A) $PS + SR = RP$
B) $PQ + SR = 0$
C) $PQ + QR = PR$
D) $PR = QS$

8 The vector equation $XZ = XY + YZ$ applies to the triangle
A)   B)   C)   D)

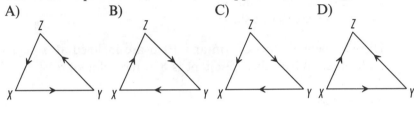

9 The diagram shows perpendicular force vectors of 6 N and 8 N together with a rectangle $WXYZ$. The resultant force is represented by:
A) WY
B) XZ
C) YW
D) ZX

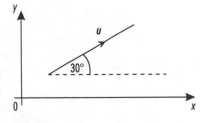

10

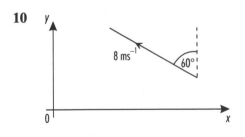

The velocity vector is drawn on a set of Cartesian axes. In order along Ox and Oy the components of this vector are:
A) $8 \sin 60°$, $8 \cos 60°$
B) $-8 \cos 60°$, $8 \sin 60°$
C) $8 \cos 60°$, $8 \sin 60°$
D) $-8 \sin 60°$, $8 \cos 60°$

11 $-3u$ has components along Ox and Oy in order of:
A) $-3|u| \cos 30°$, $-3|u| \sin 30°$
B) $-3|u| \cos 90°$, $-3|u| \sin 90°$
C) $-|u| \sin 90°$, $-|u| \cos 90°$
D) $-3|u| \sin 30°$, $-3|u| \cos 30°$

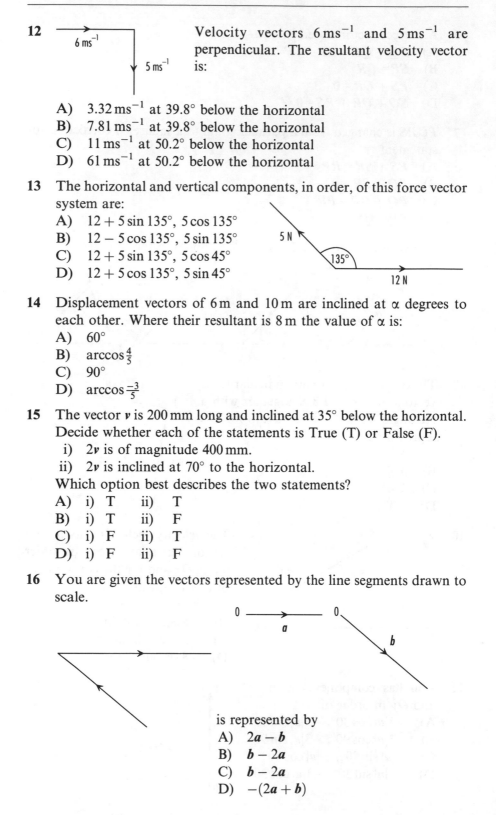

12 Velocity vectors $6\,\text{ms}^{-1}$ and $5\,\text{ms}^{-1}$ are perpendicular. The resultant velocity vector is:

A) $3.32\,\text{ms}^{-1}$ at $39.8°$ below the horizontal
B) $7.81\,\text{ms}^{-1}$ at $39.8°$ below the horizontal
C) $11\,\text{ms}^{-1}$ at $50.2°$ below the horizontal
D) $61\,\text{ms}^{-1}$ at $50.2°$ below the horizontal

13 The horizontal and vertical components, in order, of this force vector system are:
A) $12 + 5\sin 135°,\ 5\cos 135°$
B) $12 - 5\cos 135°,\ 5\sin 135°$
C) $12 + 5\sin 135°,\ 5\cos 45°$
D) $12 + 5\cos 135°,\ 5\sin 45°$

14 Displacement vectors of $6\,\text{m}$ and $10\,\text{m}$ are inclined at α degrees to each other. Where their resultant is $8\,\text{m}$ the value of α is:
A) $60°$
B) $\arccos\frac{4}{5}$
C) $90°$
D) $\arccos\frac{-3}{5}$

15 The vector v is $200\,\text{mm}$ long and inclined at $35°$ below the horizontal. Decide whether each of the statements is True (T) or False (F).
i) $2v$ is of magnitude $400\,\text{mm}$.
ii) $2v$ is inclined at $70°$ to the horizontal.
Which option best describes the two statements?
A) i) T   ii) T
B) i) T   ii) F
C) i) F   ii) T
D) i) F   ii) F

16 You are given the vectors represented by the line segments drawn to scale.

is represented by
A) $2a - b$
B) $b - 2a$
C) $b - 2a$
D) $-(2a + b)$

17 Using the standard notation the correct, equivalent, Cartesian axes are:

A)   B)

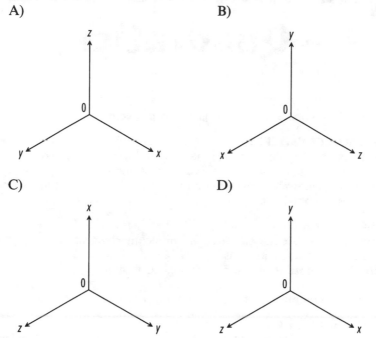

C)   D)

Questions **18–20** refer to the following information.
$a=i-3j+4k$, $b=5i-6k$, $c=-i+2j+3k$.

18 $a-c=$:
- A) $\quad -\ j+7k$
- B) $\quad -5j+\ k$
- C) $2i-\ j+\ k$
- D) $2i-5j+\ k$

19 $2b+3c=$:
- A) $7i+2j-3k$
- B) $7i+6j-3k$
- C) $7i+6j+3k$
- D) $13i-6j+9k$

20 $3a-2b+c=$:
- A) $-8i-\ 7j+27k$
- B) $-8i+\ j+13k$
- C) $-6i-11j+21k$
- D) $12i-11j+\ 3k$

8 Solving Equations: I – Quadratics

Element: Use algebra to solve engineering problems.

PERFORMANCE CRITERIA
- Formulae appropriate to the engineering problem are selected.
- The algebraic manipulation of formulae is carried out.
- Numerical values are substituted and correct solutions to engineering problems are obtained.

RANGE
Engineering problems: electrical, electronic; mechanical.
Formulae: quadratic equations.
Algebraic manipulation: transposition, substitution.

Element: Use graphs to solve engineering problems.

PERFORMANCE CRITERIA
- Coordinates and scales appropriate to the engineering problem are selected.
- Data are accurately plotted.
- The relationship between variables is identified.
- Values are accurately determined from graphs, and engineering problems are solved.

RANGE
Engineering problems: electrical, electronic; mechanical.
Coordinates: Cartesian.
Scales: linear.
Relationship: parabolic.
Values: two variables.

Introduction

In Chapter 2 we looked at linear equations and straight line graphs. Example 2.10i) used $y = 3x + 2$, the highest power of x being 1, i.e. written as x but understood to be x^1. Quadratics move one stage further with the highest power of x being 2, written as x^2, i.e. x squared. There may/may not be other terms involving x and/or pure numbers.

We need to distinguish between slightly different ideas connected with the word **quadratic**. These are **quadratic expressions** and **quadratic equations**.

ASSIGNMENT

The Assignment for this chapter looks at a conservatory to the rear of a house. Its external measurements are 6 m by 4 m. We are interested in putting a concrete path around the 3 sides and calculating the best width for that path.

Quadratic expressions

There is a difference between a **quadratic expression** and a **quadratic equation**. An expression is a list of algebraic terms, preferably written in some order as simply as possible. You cannot do anything really useful with a quadratic expression on its own. On the other hand a quadratic equation, because it is an equation, may be solved.

Example 8.1

This table distinguishes between quadratic expressions and equations. $=$ is the clue to deciding between them.

Quadratic expression	Quadratic equation
$x^2 + 5x + 6$	$x^2 + 5x + 6 = 0$
$12x^2 - 7x - 14$	$12x^2 - 7x - 14 = 0$
$3x^2 + 8$	$3x^2 + 8 = 0$
$3x^2 - 8$	$3x^2 = -8$
$5x^2 - 10x$	$5x^2 - 10x = 0$
$2x^2$	$2x^2 = 0$

At first glance it may not always be obvious that an expression is a quadratic expression. For example $(5x + 2)(4x - 7)$ is a quadratic expression but no term in x^2 is shown. Firstly let us check the meaning of each bracket.

$5x + 2$ has two terms, $5x$ and $+2$.

Also $4x - 7$ has two terms, $4x$ and -7.

Notice how the sign immediately before the pure number is attached to it. Because there are no signs written before $5x$ and $4x$ these terms are understood to be positive.

Now we will attempt to multiply out the brackets. Each term in the first bracket multiplies each term in the second bracket to give the pattern

$$(5x + 2)\ (4x - 7)$$

i.e. $5x$ multiplies $4x$ as $(+)(+) = +$

$(5)(4) = 20$

$(x)(x) = x^2$, all written as $20x^2$.

Where possible think of multiplying

SIGNS

NUMBERS

LETTERS

Also $5x$ multiplies -7 to give $-35x$,

2 multiplies $4x$ to give $8x$

and 2 multiplies -7 to give -14.

Thus $(5x + 2)(4x - 7) = 20x^2 - 35x + 8x - 14$　| Simplifying like terms.

$= 20x^2 - 27x - 14$.

Examples 8.2

Multiply out the brackets and fully simplify the algebra for

i) $(2 + x)(7 + 3x)$,　　　　ii) $(3x - 7)(9x - 5)$,

iii) $x(7 + 3x)$,　　　　iv) $8x(7 + 3x)$,

v) $-8x(7 + 3x)$.

i) For $(2 + x)(7 + 3x)$

2 multiplies 7 to give 14

2 multiplies $3x$ to give $6x$

x multiplies 7 to give $7x$

x multiplies $3x$ to give $3x^2$　| Simplifying like terms.

i.e. $(2 + x)(7 + 3x) = 14 + 13x + 3x^2$.

ii) For $(3x - 7)(9x - 5)$

$3x$ multiplies $9x$ to give $27x^2$

$3x$ multiplies -5 to give $\quad -15x$

-7 multiplies $9x$ to give $\quad -63x$

-7 multiplies -5 to give $\qquad\qquad 35$

i.e. $\quad (3x - 7)(9x - 5) = 27x^2 - 78x + 35.$

> Simplifying like terms.

iii) For $x(7 + 3x)$ the multiplication is easier because the first bracket has been reduced to a simple x term. Then x multiplies each term in the bracket to give

$$x(7 + 3x) = 7x + 3x^2.$$

iv) $8x(7 + 3x)$ follows a pattern similar to the previous example, being increased by a factor of 8 so that

$$8x(7 + 3x) = 56x + 24x^2.$$

v) $-8x(7 + 3x)$ shows the pattern continuing, with the inclusion of a minus sign, so that

$$-8x(7 + 3x) = -56x - 24x^2.$$

The order of terms **within** the brackets can be important if there are some minus signs. However, which bracket is written first does not matter. Exactly the same result of $14 - 13x + 3x^2$ will come from both $(2 - x)(7 - 3x)$ and $(7 - 3x)(2 - x)$.

▰▰▰▰ Examples 8.3 ▰▰▰▰

Multiply out the brackets and fully simplify the algebra for

i) $(2 + 3x)^2$, ii) $(2 - 3x)^2$,

iii) $(2 + 3x)(2 - 3x)$.

i) The whole bracket is squared. This means that the bracket is multiplied by itself to give

$$\begin{aligned}(2 + 3x)^2 &= (2 + 3x)(2 + 3x) \\ &= 4 + 6x + 6x + 9x^2 \\ &= 4 + 12x + 9x^2.\end{aligned}$$

ii) This example follows a style similar to the previous one,

i.e. $\begin{aligned}(2 - 3x)^2 &= (2 - 3x)(2 - 3x) \\ &= 4 - 6x - 6x + 9x^2 \\ &= 4 - 12x + 9x^2.\end{aligned}$

The answers to these two examples are very much alike, differing only by the $+12x$ and $-12x$ terms. You can expect this type of result again, but do not attempt to learn it according to any complicated rule.

iii) $(2 + 3x)(2 - 3x) = 4 - 6x + 6x - 9x^2$
$$= 4 - 9x^2.$$

Again you can expect to see this type of result again. Where the brackets look similar, differing only by the middle $+/-$ sign, the x terms disappear.

Examples 8.3 can be generalised by
$$(a + b)^2 = a^2 + 2ab + b^2$$
and $(a - b)^2 = a^2 - 2ab + b^2.$

You may find it helpful to remember:

"square the first term (a^2), square the second term (b^2), and twice the product of the terms $(+2ab$ or $-2ab)$, the results being added".

This is often shortened to:

"square the first, square the second, twice the product".

Also $(a + b)(a - b) = a^2 - b^2$
or $(a - b)(a + b) = a^2 - b^2.$

Read from the right this is the difference $(-)$ of two squares (a^2 and b^2).

ASSIGNMENT

Let us have an initial look at our Assignment. This is not a solution, more a formulation of the problem. Fig. 8.1 is a plan of our conservatory with the external dimensions. The shaded section is our path. We have no known dimension for the path so let it be of width w. There are several methods of obtaining an expression for the area of the path. One is to look at the overall area and the area of the conservatory alone. The difference in these 2 values is the area of the path.

Overall area $= (6 + w)(4 + 2w).$

Area of conservatory $= 6 \times 4$
$$= 24.$$

Path area $= (6 + w)(4 + 2w) - 24$
$$= 24 + 16w + 2w^2 - 24$$
$$=   16w + 2w^2.$$

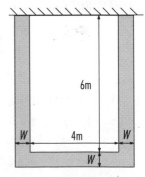

6m

4m

W W

W

Fig. 8.1

The next important question to answer is how much concrete? From our limited budget we can afford $1.5 \, \text{m}^3$ and expect a wastage of 10%. This leaves us with 90% of $1.5 \, \text{m}^3$,

i.e. $\dfrac{90}{100} \times 1.5 = 1.35 \, \text{m}^3.$

For this type of path we expect to lay the concrete to a depth of 75 mm (0.075 m) over the hardcore. This means the total path area is to be $\dfrac{1.35}{0.075} = 18\,\text{m}^2$.

We can link together our 2 calculations for path area so that

$$2w^2 + 16w = 18$$

i.e. $\quad 2w^2 + 16w - 18 = 0$

i.e. $\quad w^2 + 8w - 9 = 0.$

> Dividing each term by 2.

EXERCISE 8.1

Multiply out the following brackets, giving fully simplified answers.

1 $(x+2)(x+5)$	**26** $(4x-5)(3x+2)$
2 $(5x+1)(2x+1)$	**27** $(4-x)^2$
3 $(x+5)(x+2)$	**28** $(x-1)^2$
4 $(5x+3)(2x+4)$	**29** $(x-4)^2$
5 $(x+5)(3x+2)$	**30** $(x+5)^2$
6 $(3x+5)(4x+2)$	**31** $(2x+7)^2$
7 $(x-2)(x-5)$	**32** $(7-2x)^2$
8 $(4-x)(3-x)$	**33** $(x+2)(x-2)$
9 $(x-4)(x-10)$	**34** $(2x-13)(2x+13)$
10 $(4x-3)(3x-2)$	**35** $(2x-2)(3x+3)$
11 $(3x-2)(4x-7)$	**36** $x(7+5x)$
12 $(2x-1)(3x-1)$	**37** $4x(7-x)$
13 $(x+2)(x-5)$	**38** $-7x(2x+9)$
14 $(x+4)(x-9)$	**39** $(5x+4)(2x+3)$
15 $(2x+3)(2x-4)$	**40** $(4x-11)(2x+3)$
16 $(2x+4)(x-\frac{1}{2})$	**41** $(3+2x)(1-6x)$
17 $(7x+1)(7x-3)$	**42** $(9-x)(1-9x)$
18 $(6x+5)(11x-9)$	**43** $(2+x)x$
19 $(x-2)(x+5)$	**44** $(1-7x)(2x+9)$
20 $(x-1)(9x+10)$	**45** $(2x+\frac{1}{2})(3x-6)$
21 $(2x-2)(3x+13)$	**46** $(3+t)(5-2t)$
22 $(11x-1)(2x+8)$	**47** $(2t-7)(3t+7)$
23 $(x-2)(2x+5)$	**48** $(2-3x)4x$
24 $(3x-2)(2x+3)$	**49** $(2t-7)(7+3t)$
25 $(x+1)^2$	**50** $(7t-1)(9+2t)$

Quadratic graphs

Let us introduce these graphs by looking at a particular example of a quadratic expression, $x^2 + 3x + 2$. At this stage we do not know its likely shape. We may write the three terms $x^2 + 3x + 2$ in the form

$$y = x^2 + 3x + 2.$$

Just one letter, y, represents the sum of the three terms. y depends on their values for any given values of x.

Before attempting to plot a graph we construct a table of values, using some specimen values of x. In this case the specimen values range from $x = -4$ to $x = 2$. The body of the table between the horizontal lines has three rows of working, one for each term. The x^2 row shows each specimen value of x squared, e.g. $(-4)^2 = 16$. The $3x$ row shows each value of x multiplied by 3, e.g. $3(-4) = -12$. The row labelled 2 has the same constant value all the way along.

x	-4	-3	-2	-1	0	1	2	$\longleftrightarrow$
x^2	16	9	4	1	0	1	4	
$3x$	-12	-9	-6	-3	0	3	6	
2	2	2	2	2	2	2	2	
y	6	2	0	0	2	6	12	$\updownarrow$

Finally for each x value, the column is added to give the last value, the value of y below the second horizontal line. With the table complete we can plot a graph of y against x. Fig. 8.2 shows the points connected by a smooth curve which dips slightly below the y values of 0. This is to maintain the smoothness of the curve.

We can draw a smooth curve more easily with more points plotted closer together. In later examples we will use intervals of 0.5 rather than 1.0 for x.

Let us look at a few features of the graph. We are interested in where it crosses the axes, i.e. at $x = -2$, $x = -1$ and $y = 2$. The x

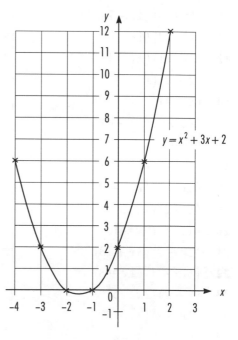

Fig. 8.2

values will be more important to us later in this chapter. Remember that
the curve crosses the y-axis when $x = 0$

so that $\quad y = x^2 + 3x + 2$

becomes $\quad y = 0^2 + 3(0) + 2 = 2$.

The shape is important. The graph of any quadratic expression is called a
parabola (Figs. 8.3) and, usually, is either

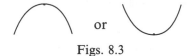

or

Figs. 8.3

Where the graph turns, at either the peak or trough, is called the **vertex**.

In our example the curve crosses the x-axis but this might not always
happen. Other curves may only touch the x-axis or may not cross it at all.

The general form of a quadratic expression can be written as

$\quad y = ax^2 + bx + c$.

The letters a, b and c are standard ones, each representing a number that
may be positive, negative or perhaps 0. We will look at them in turn.

Firstly let us look at $y = ax^2$ (a being the coefficient of x^2) and let $a = 2$,
then $a = 5$ and finally $a = -2$. Fig. 8.4 shows that as the size of a gets larger
so the curve rises more steeply, i.e. $y = 5x^2$ rises more steeply than $y = 2x^2$.
A negative value changes the shape from $\smile$ to $\frown$.

In fact $y = -2x^2$ *is a reflection of* $y = 2x^2$ *in the horizontal axis* (Fig. 8.4).

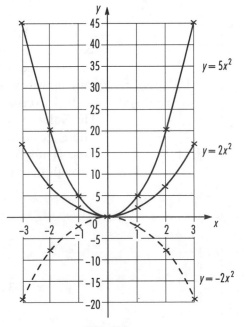

Fig. 8.4

Now let us see what happens as we change the value of b (the coefficient of x) using $y = x^2 + bx$. In turn let $b = 1$, then $b = 3$ and finally $b = -3$. As b increases the vertex moves to the left. When b is negative the vertex is to the right of the vertical axis. This is shown by the 3 examples in Fig. 8.5.

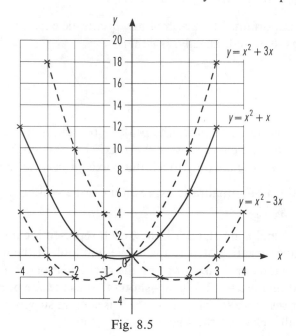

Fig. 8.5

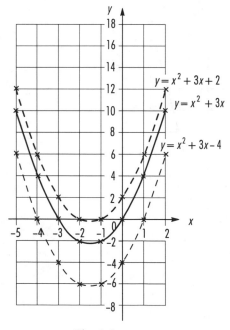

Fig. 8.6

Finally we can see that a change in c (the constant) causes the graph to shift on the vertical axis. Fig. 8.6 shows that as c increases the graph shifts vertically upwards, whilst a decrease shifts it vertically downwards. The value of c corresponds to where the curve cuts the vertical axis. Thus $y = x^2 + 3x$ passes through the origin, with $c = 0$.

$$y = x^2 + 3x + 2 \qquad \boxed{c = 2.}$$

$$y = x^2 + 3x \qquad \boxed{c = 0.}$$

$$y = x^2 + 3x - 4 \qquad \boxed{c = -4.}$$

■ EXERCISE 8.2 ■

1 Complete the following table and then plot the graph of
$y = 2x^2 + 4x - 5$.

x	-3.5	-3.0	-2.5	-2.0	-1.5	-1.0	-0.5	0	0.5	1.0
$2x^2$	24.5					2.0			0.5	
$4x$	-14.0						-2.0		2.0	
-5	-5.0		-5.0						-5.0	
y	5.5								-2.5	

2 For each quadratic expression construct a table of values and plot a graph of y against x.
   i) $y = x^2$ from $x = -4$ to $x = 3$ at intervals of 1.0.
   ii) $y = x^2 - 2x$ from $x = -2$ to $x = 5$ at intervals of 1.0.
   iii) $y = 3x^2 + 5x$ from $x = -2$ to $x = 2$ at intervals of 0.5.
   iv) $y = 3x^2 - 5x$ from $x = -2$ to $x = 2$ at intervals of 0.5.
   v) $y = x^2 - 3x + 2$ from $x = -3$ to $x = 3$ at intervals of 0.5.

3 In Question **2i)** you plotted $y = x^2$. Sketch this curve on another pair of axes. Compare the following curves by sketching them on that same set of axes

   i) $y = \frac{1}{2}x^2$,   ii) $y = -\frac{1}{2}x^2$,   iii) $y = 4x^2$,   iv) $y = -3x^2$.

4 You plotted $y = x^2 - 2x$ in Question **2ii)**. Sketch this curve on another set of axes. Compare the following curves by sketching them on that same set of axes
   i) $y = x^2 + 2x$,   ii) $y = x^2 - 5x$,   iii) $y = x^2 + 6x$.

5 Using the graph from Question **2v)**, $y = x^2 - 3x + 2$, sketch it on another set of axes. Now compare it with the graphs of
   i) $y = x^2 - 3x \div 2$,   ii) $y = x^2 - 3x$,   iii) $y = x^2 - 3x + 8$.

Quadratic equations – graphical solution

The previous section helped us to understand the shape of a parabola. Let us link together the graph and the quadratic equation. Already we have drawn the graph (Fig. 8.2) of $y = x^2 + 3x + 2$.

Suppose we wish to use this to solve the quadratic equation

$$x^2 + 3x + 2 = 0.$$

Patterns within Mathematics encourage us to compare these equations. Both of them have in common x^2, $3x$ and 2. That leaves y from the first and 0 from the second one. The only way for the two relationships to be the same is for $y = 0$, which occurs on the x-axis. Therefore if we plot the parabola, $y = x^2 + 3x + 2$, we can use it to solve the quadratic equation, $x^2 + 3x + 2 = 0$. The solutions are where the graph cuts the horizontal axis. The **solutions** of the quadratic equation are called the **roots** of the equation. Fig. 8.7 shows the roots at $x = -2$ and $x = -1$.

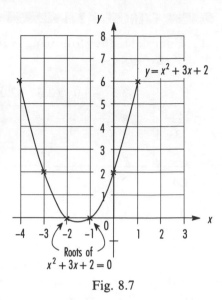

Fig. 8.7

Example 8.4

Graphically solve the quadratic equation $x^2 - 5x + 3.5 = 0$.

We start by looking at the terms of $y = x^2 - 5x + 3.5$. The coefficient of x^2 is positive which means that the basic shape of the graph is $\smile$. The constant being 3.5 means that the graph cuts the vertical axis at 3.5. At this stage this is as much as we know. At least we have some idea against which to check our plot.

We suggest that we use values of x from 0 to 3.5 at intervals of 0.5. If the curve doesn't cut the horizontal axis twice we can re-assess the situation. If necessary we might then plot some more coordinates. Now let us construct a table of values before attempting the plot.

x	0.0	0.5	1.0	1.5	2.0	2.5	3.0	3.5	$\longleftrightarrow$
x^2	0.00	0.25	1.00	2.25	4.00	6.25	9.00	12.25	
$-5x$	0.00	−2.50	−5.00	−7.50	−10.00	−12.50	−15.00	−17.50	
$+3.5$	3.50	3.50	3.50	3.50	3.50	3.50	3.50	3.50	
y	3.50	1.25	−0.50	−1.75	−2.50	−2.75	−2.50	−1.75	$\updownarrow$

At this stage we might plot y (vertically) against x (horizontally).

We know that the roots of the quadratic equation occur as the curve crosses the horizontal axis, i.e. as $y = 0$. As this happens the value of y changes from being positive to being negative, or from being negative to being positive. In the table we see that y changes from 1.25 to −0.50, meaning that there is a root somewhere in this region. The root lies between the corresponding x values of 0.5 and 1.0.

There is no other change in the sign of y so perhaps we should extend our table.

x	4.0	4.5
x^2	16.00	20.25
$-5x$	-20.00	-22.50
$+3.5$	3.50	3.50
y	-0.5	1.25

These extra values show that the sign of y does change again. Hence there is another root somewhere between the x values of 4.0 and 4.5.

From our graph in Fig. 8.8 we can read off the roots of our quadratic equation. The accuracy of these values depends on the quality of our graph. A high quality plot should get close to roots of $x = 0.84$, 4.16.

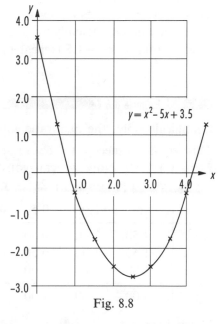

$y = x^2 - 5x + 3.5$

Fig. 8.8

<hr>

Example 8.5

Graphically solve the quadratic equation $x^2 + 3x + 2.25 = 0$.

Using the method of Example 8.4 we can plot the graph of $y = x^2 + 3x + 2.25$. In this case the specimen x values are to range from $x = -2.5$ to $x = 1.5$ at intervals of 0.5. The table shows the usual method for working out the coordinates.

x	-2.5	-2.0	-1.5	-1.0	-0.5	0.0	0.5	1.0	1.5	$\longleftrightarrow$
x^2	6.25	4.00	2.25	1.00	0.25	0.00	0.25	1.00	2.25	
$+3x$	-7.50	-6.00	-4.50	-3.00	-1.50	0.00	1.50	3.00	4.50	
$+2.25$	2.25	2.25	2.25	2.25	2.25	2.25	2.25	2.25	2.25	
y	1.00	0.25	0.00	0.25	1.00	2.25	4.00	6.25	9.00	$\updownarrow$

A glance at the table shows no change of sign for y, but a root is obvious immediately. We see $y = 0$ corresponds to the root $x = -1.5$.

Let us plot the coordinates in Fig 8.9 and draw a smooth curve through them.

This time, following through the points smoothly, the curve does not actually cross the x-axis. It touches the axis at the point where $x = 1.5$, i.e. the horizontal axis is a tangent to the curve at this point. This means that $x = -1.5$ is a root to the quadratic equation, usually termed a **repeated root** or $x = -1.5$ **(repeated).**

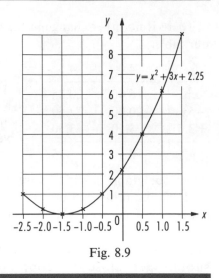

Fig. 8.9

Example 8.6

Graphically solve the quadratic equation $2x^2 = -x - 1$.

This equation needs to be re-arranged into the correct format of $2x^2 + x + 1 = 0$. Now we can construct a table of specimen values for the associated parabola $y = 2x^2 + x + 1$.

x	-1.5	-1.0	-0.5	0.0	0.5	1.0	1.5	2.0	$\longleftrightarrow$
$2x^2$	4.50	2.00	0.50	0.00	0.50	2.00	4.50	8.00	
$+x$	-1.50	-1.00	-0.50	0.00	0.50	1.00	1.50	2.00	
$+1$	1.00	1.00	1.00	1.00	1.00	1.00	1.00	1.00	
y	4.00	2.00	1.00	1.00	2.00	4.00	7.00	11.00	$\updownarrow$

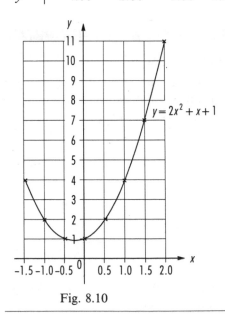

Fig. 8.10

The y values in the table show no change of sign and no hint of any change. Therefore it is unlikely that plotting more coordinates will be useful. Perhaps the curve doesn't cross the horizontal axis. In fact Fig. 8.10 shows our graph always above the horizontal axis. The $\smile$ shape has its lowest y value a little less than 1.00 with increasing values on both sides of this. Because the graph does **not** cross the horizontal axis there are **no real roots** to the quadratic equation $2x^2 + x + 1 = 0$.

We have looked at 3 possible types of quadratic equations.

  i) Curve cutting the x-axis twice,   i.e. 2 different roots;

 ii) curve touching the x-axis,   i.e. 1 repeated root;

iii) curve neither cutting nor touching the x-axis,   i.e. no real roots.

For consistency all our curves have been $\smile$ shaped, but exactly the same ideas apply to $\frown$ shaped curves.

Let us look at some features of quadratic equations for a variety of curves in Figs. 8.11 to 8.13.

    The general equation is $ax^2 + bx + c = 0$.

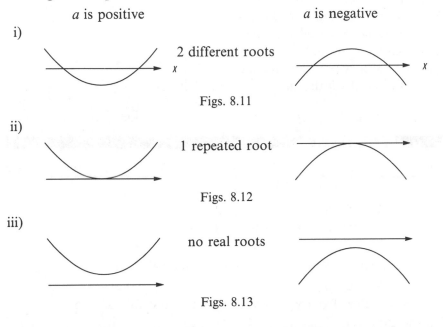

                 a is positive   *a* is negative

i)   2 different roots

Figs. 8.11

ii)   1 repeated root

Figs. 8.12

iii)   no real roots

Figs. 8.13

ASSIGNMENT

In our first look at the Assignment we deduced a quadratic equation $w^2 + 8w - 9 = 0$. w is the width of the path to be concreted around 3 sides of the conservatory. If we attempt to solve this graphically we need to plot the graph of $y = w^2 + 8w - 9$, with w horizontal and y vertical. In this practical problem negative distances have no meaning. We can try specimen values from $w = 0$ to $w = 2.5$ at intervals of 0.5 according to the table below.

w	0.0	0.5	1.0	1.5	2.0	2.5	$\longleftrightarrow$
w^2	0.00	0.25	1.00	2.25	4.00	6.25	
$+8w$	0.00	4.00	8.00	12.00	16.00	20.00	
-9	−9.00	−9.00	−9.00	−9.00	−9.00	−9.00	
y	−9.00	−4.75	0.00	5.25	11.00	17.25	$\updownarrow$

Fig. 8.14 confirms what is shown in the table, that the curve crosses the horizontal axis at $w = 1.0$.

You can see that our graph shows only 1 crossing of the horizontal axis when we might have expected to get 2 crossings. Also you can appreciate that plotting the curve for negative values of w would give that other crossing point. Remember that negative dimensions have no practical meaning. This leaves our only sensible solution to be a path around the conservatory of width 1 m.

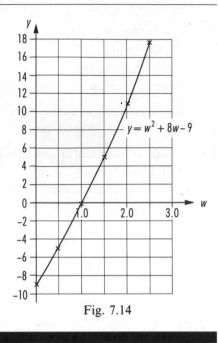

Fig. 7.14

EXERCISE 8.3

In each case plot a parabola to solve the quadratic equation. You are given ranges of specimen values for your tables.

1   $x^2 + 4x - 5 = 0$
    using values from $x = -7$ to $x = 4$ at intervals of 1.

2   $x^2 - x - 6 = 0$
    using values from $x = -4$ to $x = 6$ at intervals of 1.

3   $x^2 + 6x + 9 = 0$
    using values from $x = -4.5$ to $x = -2$ at intervals of 0.25.

4   $x^2 - 7x = 0$
    using values from $x = -2$ to $x = 9$ at intervals of 1.

5   $x^2 - 25 = 0$
    using values from $x = -6$ to $x = 6$ at intervals of 1.

6   $x^2 - 3x - 1 = 0$
    using values from $x = -1$ to $x = 4$ at intervals of 0.5.

7   $5x^2 - 9x - 6 = 0$
    using values from $x = -1$ to $x = 4$ at intervals of 0.5.

8   $2x^2 + 3x - 3 = 0$
    using values from $x = -3$ to $x = 1.5$ at intervals of 0.5.

9   $3x^2 + 5 = 8x$
    using values from $x = 0.5$ to $x = 2.0$ at intervals of 0.1.

10   $6x^2 - 2x + 1 = 0$
    using values from $x = -2$ to $x = 2$ at intervals of 0.5.

Quadratic equations – solution by factors

The first section of this chapter saw us multiplying out pairs of brackets. By way of example we started with $(x+2)(x+5)$ and reached $x^2+7x+10$. Both $(x+2)$ and $(x+5)$ are factors of the quadratic expression $x^2+7x+10$. Suppose we start with the quadratic equation

$$x^2+7x+10 = 0$$

aiming to factorise it so that

$$(x+2)(x+5) = 0.$$

> Always check back – multiply out the brackets.

These two brackets are multiplied together and their overall result is zero. Remember that any value multiplied by zero gives zero as a result. This means that one or other of the brackets is zero,

i.e. $\qquad x+2 = 0 \qquad$ or. $x+5 = 0$

i.e. $\qquad x = -2 \quad$ or $\qquad x = -5.$

Often this is shortened to $x = -2, -5.$

Both $x = -2$ and $x = -5$ are roots of the quadratic equation. Reversing the method of factorisation we can create that quadratic equation.

If $\qquad\qquad x = -2 \quad$ or $\qquad x = -5$

i.e. $\qquad\qquad x+2 = 0 \quad$ or $x+5 = 0$

and multiplying these results together we get

$$(x+2)(x+5) = 0 \times 0$$

i.e. $\quad x^2+7x+10 = 0.$

Whenever we get roots we can check that they are correct. Just substitute them into the original equation. Separately if we substitute -2 and -5 for x in $x^2+7x+10$ we should get a result of 0,

i.e. when $x = -2,$ $(-2)^2+7(-2)+10 = \quad 4-14+10 = 0$

and when $x = -5,$ $(-5)^2+7(-5)+10 = \quad 25-35+10 = 0.$

When factorising a quadratic equation (or expression) we think about the **products** we will get. They are from the first terms in the brackets and the second terms,

i.e. $(x+2)(x+5).$

These are important because $x \times x = x^2$ and $2 \times 5 = 10$ are the first and last terms of $x^2+7x+10$. Also $2+5=7$ which is the coefficient of x in the middle term.

There are other factors of 10. Because $10 \times 1 = 10$ we have 10 and 1 as those factors. However $10 + 1 = 11$ and we know that we need $7x$, not $11x$, as the middle term. There is some trial and error in discovering which pair of factors are the correct ones. The $+/-$ signs in a quadratic equation (or expression) are also important. Let us demonstrate the factors and signs using the next set of examples.

▬▬▬ Examples 8.7 ▬▬▬

By factorisation solve the quadratic equations

i) $x^2 + 10x + 21 = 0$,   ii) $x^2 - 10x + 21 = 0$,

iii) $x^2 + 4x - 21 = 0$,   iv) $x^2 - 4x - 21 = 0$.

These examples are similar to show various combinations of factors. Firstly they all start with x^2, and we know that $x \times x = x^2$, *i.e.* x and x are factors of x^2. They all finish on the left-hand side with 21. The factors of 21 are 1, 3, 7, and 21 combined as $1 \times 21 = 21$

$$\text{and } 3 \times 7 = 21.$$

i)   In $x^2 + 10x + 21 = 0$ we have $+21$. The $+$ sign means both factor brackets have the **same** sign. The $+$ sign of $+10x$ means both factor signs are $+$. So we have

$$(x+ \ )(x+ \ ) = 0.$$

The next step is to look at the addition $(+)$ of factors of 21,

i.e. $1 + 21 = 22$

and $3 + 7 = 10.$

These results and the $10x$ mean we choose 3 and 7 to give

$$(x + 3)(x + 7) = 0.$$

Then $x + 3 = 0$ or $x + 7 = 0$

i.e. $x = -3$ or $x = -7$

i.e. $x = -3, -7.$

ii)   In $x^2 - 10x + 21 = 0$ the $+$ of $+21$ means both factor brackets have the **same** sign. The $-$ of $-10x$ means the signs are $-$. So we have

$$(x- \ )(x- \ ) = 0.$$

Looking at the factors of 21 again we choose 3 and 7 because $-3 - 7 = -10$ to give

$$(x - 3)(x - 7) = 0.$$

Then $x - 3 = 0$ or $x - 7 = 0$

i.e. $x = 3$ or $x = 7$

i.e. $x = 3, 7.$

iii)   In $x^2 + 4x - 21 = 0$ the $-$ of -21 means the factor brackets have **different** signs, i.e.

$$(x+ \ )(x- \ ) = 0.$$

> We get the same result using $(x- \ )(x+ \ ).$

For the next step with factors of 21 we look at their **difference**:
$$1 - 21 = -20 \quad \text{and} \quad 21 - 1 = 20,$$
$$3 - 7 = -4 \quad \text{and} \quad 7 - 3 = 4.$$

The last option, giving 4, agrees with the middle term's coefficient so that
$$(x + 7)(x - 3) = 0.$$

Then $x + 7 = 0$ or $x - 3 = 0$

i.e. $x = -7$ or $x = 3$

i.e. $x = -7, 3.$

iv) In $x^2 - 4x - 21 = 0$ we make decisions similar to those of the previous example. We choose the third option of $3 - 7 = 4$ to agree with the middle term's coefficient so that
$$(x + 3)(x - 7) = 0.$$

Then $x + 3 = 0$ or $x - 7 = 0$

i.e. $x = -3$ or $x = 7$

i.e. $x = -3, 7.$

For each example remember to substitute each root into it's original quadratic equation. This check should give an answer of 0.

Examples 8.8

By factorisation solve the quadratic equations

i) $x^2 + 14x + 24 = 0,$ ii) $x^2 - 11x + 24 = 0,$

iii) $x^2 + 5x - 24 = 0,$ iv) $x^2 - 2x - 24 = 0.$

All these examples have 24 on the left-hand side. The factors of 24 combine as 1×24, 2×12, 3×8 and 4×6.

i)   In $x^2 + 14x + 24 = 0$ the $+$ sign of $+24$ means the same signs for the factor brackets. The $+$ sign of $+14x$ means both the factor signs are $+$.

Now $1 + 24 = 25,$

              $2 + 12 = 14,$

              $3 + 8 = 11$

and $4 + 6 = 10.$

The second option agrees with the middle term's coefficient. Combining all these decisions we get
$$(x + 2)(x + 12) = 0$$

i.e. $x + 2 = 0$ or $x + 12 = 0$

i.e. $x = -2$ or $x = -12$

i.e. $x = -2, -12.$

ii)   In $x^2 - 11x + 24 = 0$ we have the same four pairs of factor brackets based on x^2 and 24. The $+$ sign of $+24$ and the $-$ sign of $-11x$ mean the factor bracket signs are the same and negative.

Now $-1 - 24 = -25$

$-2 - 12 = -14$

$-3 - 8 = -11$

and $-4 - 6 = -10.$

This time the third option agrees with the middle term's coefficient. Combining all these decisions we get

$$(x - 3)(x - 8) = 0$$

i.e. $x - 3 = 0$ or $x - 8 = 0$

i.e. $x = 3$ or $x = 8$

i.e. $x = 3, 8.$

iii)   In $x^2 + 5x - 24 = 0$ the $-$ sign of -24 indicates different factor bracket signs. Hence we are looking at the differences of the numbers,

$1 - 24 = -23,$ $24 - 1 = 23,$

$2 - 12 = -10,$ $12 - 2 = 10,$

$3 - 8 = -5,$ $8 - 3 = 5, *$

$** \ 4 - 6 = -2,$ $6 - 4 = 2.$

The * option agrees with the middle term's coefficient. Combining all these decisions we get

$$(x + 8)(x - 3) = 0$$

i.e. $x + 8 = 0$ or $x - 3 = 0$

i.e. $x = -8$ or $x = 3$

i.e. $x = -8, 3.$

iv)   For the solution of $x^2 - 2x - 24 = 0$ we use the previous example as a basis and choose the ** option to give

$$(x + 4)(x - 6) = 0$$

i.e. $x + 4 = 0$ or $x - 6 = 0$

i.e. $x = -4$ or $x = 6$

i.e. $x = -4, 6.$

For each example remember to substitute each root into its original quadratic equation. This check should give an answer of 0.

▰▰ ASSIGNMENT ▰▰

Now we have enough factorisation skill to look at our assignment problem again. Our assignment is based on the quadratic equation $w^2 + 8w - 9 = 0$. Remember that w is the width of our concrete path around 3 sides of the conservatory. The $-$ sign of -9 means there are different signs in the factor brackets. Now the difference of those numbers is to be $+8$. Factors of 9 are 9×1 and 3×3. We choose the first option and the difference because $9 - 1 = 8$;

i.e. $(w + 9)(w - 1) = 0$

i.e. $w + 9 = 0$ or $w - 1 = 0$

i.e. $w = -9$ or $w = 1$.

   Because this is a practical problem we need to interpret our answers. Path widths can only be positive and so we need the value $w = 1$ m.

▰▰ EXERCISE 8.4 ▰▰

By factorisation solve the following quadratic equations.

1	$x^2 + 6x + 5 = 0$	16	$x^2 - 2x - 15 = 0$
2	$x^2 + 5x + 6 = 0$	17	$x^2 - 3x - 10 = 0$
3	$x^2 + 7x + 12 = 0$	18	$x^2 - 2x - 48 = 0$
4	$x^2 + 3x + 2 = 0$	19	$x^2 - x - 56 = 0$
5	$x^2 + 14x + 49 = 0$	20	$x^2 - 2x - 63 = 0$
6	$x^2 - 5x + 6 = 0$	21	$x^2 + 6x - 40 = 0$
7	$x^2 - 12x + 27 = 0$	22	$x^2 + 3x - 70 = 0$
8	$x^2 - 11x + 30 = 0$	23	$x^2 + 20x + 100 = 0$
9	$x^2 - 10x + 25 = 0$	24	$x^2 - 6x + 9 = 0$
10	$x^2 - 8x + 12 = 0$	25	$x^2 - x - 72 = 0$
11	$x^2 + 4x - 5 = 0$	26	$x^2 - x - 12 = 0$
12	$x^2 + 4x - 12 = 0$	27	$x^2 + 6x - 72 = 0$
13	$x^2 + 2x - 8 = 0$	28	$x^2 + 5x - 36 = 0$
14	$x^2 + 7x - 18 = 0$	29	$x^2 - 18x - 40 = 0$
15	$x^2 + 13x - 48 = 0$	30	$x^2 - 11x = 60$

So far in this section the coefficient of x^2 has always been 1. This might not always happen but the basic ideas about $+/-$ signs and factors continue to apply. Any coefficient change just means we have to check more combinations of factors.

▬▬▬▬ **Examples 8.9** ▬▬▬▬▬▬▬▬▬▬▬

By factorisation solve the quadratic equations
i) $2x^2 + 9x + 10 = 0$,   ii) $3x^2 - 14x + 8 = 0$,
iii) $4x^2 - 5x - 6 = 0$.

i) All our signs are $+$ which means both the factor bracket signs are $+$.
    Now we need to look at the factors of $2x^2$ and 10. Firstly, with x in
    each factor, the factors of $2x^2$ are x and $2x$. This gives us

    $(x+\ )(2x+\ ) = 0$.

    Previously at this stage the brackets were the same so order was not
    important. Different brackets now mean we must be more careful.
    The factors of 10 are

    $10 \times 1, \quad 1 \times 10,$
    $5 \times 2, \quad 2 \times 5.$

    By trial and error we test these options in turn, i.e.

    $(x + 10)(2x + 1), \qquad (x + 1)(2x + 10),$
    $(x + 5)(2x + 2), \qquad (x + 2)(2x + 5).$

    In each case we multiply out the brackets and collect together the
    terms. The first term is $2x^2$ and the last term is 10. Only the middle
    term varies. You should check for yourself that those middle terms
    are $21x$, $12x$, $12x$ and $9x$. Therefore we need the fourth option, i.e.

    $$2x^2 + 9x + 10 = 0$$

    factorises to $(x + 2)(2x + 5) = 0$

    i.e. $x + 2 = 0$ or $2x + 5 = 0$

    i.e. $x = -2$ or $2x = -5$

    i.e. $x = -2, -\dfrac{5}{2}$

    $$\text{(or } -2, -2.5).$$

ii) The signs in $3x^2 - 14x + 8 = 0$ mean both factor brackets contain $-$.
    The factors of $3x^2$ are x and $3x$ so we have

    $(x-\ )(3x-\ ) = 0$.

    The factors of 8 are 8×1, 1×8, 2×4 and 4×2. By trial and error
    we test these options in turn, i.e.

    $(x - 8)(3x - 1), \qquad (x - 1)(3x - 8),$
    $(x - 2)(3x - 4), \qquad (x - 4)(3x - 2).$

    In each case, when we multiply out the brackets and collect the terms
    the first term is $3x^2$ and the last term is 8. You should check for
    yourself that the middle terms of these options in turn are $-25x$,
    $-11x$, $-10x$ and $-14x$.
    Thus we need the fourth option, i.e.

    $$3x^2 - 14x + 8 = 0$$

    factorises to $(x - 4)(3x - 2) = 0$

i.e. $x - 4 = 0$ or $3x - 2 = 0$

i.e. $x = 4$ or $3x = 2$

i.e. $x = 4, \dfrac{2}{3}.$

iii)   In $4x^2 - 5x - 6 = 0$ the $-$ of -6 means the factor brackets have different signs. The possible factors of $4x^2$ are either $4x$ and x or $2x$ and $2x$. The possible factors of 6 are either 1 and 6 or 2 and 3. By trial and error we test all these options in turn, i.e.

$$(2x + 1)(2x - 6), \qquad (2x - 1)(2x + 6),$$
$$(2x + 2)(2x - 3), \qquad (2x - 2)(2x + 3),$$
$$(4x + 1)(x - 6), \qquad (4x - 1)(x + 6),$$
$$(4x + 2)(x - 3), \qquad (4x - 2)(x + 3),$$
$$* \ (4x + 3)(x - 2), \qquad (4x - 3)(x + 2),$$
$$(4x + 6)(x - 1), \qquad (4x - 6)(x + 1).$$

The * option, when multiplied out, gives the correct middle term of $-5x$ so that

$$4x^2 - 5x - 6 = 0$$

factorises to $(4x + 3)(x - 2) = 0$

i.e. $4x + 3 = 0$ or $x - 2 = 0$

i.e. $4x = -3$ or $x = 2$

i.e. $x = -\dfrac{3}{4}, 2$

(or $-0.75, 2$).

For each example remember to substitute each root into its original quadratic equation. This check should give an answer of 0.

Not all quadratics factorise easily. Quadratic expressions (and equations) of the type $a^2 + b^2$ will not factorise. The hint is the 2 square terms both being positive, (though both being negative would also fail). For example $x^2 + 9 = 0$ has no real roots as we will see in Examples 8.11.

We will look at other cases in the next section.

■ EXERCISE 8.5 ■

By factorisation solve the quadratic equations.

1   $3x^2 + x - 2 = 0$

2   $4x^2 - 4x + 1 = 0$

3   $6x^2 - x - 2 = 0$

4   $3x^2 - x - 2 = 0$

5   $2x^2 + 7x + 6 = 0$

6   $2x^2 + 13x + 15 = 0$

7   $3x^2 + 8x - 3 = 0$

8   $2x^2 - 7x + 6 = 0$

9   $2x^2 + 11x + 5 = 0$

10   $6x^2 + 7x - 20 = 0$

11   $6x^2 - 11x - 7 = 0$

12   $6x^2 - 13x + 6 = 0$

13	$15x^2 - x - 2 = 0$	**17**	$8x^2 + 2x - 15 = 0$
14	$3x^2 + 13x + 4 = 0$	**18**	$6x^2 + 13x + 6 = 0$
15	$4x^2 - 12x + 9 = 0$	**19**	$9x^2 + 12x + 4 = 0$
16	$25x^2 - 30x + 9 = 0$	**20**	$8x^2 - 45x - 18 = 0$

During solution by factorisation we have looked at quadratic equations with all three terms, i.e. a term in x^2, a term in x and a constant. They combine to equal zero. There may be cases where either the x term or the constant is missing. These types have easy solutions.

Examples 8.10

By factorisation solve the quadratic equations

i) $2x^2 + 5x = 0$,   ii) $5x^2 - 4x = 0$,

iii) $9x^2 - 6x = 0$.

All these examples have the same feature: only two terms. We are looking for a factor that is common to both terms. The method does **not** use division through by x. You will recall that division by 0 is **not allowed** in Mathematics. Each equation has a root $x = 0$ and so care is needed.

i) $2x^2 + 5x = 0$ | x is the factor common to both terms. |

factorises to $x(2x + 5) = 0$.

Either $x = 0$ or $2x + 5 = 0$

i.e. $x = 0$ or $2x = -5$

i.e. $x = 0, -2.5$.

ii) $5x^2 - 4x$ $= 0$

factorises to $x(5x - 4) = 0$.

Either $x = 0$ or $5x - 4 = 0$

i.e. $x = 0$ or $5x = 4$

i.e. $x = 0, 0.8$.

iii) $9x^2 - 6x = 0$ | $3x$ is the factor common to both terms. |

i.e. $3x(3x - 2) = 0$.

Either $3x = 0$ or $3x - 2 = 0$

i.e. $x = 0$ or $3x = 2$

i.e. $x = 0, \dfrac{2}{3}$.

For each example remember to substitute each root into its original quadratic equation. This check should give an answer of 0.

Examples 8.11

Solve the quadratic equations

i) $x^2 - 25 = 0$, ii) $2x^2 - 36 = 0$,
iii) $x^2 + 9 = 0$.

You will see this is an alternative method to factorisation in the first and second examples.

i) $\qquad x^2 - 25 = 0$

becomes $\quad x^2 \qquad = 25$

i.e. $\qquad\qquad x = \pm\sqrt{25}$.

> Square root of both sides.

Remember we need to include the negative solution. A negative value squared, e.g. $(-5)^2 = 25$, gives a positive result,

i.e. $\qquad x = \pm 5$,

$$\text{or} \quad x = -5, 5.$$

ii) $\qquad\qquad 2x^2 - 36 = 0$

becomes $\qquad 2x^2 = 36$

i.e. $\qquad\qquad x^2 = 18$

so that $\qquad\qquad x = \pm\sqrt{18}$

i.e. $\qquad\qquad x = \pm 4.243$ (3 decimal places)

$$\text{or} \quad x = -4.243, 4.243.$$

iii) $\qquad x^2 + 9 = 0$

starts off its solution in the same way with

$$x^2 = -9$$

i.e. $\qquad x = \pm\sqrt{(-9)}$

This is beyond our skill at the moment because we cannot find the square root of a negative number. There are no real solutions to this quadratic equation, i.e. there are no real roots.

EXERCISE 8.6

Solve the quadratic equations.

1 $x^2 + 4x = 0$

2 $3x^2 - 8x = 0$

3 $\frac{1}{2}x^2 - 2x = 0$

4 $9x^2 - 100 = 0$

5 $4x^2 - 4x = 0$

6 $9x^2 + 30x = 0$

7 $6x^2 = 18x$

8 $2x^2 + 7 = 0$

9 $(x - 3)^2 = 0$

10 $(2x + 1)^2 = 25$

Quadratic equations – solution using the formula

We know that not all quadratic equations (and expressions) factorise. Also some might not factorise easily so this alternative method uses a formula. It is easy to use, simply a matter of substituting the correct values and using a calculator.

We need to remind ourselves of a particular relation

$$(x + \alpha)^2 = x^2 + 2\alpha x + \alpha^2.$$

We will deduce the formula from the general quadratic equation

$$ax^2 + bx + c = 0.$$

Alongside we will work a numerical example so that you may compare the letters and numbers. Remember that a, b, and c represent numbers in the general equation. It is possible that a may be 1. Also if b or c is zero then an easier solution follows like Examples 8.10 or 8.11.

The method we use is called "**completing the square**".

$$ax^2 + bx + c = 0 \qquad\qquad 2x^2 - 7x - 2 = 0$$

$$ax^2 + bx = -c \qquad\qquad 2x^2 - 7x = 2$$

$$x^2 + \frac{b}{a}x = \frac{-c}{a} \qquad\qquad x^2 - \frac{7}{2}x = 1$$

We compare $x^2 + \dfrac{b}{a}x$ and $x^2 - \dfrac{7}{2}x$ with part of the relation $x^2 + 2\alpha x + \alpha^2$.

Both cases have x^2 and a term in x. However the constant term, represented by α^2, is missing. We can create this term and add it to both sides of each equation so maintaining their balance.

Now $\qquad\qquad 2\alpha = \dfrac{b}{a} \qquad$ or $\qquad 2\alpha = -\dfrac{7}{2}$

i.e $\qquad\qquad\quad \alpha = \dfrac{b}{2a} \qquad\qquad\qquad \alpha = -\dfrac{7}{4}$

Hence $\qquad\qquad \alpha^2 = \left(\dfrac{b}{2a}\right)^2 \qquad\qquad \alpha^2 = \left(-\dfrac{7}{4}\right)^2$

To create the correct format we add this to both sides of each equation. Remember that by adding to both sides we maintain the balance of each equation.

$$x^2 + \frac{bx}{a} + \left(\frac{b}{2a}\right)^2 = \left(\frac{b}{2a}\right)^2 - \frac{c}{a} \qquad\qquad x^2 - \frac{7}{2}x + \left(-\frac{7}{4}\right)^2 = \left(-\frac{7}{4}\right)^2 + 1.$$

In each case we have completed the square on the left-hand side. It will factorise into one bracket all squared to give

$$\left(x + \frac{b}{2a}\right)^2 = \frac{b^2}{4a^2} - \frac{c}{a} \qquad\qquad \left(x - \frac{7}{4}\right)^2 = \frac{49}{16} + 1$$

$$\left(x + \frac{b}{2a}\right)^2 = \frac{b^2 - 4ac}{4a^2} \qquad\qquad \left(x - \frac{7}{4}\right)^2 = \frac{49 + 16}{16}$$

$$x + \frac{b}{2a} = \pm\sqrt{\frac{b^2 - 4ac}{4a^2}} \qquad\qquad x - \frac{7}{4} = \pm\sqrt{\frac{65}{16}}$$

$$x + \frac{b}{2a} = \frac{\pm\sqrt{b^2 - 4ac}}{2a} \qquad\qquad x - \frac{7}{4} = \frac{\pm\sqrt{65}}{4}$$

There is always a pattern of 2 denominators that are the same ($2a$ in the general one and 4 in this particular example) in each equation.

$$x = \frac{-b}{2a} \pm \frac{\sqrt{b^2 - 4ac}}{2a} \qquad\qquad x = \frac{7}{4} \pm \frac{\sqrt{65}}{4}$$

$$x = \frac{-b \pm \sqrt{b^2 - 4ac}}{2a} \qquad\qquad x = \frac{7 \pm \sqrt{65}}{4}$$

$$x = \frac{7 + 8.0623}{4}, \frac{7 - 8.0623}{4}$$

$$x = \frac{15.0623}{4}, \frac{-1.0623}{4}$$

$$x = 3.766, -0.266.$$

The next set of examples shows how to use the formula. Just remember to substitute correctly for a, b and c with any relevant − signs.

Example 8.12

Use the formula to solve the quadratic equations

i) $3x^2 - 2x - 12 = 0$, ii) $x^2 + 5x - 19 = 0$.

i) We compare $3x^2 - 2x - 12 = 0$ with the general quadratic equation $ax^2 + bx + c = 0$.

This gives $a = 3$, $b = -2$, $c = -12$.

Now $x = \dfrac{-b \pm \sqrt{b^2 - 4ac}}{2a}$

becomes $x = \dfrac{-(-2) \pm \sqrt{(-2)^2 - 4(3)(-12)}}{2(3)}$

$$x = \frac{2 \pm \sqrt{4 + 144}}{6}$$

$$x = \frac{2 \pm \sqrt{148}}{6}$$

$$x = \frac{2 + 12.166}{6}, \frac{2 - 12.166}{6}$$

$$x = 2.36, -1.69.$$

ii) We compare $x^2 + 5x - 19 = 0$ with the general quadratic equation $ax^2 + bx + c = 0$.

This gives $a = 1, b = 5, c = -19$.

Now $x = \dfrac{-b \pm \sqrt{b^2 - 4ac}}{2a}$

becomes $x = \dfrac{-5 \pm \sqrt{5^2 - 4(1)(-19)}}{2(1)}$

$$x = \frac{-5 \pm \sqrt{25 + 76}}{2}$$

$$x = \frac{-5 \pm \sqrt{101}}{2}$$

$$x = \frac{-5 + 10.0499}{2}, \frac{-5 - 10.0499}{2}$$

$$x = 2.52, -7.52.$$

For each example remember to substitute each root into its original quadratic equation. This check should give an answer of 0. More decimal places used means a more accurate check.

▆▆▆▆ Examples 8.13 ▆▆▆▆

Use the formula to solve the quadratic equations

i) $x^2 + 3x + 2.25 = 0$,   ii) $x^2 + 2x + 5.5 = 0$.

i) The values for substitution are $a = 1, b = 3, c = 2.25$,

so that $x = \dfrac{-b \pm \sqrt{b^2 - 4ac}}{2a}$

becomes $x = \dfrac{-3 \pm \sqrt{3^2 - 4(1)(2.25)}}{2(1)}$

$$x = \frac{-9 \pm \sqrt{9 - 9}}{2}$$

$$x = \frac{-3}{2}$$

i.e. $x = -1.5$.

You can see this solution leads to $\sqrt{0}$ and then only gives one solution for x. This is the case of a repeated root, i.e. $x = -1.5$ repeated.

ii) The values for substitution are $a=1$, $b=2$, $c=5.5$

so that $\qquad x = \dfrac{-b \pm \sqrt{b^2 - 4ac}}{2a}$

becomes $\qquad x = \dfrac{-2 \pm \sqrt{2^2 - 4(1)\,(5.5)}}{2(1)}$

$\qquad\qquad x = \dfrac{-2 \pm \sqrt{-18}}{2}\,.$

We can go no further with this solution because there are no real answers to the square root of a negative number. This means this quadratic equation has no real roots.

We can look a little more closely under the $\sqrt{}$ in the general formula

$$x = \dfrac{-b \pm \sqrt{b^2 - 4ac}}{2a}$$

It is possible to look at 3 solution types which we can link with our earlier graph work.

i) $b^2 - 4ac > 0$, $\qquad$ i.e. $b^2 > 4ac$

is the usual situation leading to 2 different roots. The graph crosses the horizontal axis twice (Figs. 8.15).

Figs. 8.15

ii) $b^2 - 4ac = 0$, $\qquad$ i.e. $b^2 = 4ac$

leads to one repeated root. The graph touches the horizontal axis, i.e. that axis acts as a tangent (Figs. 8.16).

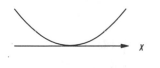

Figs. 8.16

iii) $b^2 - 4ac < 0$, $\qquad$ i.e. $b^2 < 4ac$

leads to no real roots. The graph does not cross the horizontal axis, i.e. the graph and axis do not intersect (Figs. 8.17).

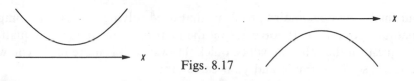

Figs. 8.17

■■■■■ **ASSIGNMENT** ■■■■■

We return to our conservatory project and the concrete path going around the 3 sides. Suppose the hardcore is rather uneven meaning that in total we waste slightly more concrete than before. If now we lose 15% this leaves us with 85% of our $1.5 \, m^3$ delivery, i.e. $1.5 \times \dfrac{85}{100} = 1.275 \, m^3$.

Remember that the depth of concrete is to be 75 mm which gives us a total area of $\dfrac{1.275 \, m^3}{75 \, mm} = 17 \, m^2$.

In the quadratic equation we replace our original 18 with 17

so that $\qquad 2w^2 + 16w = 17$

i.e $\qquad 2w^2 + 16w - 17 = 0$

$$w = \frac{-16 \pm \sqrt{16^2 - 4(2)(-17)}}{2(2)}$$

$$w = \frac{-16 \pm \sqrt{392}}{4}$$

$$w = \frac{-16 + 19.799}{4}, \quad \frac{-16 - 19.799}{4}$$

$$w = 0.95 \text{ only.}$$

The second answer, being negative, has no physical meaning relevant to our concrete path. We can have only positive widths, in this case 0.95 m.

■■■■■ **EXERCISE 8.7** ■■■■■

Use the formula to solve the quadratic equations.

1	$x^2 + 20x - 34 = 0$	6	$x^2 - 6x - 3 = 0$
2	$x^2 - x - 1 = 0$	7	$x^2 - 12x + 36 = 0$
3	$x^2 - 5x - 12 = 0$	8	$3x^2 + 14x + 2 = 0$
4	$x^2 - 28x - 18 = 0$	9	$2x^2 + 4x + 5 = 0$
5	$x^2 + 3x - 2 = 0$	10	$15x^2 - 9x + 1$

Our final exercise contains problems that need solving, rather than simply quadratic equations. If you can spot the factors of the quadratic equation in a question the solution will be quick. However, it is more likely you will need to use the formula and your calculator.

EXERCISE 8.8

1 The power dissipated by a circuit component is 6 W. It is connected in series with a resistor of 15Ω and an emf of 25 V, as shown in the diagram.

Calculate the current, I, if $15I^2 - 25I + 6 = 0$.

```
       I
  o─────►┌──────┐
         │ 15Ω  │
  25v     └──────┘
  o──────┌──────────┐
         │ COMPONENT│
         └──────────┘
```

2 $s = ut + \frac{1}{2}at^2$ refers to the motion of a vehicle.

s is the displacement,
u is the initial velocity,
a is the acceleration

and t is the time for the motion.

Calculate the time for the motion over a displacement of 70 m if the initial velocity is 1.56 ms^{-1} and the acceleration is 2.12 ms^{-2}.

3 Apart from the units, the area of a circle $(A = \pi r^2)$ and its circumference $(C = 2\pi r)$ are equal. What is the length of the radius, r? The area of the circle is now increased so that, apart from the units again, the area is 3 times the circumference; i.e. $A = 3C$. Calculate the length of the new radius.

4 Your company needs some new equipment and you decide to borrow £75 000 repayable over 2 years. The bank tells you the total repayments will be £105 000. The relevant compound interest formula is

$$105\,000 = 75\,000\left(1 + \frac{R}{100}\right)^2$$

where R is the rate of interest you will be paying.
Calculate this value of R.

5 The voltage (V volts) and current (I amps) of a non-linear resistor are related according to $I = 0.025V + 0.005V^2$. If $I = 0.91$ amp calculate the voltage, V.

6 The acceleration, f, of a vehicle over a distance, s, changes its velocity from u to v. These are related by $v^2 = u^2 + 2fs$. You are given $f = 2\,\text{ms}^{-2}$ and $s = 150\,\text{m}$. Find the initial velocity, u, if the final velocity is 4 times larger than it.

7 The diagram shows a floor area of 120 m² to be tiled with new quarry tiles. Calculate the value of x.

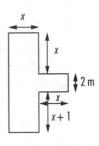

8   A rigid beam of length 10 m carries a uniformly distributed total load of 10 000 N. The bending moment, M, is related to its distance from one end of the beam, x, according to $M = -10\,000 + 5000x - 500x^2$

    i) Calculate the value of x when $M = 2000$ Nm.

    ii) Where along the beam is the bending moment 0?

9   The diagram shows a cone of height h, radius r and slant height l. They are related according to Pythagoras' Theorem. You are given that $l = 0.4$ m.

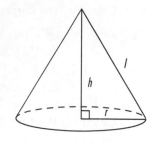

    i) If the height and radius are equal calculate their values.

    ii) The curved surface area is $\pi r l$ and the area of the circular base is πr^2. Their total area is 10 m^2, i.e. $\pi r l + \pi r^2 = 10$. Calculate the radius of the cone.

10   2 resistors, R_1 and R_2, are in parallel with a combined resistance of 4Ω according to $\dfrac{1}{4} = \dfrac{1}{R_1} + \dfrac{1}{R_2}$.

When they are in series we are given their combined resistance to be 21Ω according to $21 = R_1 + R_2$.

We may make R_2 the subject of the second formula so that
$$21 - R_1 = R_2.$$

Substituting into the first equation we get
$$\frac{1}{4} = \frac{1}{R_1} + \frac{1}{21 - R_1}.$$

Re-arrange this relation into a more usual quadratic equation. Solve your equation for R_1.

■■■■■ MULTI-CHOICE TEST 8 ■■■■■

1   An example of a quadratic equation is

    A) $x^2 + 2x + 3$

    B) $x^3 + x^2 + 3$

    C) $y = x^2 + 2x - 2$

    D) $5 - x^2 - x = 0$

2 $(x - 8)(x + 3) =$

    A) $x^2 - 5x - 11$

    B) $x^2 - 5x - 24$

    C) $x^2 - 11x - 5$

    D) $x^2 - 11x - 24$

3 $(x+3)^2 =$

    A) $x^2 + 6$

    B) $x^2 + 9$

    C) $x^2 + 6x + 9$

    D) $x^2 + 9x + 6$

Questions **4** and **5** share the accompanying diagram with the labelled graphs.

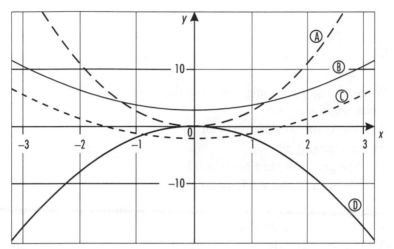

4   Which graph represents $y = 3x^2$?

5   Which graph represents $y = -2x^2$?

Questions **6** and **7** share the accompanying diagram with the labelled graphs.

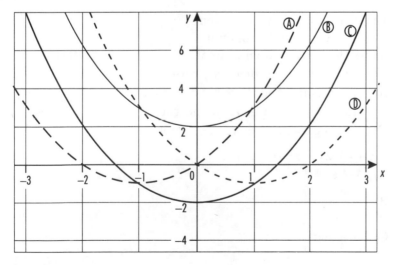

6   What is the label for the graph $y = 2x + x^2$?

7   What is the label for the graph $y = x^2 - 2$?

8   The general form of a quadratic expression may be written as $y = ax^2 + bx + c$ where a, b and c represent numbers. When a is negative the general shape is

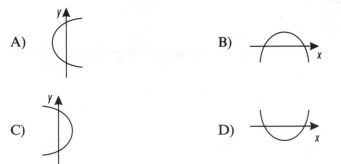

A)

B)

C)

D)

9   When plotting y against x to give the general graph $y = ax^2 + bx + c$ a change in a causes

A)   the graph to shift vertically

B)   nothing at all

C)   the graph to shift horizontally

D)   the sign of the gradient to change

10   The graph shows $y = x^2 - x - 2$. The solutions of the quadratic equation $x^2 - x - 2 = 0$ are at

A)   (i) and (ii) only

B)   (i) and (iii) only

C)   (ii) and (iii) only

D)   (i), (ii) and (iii) only

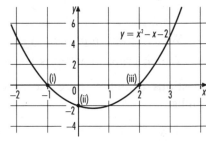

11   The general quadratic equation is $ax^2 + bx + c = 0$. The diagram shows the x-axis and an example of a parabola. In this case

A) a is positive and there is 1 root

B) a is negative and there are no roots

C) a is positive and there are no roots

D) a is negative and there are 2 roots

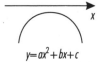

12 $(3x + 4)(2x - 7) = 0$ has roots $x =$

A) -4 and 7

B) $-\dfrac{4}{3}$ and $\dfrac{7}{2}$

C) $-\dfrac{3}{4}$ and $\dfrac{2}{7}$

D) $\dfrac{4}{3}$ and $\dfrac{2}{7}$

13 The simplest quadratic equation with roots of 1 and −3 is
A) $x^2 + 2x − 3 = 0$
B) $x^2 − 2x − 3 = 0$
C) $x^2 − 2x + 3 = 0$
D) $x^2 + 2x + 3 = 0$

14 The roots of the quadratic equation $x^2 − 3x = 0$ are
A) −3 and −3
B) −3 and 0
C) −3 and 3
D) 0 and 3

15 The graph represents $y =$
A) $x^2 − x − 6$
B) $x^2 + x − 6$
C) $x^2 − 6x + 1$
D) $x^2 + 6x − 1$

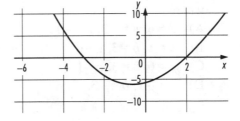

16 If $\dfrac{2}{x+1} = 3 − \dfrac{x}{2−x}$ then
A) $x^2 − x − 1 = 0$
B) $x^2 − 4x + 1 = 0$
C) $2x^2 − 2x − 1 = 0$
D) $5x − 4 = 0$

17 If $x^2 + 9x + 7 = 0$ then $x =$
A) $\dfrac{9 \pm \sqrt{109}}{2}$

B) $9 \pm \dfrac{\sqrt{53}}{2}$

C) $-9 \pm \sqrt{\dfrac{109}{2}}$

D) $\dfrac{-9 \pm \sqrt{53}}{2}$

18 The solutions to the equation $2x^2 + 5x + 3 = 0$ are $x =$
A) -3 and $-\dfrac{1}{2}$

B) $-\dfrac{3}{2}$ and -1

C) $\dfrac{1}{2}$ and 3

D) 1 and $\dfrac{3}{2}$

19 Completing the square upon $x^2 + x - 1 = 0$ gives

A) $(x + 2)^2 - 5 = 0$

B) $(x - 1)^2 - 1 = 0$

C) $\left(x + \dfrac{1}{2}\right)^2 - \dfrac{5}{4} = 0$

D) $\left(x - \dfrac{1}{2}\right)^2 - 1 = 0$

20 $6 - 3x - x^2 = 0$ may be solved using the formula $x = \dfrac{-b \pm \sqrt{b^2 - 4ac}}{2a}$
where

A) $a = 6,\ b = -3,\ c = -1$

B) $a = 1,\ b = 3,\ c = 6$

C) $a = 6,\ b = 3,\ c = 1$

D) $a = -1,\ b = -3,\ c = 6$

9 Solving Equations: II – Simultaneously

Element: Use algebra to solve engineering problems.

PERFORMANCE CRITERIA
- Formulae appropriate to the engineering problem are selected.
- The algebraic manipulation of formulae is carried out.
- Numerical values are substituted and correct solutions to engineering problems are obtained.

RANGE
Engineering problems: electrical, electronic; mechanical.
Formulae: pairs of simultaneous equations.
Algebraic manipulation: transposition, substitution.

Element: Use graphs to solve engineering problems.

PERFORMANCE CRITERIA
- Coordinates and scales appropriate to the engineering problem are selected.
- Data are accurately plotted.
- The relationship between variables is identified.
- Values are accurately determined from graphs, and engineering problems are solved.

RANGE
Engineering problems: electrical, electronic; mechanical.
Coordinates: Cartesian.
Scales: linear.
Relationship: linear, parabolic.
Values: two variables.

Introduction

We know equations are for solving in Mathematics. In fact we have already attempted to solve separately linear and quadratic equations.

First let us consider our chapter title. Things that occur **simultaneously** are things that occur at the **same time**. For example, if 2 people join the end of a queue simultaneously then they join it at the same place and time. When we solve equations simultaneously we solve them together: the solution for one of them must apply to them all. Most of our examples and exercises will consider a pair of simultaneous equations. The same principles in fact apply to more than just 2 equations.

ASSIGNMENT

This Assignment looks at a production company. It produces 3 types of interior car trim for the domestic automotive industry. These are Standard, Deluxe and Prestige styles. The company has a split site operation over two factories. Later in this chapter we will look at its attempts to meet a production target.

The company's sales force has been particularly active. Many of these new smaller orders have to be squeezed into the production schedules. However, there is a very important order for a longstanding customer. It is for 700 Standards, 900 Deluxes and 500 Prestiges.

The 2 factories are differently equipped. This means their daily production capacities are different. The following table shows the capacity figures for each type of trim.

Daily Production	Standard	Deluxe	Prestige
First Site	100	300	100
Second Site	300	100	100

We want to know how to schedule this order using both sites. Suppose the order will use x days of production at the first site and y days at the second site. Combining the daily production capacities with these days gives

Order Production	Standard	Deluxe	Prestige
First Site	$100x$	$300x$	$100x$
Second Site	$300y$	$100y$	$100y$
Total	700	900	500

Having set out the problem we will return to it once we have looked at some relevant Mathematics.

Coordinates and graphs

Before we attempt to solve any equations let us look at points lying on graphs. Whenever we plot a graph we plot only a **selection of points**, usually (x, y). Then we draw either a straight line or a smooth curve through them. The graph passes through many more points than our selected ones. All such points satisfy the particular relationship between x and y. Our first set of examples demonstrates whether points lie or do not lie on a graph.

■■■■ Examples 9.1 ■■■■

For the quadratic function $y = x^2 + 3x + 2$ decide if the following points lie on the graph

i) $(0, 2)$, ii) $(3, 20)$, iii) $(2, 0)$, iv) $(0.50, 3.75)$, v) $(-0.4, 0.9)$.

The right-hand side of $y = x^2 + 3x + 2$ is the more complicated side. Hence we will start with that side, substituting the values of x in each case.

i) Using $(0, 2)$ we have $x = 0$ to substitute in $x^2 + 3x + 2$ to give
$$0^2 + 3(0) + 2 = 0 + 0 + 2 = 2.$$
This answer agrees with the y value of 2 in $(0, 2)$. Hence $(0, 2)$ lies on the curve.

ii) Using $(3, 20)$ we have $x = 3$ to substitute in $x^2 + 3x + 2$ to give
$$3^2 + 3(3) + 2 = 9 + 9 + 2 = 20.$$
This answer agrees with the y value of 20 in $(3, 20)$. Hence $(3, 20)$ lies on the curve.

iii) Using $(2, 0)$ we have $x = 2$ to substitute in $x^2 + 3x + 2$ to give
$$2^2 + 3(2) + 2 = 4 + 6 + 2 = 12.$$
This answer *differs* from the y value in $(2, 0)$. Hence $(2, 0)$ does *not* lie on the curve.

iv)   Using $(0.50, 3.75)$ we have $x = 0.50$ to substitute in $x^2 + 3x + 2$ to give

$$(0.5)^2 + 3(0.5) + 2 = 0.25 + 1.50 + 2 = 3.75.$$

This answer agrees with the y value in $(0.50, 3.75)$. Hence $(0.50, 3.75)$ lies on the curve.

v)   Using $(-0.4, 0.9)$ we have $x = -0.4$ to substitute in $x^2 + 3x + 2$ to give

$$(-0.4)^2 + 3(-0.4) + 2 = 0.16 - 1.20 + 2 = 0.96.$$

This answer *differs* from the y value in $(-0.4, 0.9)$. Hence $(-0.4, 0.9)$ does *not* lie on the curve.

Fig. 9.1 shows the curve together with these 5 specimen pairs of coordinates. It confirms our "lies" and "not lie" decisions.

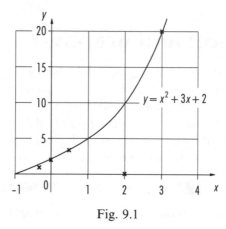

Fig. 9.1

Simultaneous linear equations – graphical solution

Let us start with 2 straight line (linear) graphs. We know that every point lying on a straight line satisfies the x and y relation. If possible, we need to find a point that satisfies **both relations**. This can only happen if the point lies on both lines, i.e. if those lines **cross (intersect)**.

Example 9.2

Graphically solve the pair of simultaneous equations

$$2x + 3y = 24$$

and $y - 4x = 1.$

We know the general equation of a straight line is $y = mx + c$. It might be easier if we re-arrange our equations into this form.

$$2x + 3y = 24$$

becomes $3y = 24 - 2x$

i.e. $y = 8 - \dfrac{2}{3}x.$

> Subtracting $2x$ from both sides.
> Dividing by 3.

We know this is a straight line of gradient $-\dfrac{2}{3}$ and vertical intercept 8.

Also $y - 4x = 1$

becomes $y = 4x + 1.$

We know this is a straight line of gradient 4 and vertical intercept 1.

Because these are known straight lines our tables need only 3 values for x: 2 and 1 as a check. We can choose those values of x at random.

$y = 8 - \dfrac{2x}{3}$

x	-6	0	3
8	8	8	8
$-\dfrac{2x}{3}$	4	0	-2
y	12	8	6

$y = 4x + 1$

x	-5	0	4
$4x$	-20	0	16
$+1$	1	1	1
y	-19	1	17

The graphs are plotted in Fig. 9.2. We see that the graphs do intersect and can read off their point of intersection as $(1.5, 7)$. This means the solution to the pair of simultaneous equations is $x = 1.5$ and $y = 7$.

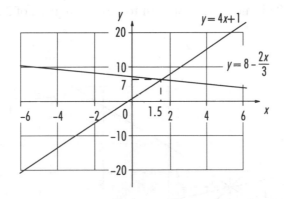

Fig. 9.2

The pair of equations $2x + 3y = 24$ and $y - 4x = 1$ of Example 9.2 have 2 **unknowns (variables)**. Whenever we attempt to solve any simultaneous equations we need at least as many equations as unknowns.

▆▆▆▆▆ **Example 9.3** ▆▆▆▆▆▆▆▆▆▆▆▆▆▆▆▆▆▆▆▆▆

Graphically solve the simultaneous equations

$$2x + y + 3 = 0$$
$$2y + x = 0$$
$$3y - 4x = 11.$$

Like our method in Example 9.2 we can re-arrange these 3 equations so y is the subject,

i.e. $\qquad y = -2x - 3,$

$$y = -\frac{x}{2}$$

$$y = \frac{4x}{3} + \frac{11}{3}.$$

As an exercise for yourself you might like to check these re-arrangements are correct.

We recognise these equations represent straight lines and so need to plot just 3 points. The method of table construction in the previous example will have refreshed your memory. So the ones for this example show just the first (x) and last (y) rows.

$y = -2x - 3$

x	-2	0	2
y	1	-3	-7

$y = -\dfrac{x}{2}$

x	-4	2	6
y	2	-1	-3

$y = \dfrac{4x}{3} + \dfrac{11}{3}$

x	-2	1	4
y	1	5	9

The graphs are plotted in Fig. 9.3. We see all 3 graphs do intersect at one point, $(-2, 1)$. This means the solution to our original set of 3 equations is $x = -2$ and $y = 1$.

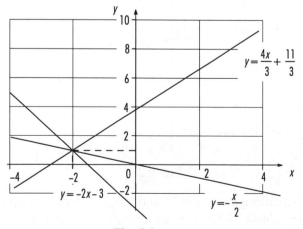

Fig. 9.3

Example 9.4

Graphically solve the simultaneous equations

$$2x + y + 3 = 0$$
$$2y + x = 0$$
$$3y - 2x = 12.$$

Like our method in Example 9.2 we can re-arrange these 3 equations so y is the subject,

i.e.
$$y = -2x - 3$$
$$y = -\frac{x}{2}$$
$$y = \frac{2x}{3} + 4.$$

As an exercise for yourself you might like to check these re-arrangements are correct.

We recognise these equations represent straight lines and so will plot just 3 points. Again the tables show just the first (x) and last (y) rows.

$y = -2x - 3$

x	-2	0	2
y	1	-3	-7

$y = -\dfrac{x}{2}$

x	-4	2	6
y	2	-1	-3

$y = \dfrac{2x}{3} + 4$

x	-3	3	6
y	2	6	8

The graphs are plotted in Fig. 9.4. We see all 3 graphs do *not* pass through one point simultaneously. This means there is no one solution. We do have solutions to pairs of equations.

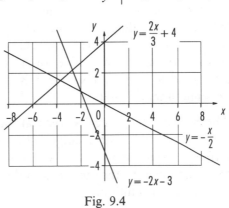

Fig. 9.4

$$2x + y + 3 = 0$$
and $\quad 2y + x = 0 \quad$ both pass through $(-2, 1)$

$$2x + y + 3 = 0$$
and $\quad 3y - 2x = 12 \quad$ both pass through $(-2.625, 2.25)$

$$2y + x = 0$$
and $\quad 3y - 2x = 12 \quad$ both pass through $(-3.43, 1.71)$.

Only as pairs, but not as all 3 together, can we solve them simultaneously.

████████ **Example 9.5** ████████

Graphically solve the simultaneous equations

$$y = 2x + 3$$

and $y = 2x - 4$.

We can plot these straight lines as usual (Fig. 9.5).

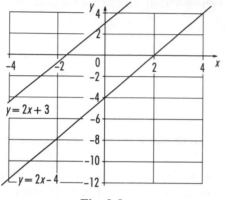

Fig. 9.5

Comparing the original equations with the general equation $y = mx + c$ we see they both have a gradient of 2. This means they are **parallel** and so **cannot intersect**, i.e. there is **no solution**.

There are 2 other types of parallel lines that you may meet. They may be horizontal lines like those in Fig. 9.6. The examples show totally different values of y as $\frac{1}{2}$ and 3. Whatever the values of x separately these values of y never change. Never can the same value of y satisfy both lines and so there is no simultaneous solution.

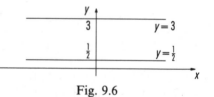

Fig. 9.6

Alternatively they may be vertical lines like those in Fig. 9.7. The examples show totally different values of x as -1.0 and 1.5 . Again these can never be the same and so there is no simultaneous solution.

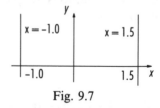

Fig. 9.7

Thus for any parallel lines there can be no simultaneous solution. Do not be disappointed. Mathematics is a tool to aid your decision making. We must move away from early ideas of absolute truth into the real world.

There is one more case where you will get no solution. Sometimes you *think* you have a pair of equations. Closer inspection reveals that one is just a multiple of the other.

Examples 9.6

We look at pairs of equations in this set of examples.

i) $\quad\quad y - 2x = 3$

and $2y - 4x = 6.$

The second equation is twice the first equation, i.e. in the second equation we can divide through by a common factor of 2 to create the first equation.

ii) $\quad\quad 2y - 4x \quad\quad = 6$

and $-3y + 6x + 9 = 0.$

In this case the second equation is -1.5 times the first equation. Do not be misled by the slightly different forms. In the second equation, if you prefer, move the 9 to the right-hand side. This gives $-3y + 6x = -9$. Now you will be able to compare them both more easily and spot the factor of -1.5.

ASSIGNMENT

Earlier in this chapter we set out the order production figures in total and for each site. We may now combine them to create 3 linear equations

$$100x + 300y = 700$$
$$300x + 100y = 900$$
$$100x + 100y = 500.$$

Each term in each equation is a multiple of 100. We can cancel through by this figure and still preserve the balances. Also we may re-arrange each one to make y the subject,

i.e. $\quad\quad\quad y = \dfrac{7}{3} - \dfrac{x}{3}$

$$y = 9 - 3x$$
$$y = 5 - x.$$

The next step is to construct tables of values. Each equation represents a straight line so we need just 3 specimen values. Because production has to be positive the values of x and y have to be positive too.

$y = \dfrac{7}{3} - \dfrac{x}{3}$ $\quad\quad\quad\quad y = 9 - 3x$ $\quad\quad\quad\quad y = 5 - x$

x	1	2.5	4
y	2	1.5	1

x	0	2	3
y	9	3	0

x	0	1	3
y	5	4	2

The graphs are plotted in Fig. 9.8. We see that all 3 straight lines do not pass through the same point. This means there is no one solution. There is no optimum solution. Instead we have 3 solutions because we have 3 intersections. These are $x=2$, $y=3$; $x=2.5$, $y=1.5$ and $x=4$, $y=1$. The points are marked as A, B and C on Fig. 9.8. We will look at the meaning of these results at the end of the next section.

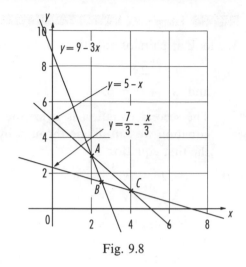

Fig. 9.8

■ EXERCISE 9.1 ■

Graphically solve the following simultaneous equations. In each case choose 3 specimen values of x and plot the straight lines.

1 $x+y=11$ and
   $x-y=1$

2 $x+3y=8$ and
   $x-2y=3$

3 $3y=x+1$ and
   $5y=20-2x$

4 $2x-y=0$ and
   $4x-5y=-3$

5 $3y-5x=4$ and
   $y=1$

6 $x+2y=5$ and
   $2x-y=7$

7 $2x+3y=3$ and
   $x=0$

8 $7x-6y=18$ and
   $6y=7x-12$

9 $3x+2y-4=0$ and
   $x+3y-11=0$

10 $2x-5y-23=0$ and
   $15y=69-6x$

Simultaneous linear equations – solution by elimination

For our second method we can look again at some of our earlier examples. We will have 2 equations with 2 unknowns, x and y. The aim is to make both x terms the same, or both y terms the same. Often we can achieve this by multiplying one or both equations by selected values. Remember that any multiplication must be **consistent throughout an equation**. Also, equations must balance themselves about the "$=$" sign. We will maintain this balance if we act consistently with the left-hand sides and right-hand sides.

Examples 9.7

Solve the pairs of simultaneous equations

i) $\qquad x + y = 9$

and $x - y = 3$;

ii) $\qquad 4x + y = 14$

and $\quad x + y = 8$.

We can choose to eliminate either x or y in each case. Firstly let us eliminate y.

i) In $\qquad x + y = 9$

and $\qquad x - y = 3$

we add these equations knowing that $(+y) + (-y) = y - y = 0$ so the y terms disappear,

i.e. $\qquad x + x = 9 + 3$

$$2x = 12$$

$$\frac{2x}{2} = \frac{12}{2}$$

$$x = 6.$$

To find y we substitute this value into one or other of our original equations. Let us substitute into the first equation so that

$$6 + y = 9$$

i.e. $\qquad y = 9 - 6$

$$y = 3.$$

Our complete solution is $x = 6$, $y = 3$.

We should check our solution with the other equation. Substitute both values into the left-hand side to get

$$x - y = 6 - 3 = 3$$

This is consistent with our second equation so confirming our solution.

ii) In $\qquad 4x + y = 14$

and $\qquad x + y = 8$

we subtract these equations knowing that $(+y) - (+y) = y - y = 0$ so the y terms disappear again,

i.e $\qquad 4x - x = 14 - 8$

$$3x = 6$$

$$\frac{3x}{3} = \frac{6}{3}$$

$$x = 2.$$

To find y we substitute this value into one or other of our original equations. Let us substitute into the first equation to get

$$4(2) + y = 14$$
$$8 + y = 14$$
$$y = 14 - 8$$
$$y = 6.$$

Our complete solution is $x = 2$, $y = 6$.

We should check our solution with the other equation. Substitute both values into the left-hand side to get

$$x + y = 2 + 6 = 8.$$

Again this is consistent with our second equation so confirming our solution.

ii) Again. Let us repeat the second example, this time eliminating x.

In $4x + y = 14$

and $x + y = 8$ we need both x terms the same.

It is easier numerically if they become $4x$. We achieve this by multiplying throughout the second equation by 4,

i.e. $4x + y = 14$

and $4x + 4y = 32.$

We subtract the equations since the signs for both $4x$ terms are the same (i.e. $4x - 4x = 0$).

Then $y - 4y = 14 - 32$

i.e. $-3y = -18$

$$\frac{-3y}{-3} = \frac{-18}{-3}$$
$$y = 6.$$

To find x we substitute this value into one or other of our original equations. Let us substitute into the first equation to get

$$4x + 6 = 14$$

i.e. $4x = 14 - 6$

$$4x = 8$$
$$\frac{4x}{4} = \frac{8}{4}$$
$$x = 2.$$

Just as before we can check our complete solution of $x = 2$, $y = 6$ by substitution.

■■■■■■■ **Examples 9.8** ■■■■■■■

Solve the pairs of simultaneous equations

i) $\quad\quad 5x - 3y = 21$

$\quad$ and $\quad 7x + 8y = 5$;

ii) $\quad\quad 2x + 3y = 7$

$\quad$ and $\quad x + 4y = 11.$

We can choose to eliminate either x or y in each case. Our choice is to eliminate y.

i) In $\quad 5x - 3y = 21$

$\quad$ and $\quad 7x + 8y = 5$

the y terms are not the same. To make them the same we multiply the first equation by 8 and the second equation by 3. Then they will be $24y$, 24 being the lowest common multiple (LCM) of 3 and 8.

$$5x - 3y = 21 \quad\quad \times 8$$
$$\text{and} \quad 7x + 8y = 5 \quad\quad \times 3$$
$$\text{become} \quad 40x - 24y = 168$$
$$\text{and} \quad 21x + 24y = 15$$
$$\text{Add} \quad 40x + 21x = 168 + 15$$
$$\text{i.e.} \quad\quad 61x = 183$$
$$\frac{61x}{61} = \frac{183}{61}$$
$$x = 3.$$

To find y we substitute this value into one or other of our original equations. Let us substitute into the second equation for a change to get

$$7(3) + 8y = 5$$
$$\text{i.e.} \quad\quad 21 + 8y = 5$$
$$8y = 5 - 21$$
$$8y = -16$$
$$\frac{8y}{8} = \frac{-16}{8}$$
$$y = -2.$$

Our complete solution is $x = 3$, $y = -2$.

We check our solution with the other equation. Substitute both values into the left-hand side so that

$$5x - 3y = 5(3) - 3(-2) = 15 + 6 = 21$$

This is consistent with our second equation so confirming our solution.

ii)   In $2x + 3y = 7$

and $x + 4y = 11$

the y terms are not the same. To make them the same we multiply the first equation by 4 and the second equation by 3. Then they will be $12y$, 12 being the lowest common multiple (LCM) of 4 and 3.

	$2x + 3y = 7$	$\times 4$
and	$x + 4y = 11$	$\times 3$
become	$8x + 12y = 28$	
and	$3x + 12y = 33$	
Subtract	$8x - 3x = 28 - 33$	
i.e.	$5x = -5$	
	$\dfrac{5x}{5} = \dfrac{-5}{5}$	
	$x = -1.$	

To find y we substitute this value into one or other of our original equations. Let us substitute into the first equation to get

$$2(-1) + 3y = 7$$

i.e. $-2 + 3y = 7$

$$3y = 7 + 2$$
$$3y = 9$$
$$y = 3.$$

We check our solution with the other equation. Substitute both values into the left-hand side so that

$$x + 4y = -1 + 4(3) = -1 + 12 = 11.$$

--- **Examples 9.9** ---

Solve the pairs of simultaneous equations

i) $4x + y = 14$

and $4x + y = 9$;

ii) $4x + y = 14$

and $8x + 2y = 28.$

i) This first pair of equations has the same terms on the left-hand sides yet different ones on the right. Remember that the solutions for x and y must apply to **both** equations simultaneously. How can $4x + y$ give different answers on the right? This cannot happen. These equations are inconsistent. they have no simultaneous solution. As we saw earlier in the chapter graphs show them to be a pair of parallel lines.

ii) Look closely at the second pair of equations. Really we have only
one equation because $8x + 2y = 28$ is just twice $4x + y = 14$. As we
have only one equation repeated and not a pair of them we are
unable to solve them simultaneously. Each equation is **dependent** on
the other. Remember because we have 2 unknowns, x and y, we need
at least 2 different equations.

■■■■■■ ASSIGNMENT ■■■■■■

Let us return to our Assignment. We will attempt to solve our 3 equations
simultaneously. Using the first pair

$$y = \frac{7}{3} - \frac{x}{3}$$

and $y = 9 - 3x$

we subtract them, so eliminating y,

i.e. $0 = \dfrac{7}{3} - 9 - \dfrac{x}{3} + 3x$ $\boxed{-(-3x) = +3x.}$

$$0 = -\frac{20}{3} + \frac{8x}{3}$$

$$\frac{20}{3} = \frac{8x}{3}$$ $\boxed{\text{Dividing by } \dfrac{8}{3}}$

$$2.5 = x.$$

To find y we can substitute for x in the second equation,

i.e. $y = 9 - 3\,(2.5)$

$y = 9 - 7.5$

$y = 1.5.$

As an exercise you should check this solution, $x = 2.5$ and $y = 1.5$, by
substituting into the first equation.

We can check to see whether it satisfies the third equation, $y = 5 - x$. In
fact $x = 2.5$ gives $y = 5 - 2.5 = 2.5$. We need $y = 1.5$, meaning $y = 5 - x$ does
not pass through our solution point of $(2.5, 1.5)$.

Now we know the 3 equations do not pass through the same point. Our
method continues by taking the equations in pairs.

Using the first and last equations

$$y = \frac{7}{3} - \frac{x}{3}$$

and $y = 5 - x$

we subtract them, so eliminating y,

i.e. $0 = \dfrac{7}{3} - 5 - \dfrac{x}{3} + x$

$$0 = -\frac{8}{3} + \frac{2x}{3}$$

$$\frac{8}{3} = \frac{2x}{3}$$

> Dividing by $\frac{2}{3}$

$$4 = x.$$

To find y we can substitute for x in the last equation,

i.e. $y = 5 - 4$

$y = 1.$

As an exercise you should check this solution, $x=4$ and $y=1$, by substitution in the usual way.

To find the final point of intersection we use the remaining equation pairing of

$$y = 9 - 3x$$

and $y = 5 - x.$

We subtract them to eliminate y,

i.e. $0 = 9 - 5 - 3x + x$

$0 = 4 - 2x$

$2x = 4$

$x = 2.$

To find y we can substitute for x in the last equation,

i.e. $y = 5 - 2$

$y = 3.$

Having solved our problem by 2 different methods let us interpret our answers. x is the number of days' production at the First Site and y is the number of days' production at the Second Site. We have the following table from earlier in the chapter

Order Production	Standard	Deluxe	Prestige
First Site	$100x$	$300x$	$100x$
Second Site	$300y$	$100y$	$100y$
Total	700	900	500

and can substitute for x and y in turn.

$x=2.5$ and $y=1.5$

Actual Production	Standard	Deluxe	Prestige
First Site	250	750	250
Second Site	450	150	150
Total	700	900	400

These production totals show that the 500 Prestige trim part of the order would not be met. Hence we must reject this solution. However, it is possible that, unknown to us, there may be stock to make up the shortfall.

$x = 4$ and $y = 1$

Actual Production	Standard	Deluxe	Prestige
First Site	400	1200	400
Second Site	300	100	100
Total	700	1300	500

These production totals show all the order being met. We would have a surplus of 400 Deluxe trims. If we used this solution we would need to be able to sell this surplus to another customer.

$x = 2$ and $y = 3$

Actual Production	Standard	Deluxe	Prestige
First Site	200	600	200
Second Site	900	300	300
Total	1100	900	500

All the order is met with a surplus of 400 Standard trims. Again we ask whether this surplus might be sold to another customer.

The Mathematics has helped us with some possible production plans. However there are many queries yet to be resolved. For example, here are a few of them.

i)   Do we hold stock to meet any production shortfalls?
ii) Can we easily sell production surpluses?
iii) With different numbers of production days can we balance up the schedule with other orders?
iv) What are the production costs between plants?

EXERCISE 9.2

Solve the following pairs of simultaneous equations.

1   $x - y = 19$ and
    $x + y = 1$

2   $3x + y = 7$ and
    $x + y = 11$

3   $2x + y = 9$ and
    $x - y = 0$

4   $x - 2y = -4$ and
    $x + 2y = 0$

5   $3x - y = 2$ and
    $5x + 2y = 40$

6   $-3x + 8y = 5$ and
    $x - 5y = 3$

7 $3x + 2y = 2$ and
    $x + 3y = 6.5$

8 $x + \dfrac{1}{2}y = 26$ and

    $\dfrac{1}{3}x - y = 4$

9 $2x - 5y = 9$ and
    $7x - y = 15$

10 $\dfrac{1}{2}x - 2y = 2.5$ and

    $\dfrac{1}{3}x + y = 0.5$

Linear and quadratic equations – graphical solution

In Chapter 8 we looked at the various shapes associated with quadratics. Now we aim to link together our knowledge and skills concerning straight lines and quadratic curves. The next comment applies to all curves: the accuracy of your plot will affect your answers. Do not expect accuracy beyond 1 or 2 decimal places.

■■■■■■■ **Example 9.10** ■■■■■■■

Solve the simultaneous equations $y = x^2 + 3x + 2$
                    and $y = x + 2$.

We construct 2 tables. Because the first equation represents a curve we need to use quite a few values of x. The specimen values of x are from $x = -3.0$ to $x = 1.0$ at intervals of 0.5. If we need more values we just extend our table.

$y = x^2 + 3x + 2$

x	-3.0	-2.5	-2.0	-1.5	-1.0	-0.5	0.0	0.5	1.0 ↔
x^2	9.00	6.25	4.00	2.25	1.00	0.25	0.00	0.25	1.00
$+3x$	-9.00	-7.50	-6.00	-4.50	-3.00	-1.50	0.00	1.50	3.00
$+2$	2.00	2.00	2.00	2.00	2.00	2.00	2.00	2.00	2.00
y	2.00	0.75	0.00	-0.25	0.00	0.75	2.00	3.75	6.00 ↕

The second equation can be compared with the general $y = mx + c$. It is a straight line of gradient 1 and intercept 2. We need to plot only 3 points for a straight line. For our table we have chosen these at random.

$y = x + 2$

x	-3	0	3
y	-1	2	5

The graphs are plotted in Fig. 9.9. We can read off the solutions where the graphs intersect. The pairs of solutions are $x = -2$, $y = 0$ and $x = 0$, $y = 2$. You will notice the solutions lie on the axes. This is neither significant nor usual.

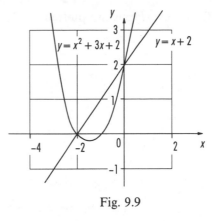

Fig. 9.9

Suppose we look again at this pair of equations

$$y = x^2 + 3x + 2$$
$$\text{and} \quad y = \quad\quad x + 2.$$

We can eliminate y by subtracting the second equation from the first one,

i.e. $\quad 0 = x^2 + 2x$.

This gives us a quadratic equation in x which we can solve by factorisation

i.e. $\quad 0 = x(x + 2)$

$\therefore \quad x = 0 \quad \text{or} \quad x + 2 = 0$

i.e. $\quad x = 0, -2$.

To find y now we may substitute these values in turn into one of our original equations. We choose the second equation, $y = x + 2$, because it is the simpler one.

$$x = 0 \quad \text{gives } y = 0 + 2 \quad = 2$$
$$\text{and} \quad x = -2 \quad \text{gives } y = -2 + 2 = 0.$$

This alternative method gives the same pair of solutions, $x = -2$, $y = 0$ and $x = 0$, $y = 2$, as the graphical method.

We can use the other equation, $y = x^2 + 3x + 2$, to check our results,

i.e. $\quad x = 0 \quad \text{gives } y = 0^2 + 3(0) + 2 \quad\quad = 2$

and $\quad x = -2 \quad \text{gives } y = (-2)^2 + 3(-2) + 2 = 0.$

This confirms our solutions to be correct.

Example 9.11

Solve the simultaneous equations $\quad y = x^2 + 3x + 2$
$$\text{and} \quad y + 3x + 7 = 0.$$

We have the table for $y = x^2 + 3x + 2$ in Example 9.10. $y + 3x + 7 = 0$ can be re-arranged to $y = -3x - 7$. This is known to be a straight line with a gradient of -3 and an intercept of -7. Choosing 3 specimen values for x we have the table below.

$y = -3x - 7$

x	-4	-2	-1
y	5	-1	-4

Both our graphs are plotted in Fig. 9.10.

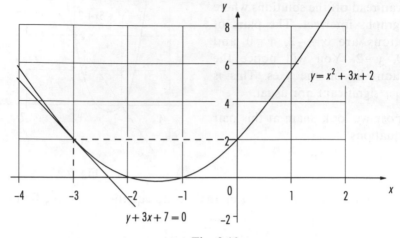

Fig. 9.10

We see the straight line is a tangent to the curve at the point where $x = -3$. The coordinates of this point are $(-3, 2)$ meaning the simultaneous solution is $x = -3$, $y = 2$. The **tangential** aspect means this is the **only solution**. you can think of it as a repeated solution, similar to the idea we saw in Chapter 8 on quadratic equations.

Suppose we look again at this pair of simultaneous equations

$$y = x^2 + 3x + 2$$

and $y = -3x - 7$.

We can eliminate y by subtracting the second equation from the first one,

i.e. $0 = x^2 + 6x + 9$.

This is a quadratic equation in x which we can solve by factorisation,

i.e. $0 = (x + 3)^2$

$\therefore$ $x + 3 = 0$

i.e. $x = -3$ repeated.

To find y we substitute this value into one of our original equations. We choose the second equation, $y + 3x + 7 = 0$, because it is the simpler one.

$x = -3$ gives $y + 3(-3) + 7 = 0$

i.e. $y - 9 + 7 = 0$

$y - 2 = 0$

$y = 2$.

The complete solution is $x = -3$, $y = 2$ repeated.

We can use the other equation, $y = x^2 + 3x + 2$, to check our result.

$x = -3$ gives $y = (-3)^2 + 3(-3) + 2 = 9 - 9 + 2 = 2$.

This confirms our solution to be correct.

Example 9.12

Solve the simultaneous equations $y = x^2 + 3x + 2$
and $y = x - 1$.

We have the table for $y = x^2 + 3x + 2$ in Example 9.10. Because $y = x - 1$ is a known straight line we need to plot only 3 points. Our table shows just 3 specimen values.

$y = x - 1$

x	-1	0	1.5
y	-2	-1	0.5

Both graphs are plotted in Fig. 9.11. We see there is **no intersection** of the curve and the straight line. This means there is **no simultaneous solution**. No pair of coordinates satisfies both equations.

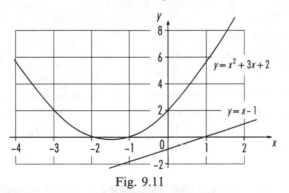

Fig. 9.11

As we have done previously we may look at another method. Again we eliminate y by subtraction,

i.e. $\quad y = x^2 + 3x + 2$

and $\quad y = \qquad x - 1$

give $\quad 0 = x^2 + 2x + 3$

This quadratic equation will not factorise. The quadratic equation formula will not work because it involves the **square root of a negative number**. This is consistent with the curve not crossing the horizontal axis. Fig. 9.12 provides us with that confirmation.

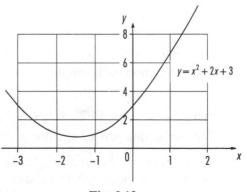

Fig. 9.12

▬▬▬ EXERCISE 9.3 ▬▬▬

Graphically solve the following pairs of simultaneous equations. In each case you will see they are a quadratic equation and a linear equation.

1 $y = x^2 + x - 2$ and
$y = 4x + 2$.
In your tables use values of
x from $x = -2$ to $x = 5$ at
intervals to 0.5.

2 $y = 3x^2 + x + 1$ and
$y = -1.5x + 6$.
In your tables use values of
x from $x = -2$ to $x = 2$ at
intervals of 0.5.

3 $y = 2x^2 + 11x + 5$ and
$y = 3x - 3$.
In your tables use values of
x from $x = -6$ to $x = 1$ at
intervals of 0.5.

4 $y = 4x^2 - 4x + 1$ and
$y + x + 2 = 0$.
In your tables use values of
x from $x = -1$ to $x = 3$ at
intervals of 0.5.

5 $y = 2x^2 - 7x + 6$ and
$2y + x = 6$.
In your tables use values of
x from $x = 0$ to $x = 4$ at
intervals of 0.5.

The theory is complete. Now you are in a position to try a mixed set of questions in this last exercise. In each question the equations have been formed for you.

▬▬▬ EXERCISE 9.4 ▬▬▬

1 The electrical circuit shows 2 emfs together with various resistors. The sums of the relevant voltages produce the simultaneous equations. They are in terms of the currents, I_1 and I_2 amps.

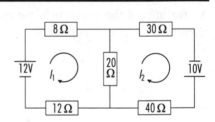

$$8I_1 + 20(I_1 - I_2) + 12I_1 = 12$$
and $40I_2 + 20(I_2 - I_1) + 30I_2 = 10$.

These equations will simplify. Show that they may be reduced to

$$10I_1 - 5I_2 = 3$$
and $$-2I_1 + 9I_2 = 1.$$

Now solve this pair of equations for I_1 and I_2.

2 A uniform heavy beam of length 20 m has a bending moment, M. x is the distance measured from one end and M is related to x by $M = 2x(20 - x)$. Construct a table of values of x and M using intervals of 2 m all along the beam. Plot a graph of M against x. From your graph, what is the maximum bending moment and where does it occur?

Another beam is shorter at 15 m. Its bending moment is given by $M = 4x(15 - x)$. Again, along the length of the beam construct a table of values for M and x. On the same set of axes hence plot a graph of M against x. What is the maximum bending moment for this second beam and where does it occur?

Where are the 2 bending moments equal? What is that common value of M?

3   The resistance, $R\,\text{k}\Omega$, in an electrical circuit is related to its temperature, $T°$ Celsius.

At first this is thought to be $R = 200T - 450$. Using 3 specimen values of T from 20 to 150 construct a simple table. Hence plot a graph of R against T.

An improved relation is found to be $R = 2T^2 + 3T - 500$. Using values of T from 20 to 150 at intervals of 10 construct a table for this new relationship. On the same pair of axes plot the new graph of R against T. Use your graphs to find at what temperature(s) the predicted resistances might be the same. Also find the values of
   i) T when $R = 24\,\text{k}\Omega$,
   ii) R when $T = 135°$ Celsius.

4   There are two vehicles, a van and a car. The van is travelling at a constant speed of $18\,\text{ms}^{-1}$. The distance it travels, s metres, is related to time, t seconds, by $s = 18t$. Choose 3 specimen values of t up to 20 seconds. Construct a table and then plot a graph of s against t.

The car, from a standing start, attempts to catch the van. The distance it travels, s, is related to time, t, by $s = \dfrac{4}{3}t^2$. Using values of t from $t = 0$ to $t = 20$ seconds at intervals of 2 seconds construct a table of values. On the same set of axes plot a graph of s against t. When does the car catch the van? By what distance is it ahead 20 seconds after the start?

Suppose the car had been travelling at $5\,\text{ms}^{-1}$ initially so that $s = 5t + \dfrac{4}{3}t^2$.

Construct another table and plot a third graph on the axes. From your graphs find the new time when the overtaking now occurs.

5   An engineering company's position in the market is oriented to the price of its product, £x. It would like to supply more at a higher price, but as the price rises so demand falls away. The market is in equilibrium when supply equals demand. For the following straight line laws one represents supply and one represents demand. They are $y = 100 + 10x$ and $y = 350 - 8x$.

Choose 3 specimen values of x between 5 and 40. Construct tables for these relations and plot the graphs on one set of axes.

Decide which graph represents demand and which represents supply. Decide on the equilibrium price.

Improved research shows that $y = 400 - 1.5x^2$ and $y = 50 + 0.5x^2$ are more likely supply and demand curves. Use values of x from 5 to 25 at intervals of 2.5 to construct the necessary tables. Hence plot these graphs on a fresh pair of axes. Read off from your graph the improved estimate of the equilibrium price.

■ MULTI-CHOICE TEST 9 ■

The diagram for Questions **1** and **2** shows the graphs of $y = 3x + 2$ and $y = 4x - 1$.

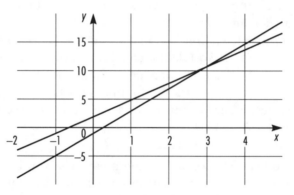

On the diagram we could plot the pairs of coordinates labelled

A) $(-1, 0)$

B) $(2, 0)$

C) $(3, 11)$

D) $(11, 3)$

1   Which pair of coordinates lies on the line $y = 3x + 2$?

2   Which pair of coordinates lies on the line $y = 4x - 1$?

3   The diagram shows the graphs of $y = 1 - 2x$, $y = 2x + 9$ and $7y = -2x - 17$. Also there are the points labelled A, B, and C and D. Which point is the simultaneous solution for $y = 2x + 9$ and $7y = -2x - 17$?

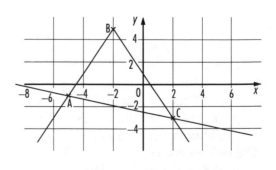

4 The diagram shows 2 graphs with no simultaneous solution. This is because they have

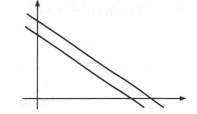

A) the same intercept
B) different intercepts
C) the same gradient
D) different gradients

5 Into the equation $2x - y = 24$ substitute $y = 6 - x$ to get $x =$

A) 6
B) 10
C) 18
D) 30

6 In the pair of simultaneous equations $2x + 3y = 17$ and $3x + 4y = 24$ you are given $x = 4$. Hence the value of y is

A) 3
B) $\dfrac{17}{3}$
C) 6
D) $\dfrac{25}{3}$

7 The solution to the simultaneous equations $2x - 5y = 3$ and $x - 3y = 1$ is

A) $x = 1$ and $y = 4$
B) $x = 4$ and $y = 1$
C) $x = 4$ and $y = 3$
D) $x = 13$ and $y = 4$

8 Two electrical currents, I_1 and I_2, are connected by the equations $4I_1 - 5I_2 = 7$ and $12I_2 - 3I_1 = 3$. In order the values of I_1 and I_2 are

A) -1 and 4
B) $\dfrac{1}{2}$ and -2
C) 1 and 3
D) 3 and 1

9 A motorist travels p km at $50\,\text{kmh}^{-1}$ and q km at $60\,\text{kmh}^{-1}$. The total time taken is 5 hours and the average speed is $56\,\text{kmh}^{-1}$. This scenario is modelled by

A) $50p + 60q = 5$ and $\dfrac{p}{50} + \dfrac{q}{60} = \dfrac{56}{5}$

B) $\dfrac{p}{50} + \dfrac{q}{60} = 5$ and $\dfrac{p+q}{5} = 56$

C) $6p + 5q = 1500$ and $p + q = 280$

D) $50p + 60q = 5$ and $p + q = 280$

10 Given the pair of equations $3x - 2y = 4.5$ and $-4.5x + 3y = -6.25$ the solution is

A) $x = 1.5$ and $y = -2.25$

B)   no solution because they are inconsistent

C) $x = -1.25$ and $y = 2.5$

D)   no solution because they are dependent

11 You are given two numbers P and Q. Their sum is 18 and their difference is 12. The equations modelling these statements are

A) $P + Q = 18$ and $P/Q = 12$

B) $P + Q = 18$ and $P - Q = 12$

C) $P/Q = 18$ and $P + Q = 12$

D) $P - Q = 18$ and $P + Q = 12$

12 A set of three simultaneous linear equations has a solution. This means the graphs must intersect

A)   once

B)   twice

C)   three times

D)   at least four times

Questions **13, 14, 15** and **16** use the following information. An engineering company produces goods, G, each at a price, P. In an unrestricted market the laws of supply and demand apply.

For the law of supply: the company wishes to supply more as the price rises.

For the law of demand: the customer wishes to buy less as the price rises.

Where the graphs of these laws intersect is the market equilibrium price.

The equations for these graphs are $G = 200 + 7P$ ...... ①

and $G = 640 - 4P$ ...... ②

13 Decide whether each of these statements is True (T) or False (F).

i)   The law of supply is $G = 200 + 7P$.

ii)   The law of demand is $G = 640 - 4P$.

Which option best describes the two statements?

A)   i) T   ii)   T

B)   i) T   ii)   F

C)   i) F   ii)   T

D)   i) F   ii)   F

14 Multiplying equation ① by 4 and equation ② by 7 gives

A) $4G = 800 + 11P$ and $7G = 4480 - 11P$

B) $G = 200 + 28P$ and $G = 640 - 28P$

C) $4G = 800 + 28P$ and $G = 4480 - 28P$

D) $4G = 800 + 28P$ and $7G = 4480 - 28P$

15 Subtracting equation ② from equation ① gives
  A) $0 = -440 + 3P$
  B) $0 = 440 + 11P$
  C) $0 = -440 + 11P$
  D) $0 = -840 + 11P$

16 The complete solution for P and G in order is
  A) -40 and -80
  B) -40 and 800
  C) 40 and 280
  D) 40 and 480

Questions **17**, **18**, **19** and **20** use the following diagram and information. The diagram shows the graphs of $y = x^2 - 3x + 2$ and $y = 3x + 6$. Also labelled are the points (i), (ii), (iii), (iv) and (v).

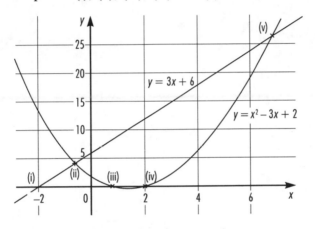

17 The simultaneous solutions for these graphs are at
  A) (i) and (ii)
  B) (ii) and (iii)
  C) (ii) and (v)
  D) (iii) and (iv)

18 The solutions of the quadratic equation $x^2 - 3x + 2 = 0$ occur at
  A) (i) and (ii)
  B) (ii) and (iii)
  C) (ii) and (v)
  D) (iii) and (iv)

19 The solutions of the quadratic equation $x^2 - 6x - 4 = 0$
  A) occur at (i), (ii) and (iii)
  B) cannot be found from this diagram
  C) (ii) and (v)
  D) (iii) and (iv)

20   The solutions of the quadratic equation $x^2 - 3x + 8 = 0$
  A)   occur at (i), (ii) and (iii)
  B)   cannot be found from this diagram
  C)   (ii) and (v)
  D)   (iii) and (iv)

10 Introducing Indices and Logarithms

Element: Use algebra to solve engineering problems.

PERFORMANCE CRITERIA
- Formulae appropriate to the engineering problem are selected.
- The algebraic manipulation of formulae is carried out.
- Numerical values are substituted and correct solutions to engineering problems are obtained.

RANGE
Engineering problems: electrical, electronic; mechanical.
Formulae: logarithmic expressions, exponential expressions.
Algebraic manipulation: substitution, transformation.

Introduction

This chapter looks at 3 ideas. They are indices (which you will have seen before), common logarithms and natural logarithms. They are all related and you will see the relationships throughout the chapter.

ASSIGNMENT

This Assignment looks at the power gain of electrical components and systems. There is a formula for power gain. It is a logarithmic ratio of output power, P_0, and input power, P_i.

Power gain $= 10 \log_{10} \left(\dfrac{P_0}{P_i} \right)$.

The units for power gain are decibels, dB.

The same formula works for a power loss. We know that $10 - 35$ is the same as $10 + (-35)$, i.e. subtracting 35 is the same as adding -35. Similarly power loss $(-)$ is a negative power gain $(+(-))$.

The Assignment is in 2 parts. In the first part there is a calculation based on an amplifier. The second part looks at a simple system involving a series of transmission lines and amplifiers.

The laws of indices

We can write numbers in many forms. One of these is

$$\textbf{NUMBER} = \textbf{BASE}^{\textbf{index}}$$

e.g. $2 = 10^{0.3010}$.

The plural of **index** is **indices**. An alternative word for **index** is **power**. It is different from electrical power. We raise the base to a power (or index). When we look at this example the index is rather awkward as a decimal. We are more used to whole numbers or simple fractions like $\frac{1}{2}$ or $\frac{3}{4}$.

As a first step let us recall the basic laws of indices. Whenever we simplify indices they must relate to the **same base**. If the bases are different there is no easy simplification.

We have a look at 6 laws of indices to jog your memory. Each law is followed by a simple example.

LAW 1 $b^m \times b^n = b^{m+n}$

multiplying: add the indices

e.g. $2^4 \times 2^5 = 2^{4+5} = 2^9$ or 512.

LAW 2 $\qquad b^m \div b^n = b^{m-n}$

dividing: subtract the indices

e.g. $\qquad\qquad \dfrac{2^8}{2^5} = 2^{8-5} = 2^3 \qquad$ or 8.

LAW 3 $\qquad (b^m)^n = b^{mn}$

raise to a power: multiply the indices

e.g. $\qquad\qquad (2^3)^2 = 2^{3 \times 2} = 2^6 \qquad$ or 64.

LAW 4 $\qquad b^0 = 1$

raise to a power 0:   answer is always 1.

This may seem strange at first glance, but the demonstration is easy.

Suppose we have $\dfrac{2^6}{2^6}$. We can look at this in two different ways.

$$\frac{2^6}{2^6} = 2^{6-6} = 2^0.$$

Also $\qquad \dfrac{2^6}{2^6} = \dfrac{2 \times 2 \times 2 \times 2 \times 2 \times 2}{2 \times 2 \times 2 \times 2 \times 2 \times 2}$ | Cancelling 2 six times. |

$$= \frac{1}{1}$$

$$= 1.$$

In each case we started with $\dfrac{2^6}{2^6}$ yet reached answers in different forms.

Equating our answers shows that $2^0 = 1$.

LAW 5 $\qquad b^{-p} = \dfrac{1}{b^p}$

e.g. $2^{-4} = 2^{0-4}$ $2^7 = 2^{--7} = 2^{0--7}$

$\qquad\qquad\quad = \dfrac{2^0}{2^4} \qquad\qquad\qquad = \dfrac{2^0}{2^{-7}}$ | Using the second law. |

$\qquad\qquad\quad = \dfrac{1}{2^4}. \qquad\qquad\qquad = \dfrac{1}{2^{-7}}.$ | Using the fourth law. |

LAW 6 $\qquad\qquad b^{m/n} = \sqrt[n]{b^m} \text{ or } (\sqrt[n]{b})^m$

e.g. $2^{5/3}$ may be written as $\sqrt[3]{2^5}$ or $(\sqrt[3]{2})^5$.

We can use the calculator and sequence of buttons learned in Chapter 1.

$\sqrt[3]{2^5}$ has a cube root (i.e. the power $\frac{1}{3}$) and uses

$\boxed{2} \quad \boxed{x^y} \quad \boxed{5} \quad \boxed{=}$

to display 32 and then

$\boxed{x^{1/y}} \quad \boxed{3} \quad \boxed{=}$

to give an answer of 3.1748 . . .

Alternatively $(\sqrt[3]{2})^5$ uses

$$\underline{2|} \quad \underline{x^{1/y}|} \quad \underline{3|} \quad \underline{=|}$$

to display 1.2599... and then

$$\underline{x^y|} \quad \underline{5|} \quad \underline{=|}$$

to give the same answer of 3.1748...

You will see some similarities between these laws of indices and the laws of logarithms.

Logarithms

We know NUMBER = BASE^{index}. If a number, N, is written as b^x then the index, x, is called the **logarithm**. x is the logarithm of N to the base b.
   This is written as $N = b^x$ or $\log_b N = x$.
   log is the shortened form of logarithm.
   It is important to know that **logarithms are defined only for positive numbers**.
   In $\log_b N = x$ the base, b, can take many values. You will find the two important and usual ones on your calculator. These are the bases 10 and e. All the laws of logarithms apply to any base, but we will concentrate on these particular bases.

Common logarithms

Logarithms to the base 10 are called **common logarithms**. Before cheap electronic calculators people used tables of values of common logarithms. Now we can find values by simply pressing buttons on a calculator. We replace b with 10 in our basic definitions so that $N = 10^x$ and $\log_{10} N = x$. The notation can be shortened by omitting 10 from $\log_{10} N$. It is understood that $\log N$ means the common logarithm of N.

■■■■ Examples 10.1 ■■■■

Find the values of

i) $\log 2$,   ii) $\log 4$,   iii) $\log 10$,   iv) $\log 100$,   v) $\log 0.1$,   vi) $\log 1$.

The calculator button we need is $\underline{\log|}$. Just input the number and press $\underline{\log|}$ to display each answer.

   i) $\log 2$ uses the buttons $\underline{2|} \underline{\log|}$ to give 0.3010 (4 dp).
   ii) $\log 4 = 0.6021$ (4 dp).
   iii) $\log 10 = 1$.
   iv) $\log 100 = 2$.
   v) $\log 0.1 = -1$.
   vi) $\log 1 = 0$.

In place of log $N = x$ we can think of the alternative version, $N = 10^x$. Then we can re-write each question part to show the logarithm in the index position.

i) $2 = 10^{0.3010}$.

ii) $4 = 10^{0.6021}$.

iii) $10 = 10^1$.

iv) $100 = 10^2$.

v) $0.1 = 10^{-1}$.

vi) $1 = 10^0$.

All these examples started with a number, N, and gave a logarithm, x. Suppose we start with x and attempt to find the number, N, according to $N = 10^x$. Above the log| button on your calculator you should find the inverse function, 10^x. This used to be called the **antilog**. The next set of examples uses this inverse function.

■■■■■■■■ **Examples 10.2** ■■■■■■■■■■■■■■■■■

Find the numbers whose common logarithms are

i) 0.4771, ii) −0.3010, iii) 3.7000.

The order of calculator buttons involves inputting the value and then pressing inv| log|. The effect of inv| log| is to apply 10^x.

i) log $N = 0.4771$ is the same as $N = 10^{0.4771}$.
   The order of calculator buttons is

   0.4771| inv| log|

   to display 2.99985...
   Correct to 3 decimal places this is $N = 3$.

ii) log $N = -0.3010$ is the same as $N = 10^{-0.3010}$ or $\dfrac{1}{10^{0.3010}}$.

   Using the same calculator order we get $N = 0.500$ (3 dp).

iii) log $N = 3.700$ is the same as $N = 10^{3.7000}$ so $N = 5012$ (4sf).

■■■■■■■ **EXERCISE 10.1** ■■■■■■■■■■■■■■■■■■■■■

Find the values of

1	log 5.75	6	log 0.50
2	log 13.40	7	log 10^6
3	log 1000	8	log 0.25
4	log 26.80	9	log −0.25
5	log 200	10	log 1.76

Find the numbers with the following common logarithms. Give your answers to 3 significant figures where appropriate.

11	1.23		**16**	0.699
12	−0.55		**17**	−1.25
13	0.6021		**18**	1.658
14	0.1761		**19**	3.69
15	2.46		**20**	1.949

Common logarithmic graph

It is always interesting to see a diagram of a function. We can use the calculator's log| button to create a table.

N	0.05	0.1	0.5	1.0	10.0	50.0	100.0
$\log N$	−1.301	−1.000	−0.301	0.000	1.000	1.699	2.000

The specimen values of N have not been chosen for regularity. Instead they have been chosen so you can see the tremendous change in N and how this affects $\log N$. The graph of $y = \log N$ is shown in Fig. 10.1.

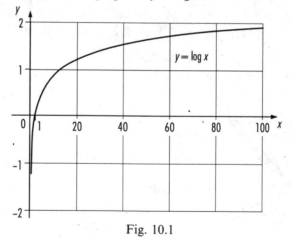

Fig. 10.1

Notice the graph crosses the horizontal axis at 1, i.e. $\log 1 = 0$. Another important feature is that $\log 10 = 1$. Generally we find when the number, N, and the base, b, are the same then the logarithm is 1, i.e. $\log_b b = 1$.

As N increases the gradient becomes less steep, tending to flatten a little. As N decreases, approaching 0, the graph tends to $-\infty$. We write this mathematically; as $x \to 0$ then $\log_b x \to -\infty$. The vertical axis acts as an asymptote. An asympote is a line a graph approaches but never quite touches. In this case the log graph approaches the vertical axis but never quite touches it.

■ ASSIGNMENT ■

1.   We are going to look at an amplifier's power gain. Suppose its input power is 175 mW and its output power is 10 W. When we introduced our Assignment we had an output power, P_0, and input power, P_i.

Using our values we have $P_0 = 10$ W and $P_i = 175$ mW $= 0.175$ W. Notice how we amend the input power so that the units are consistently watts, W. We can re-write our formula too. We have learned that for common logarithms we do not need to include the base, 10.

Now   power gain $= 10\log\left(\dfrac{P_0}{P_i}\right)$

becomes   power gain $= 10\log\left(\dfrac{10}{0.175}\right)$

$$= 10\log(57.143)$$
$$= 10 \times 1.757$$
$$= 17.6 \text{ dB}.$$

2.   In this second part we are going to combine 4 different sections, 2 transmission lines and 2 amplifiers. We will input a signal of 600 mW into the first section of transmission line. The question is what happens to that power input by the end of the system? The output from one section of the system is the input into the next section.

Firstly let us be consistent with the units, 600 mW $= 0.6$ W. Suppose we have the following power gains/losses:

Transmission line 1,   power loss $= 35$ dB
                              (i.e. power gain $= -35$ dB)
Amplifier 1,   power gain $= 25$ dB
Transmission line 2,   power loss $= 45$ dB
                              (i.e. power gain $= -45$ dB)
Amplifier 2,   power gain $= 50$ dB

We can look at each section applying our formula in turn.

$$\text{power gain} = 10\log\left(\frac{P_2}{P_1}\right)$$

$$-35 = 10\log\left(\frac{P_2}{0.6}\right)$$

i.e. $$-3.5 = \log\left(\frac{P_2}{0.6}\right)$$

i.e. $$\frac{P_2}{0.6} = 10^{-3.5}$$

$$P_2 = 0.6 \times 10^{-3.5} \text{ W}.$$

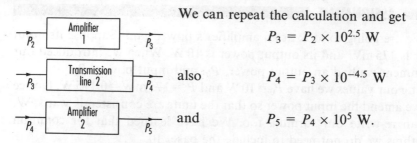

We can repeat the calculation and get

$$P_3 = P_2 \times 10^{2.5} \text{ W}$$

also

$$P_4 = P_3 \times 10^{-4.5} \text{ W}$$

and

$$P_5 = P_4 \times 10^5 \text{ W}.$$

As an exercise you should check these calculations for yourself. We can link together all the calculations:

$$
\begin{aligned}
P_5 &= P_4 \times 10^5 \\
&= (P_3 \times 10^{-4.5}) \times 10^5 \qquad \text{Substituting for } P_4. \\
&= (P_2 \times 10^{2.5}) \times 10^{-4.5} \times 10^5 \qquad \text{Substituting for } P_3. \\
&= (0.6 \times 10^{-3.5}) \times 10^{2.5} \times 10^{-4.5} \times 10^5 \\
&= 0.6 \times 10^{-3.5+2.5-4.5+5} \\
&= 0.6 \times 10^{-0.5} \\
&= 0.6 \times 0.316 \\
&= 0.190 \text{ W} \qquad \text{or } 190 \text{ mW.}
\end{aligned}
$$

You can see that the initial input of 600 mW has been reduced to a final output of 190 mW. In the combination of all the calculations you can spot the power gains/losses in each index.

For an extension to this problem you might like to replace an amplifier so you achieve neither a power gain nor loss.

Natural logarithms

Logarithms to the base e are called **natural** or **Naperian** logarithms. Again before cheap electronic calculators people used tables of values of natural logarithms. Now we can find values by simply pressing buttons on a calculator. We replace b with e in our basic definitions so that

$$N = e^x \text{ and } \log_e N = x.$$

The notation can be shortened from $\log_e N$. It is understood that $\ln N$ means the natural logarithm of N.

e^x is the **exponential** function. We will look at this in detail in Chapter 11.

Numerical calculations can be performed using any type of logarithms. However, natural logarithms are the more widely used type in mathematics, science and technology. We can check out the value of e (i.e. e^1) by using a calculator.

Input 1| inv| ln| to display 2.71828 ... Often this is quoted correct to 3 decimal places as 2.718.

We look at some values in the next set of examples.

Examples 10.3

Find the values of

i) ln 2,   ii) ln 4,   iii) ln 2.718,   iv) ln 7.389,   v) ln 0.35,   vi) ln 1.

The calculator button we need is ln|. Just input the number and press ln| to display each answer.

i)   ln 2 uses the buttons 2| ln| to give 0.693 (3 dp).

ii)   ln 4 = 1.386 (4 sf).

iii)   ln 2.718 = 1 (3 sf).

iv)   ln 7.389 = 2 (3 sf).

v)   ln 0.35 = −1.0498 (4 dp).

vi)   ln 1 = 0.

In place of $\ln N = x$ we can think of the alternative version, $N = e^x$. Now for each example part we give the alternative version, showing the logarithm in the index position.

i) $2 = e^{0.693}$.

ii) $4 = e^{1.386}$.

iii) $2.718 = e^1$.

iv) $7.389 = e^2$.

v) $0.35 = e^{-1.0498}$.

vi) $1 = e^0$.

All these examples started with a number, N, and gave a logarithm, x. Suppose we start with x and attempt to find the number, N, according to $N = e^x$. Above the ln| button on your calculator you should find the inverse function, e^x. This used to be called the **antilog**. The next set of examples uses this inverse function.

Examples 10.4

Find the numbers whose natural logarithms are

i) 0.4771,   ii) −0.3010,   iii) 3.7000.

The order of calculator buttons involves inputting the value and then pressing inv| ln|. The effect of inv| ln| is to apply e^x.

i)   ln $N = 0.4771$ is the same as $N = e^{0.4771}$.

The order of calculator buttons is

0.4771| inv| ln|

to display 1.61139 ...

Correct to 3 decimal places this is $N = 1.611$.

ii) $\ln N = -0.3010$ is the same as $N = e^{-0.3010}$ or $\dfrac{1}{e^{0.3010}}$.

Using the same calculator order we get $N = 0.740$ (3 dp).

iii) $\ln N = 3.700$ is the same as $N = e^{3.7000}$ so $N = 40.45$ (4 sf).

■■■■■ EXERCISE 10.2 ■■■■■■

Find the values of

1	$\ln 15.75$		**6**	$\ln 0.15$
2	$\ln 3.40$		**7**	$\ln e^6$
3	$\ln 1000$		**8**	$\ln 0.3679$
4	$\ln 6.80$		**9**	$\ln -0.75$
5	$\ln 250$		**10**	$\ln 175$

Find the numbers with the following natural logarithms. Give your answers to 3 significant figures where appropriate.

11	1.23		**16**	0.699
12	−0.55		**17**	−1.25
13	0.6021		**18**	1.658
14	0.1761		**19**	3.69
15	2.46		**20**	1.949

Natural logarithmic graph

Again, a diagram of this function should be interesting. We can use the calculator's ln ⌋ button to create a table.

N	0.05	0.1	0.5	1.0	10.0	50.0	100.0
$\ln N$	−2.996	−2.303	−0.693	0.000	2.303	3.912	4.605

From the above table of values you can see that for very large changes in N there are relatively small changes in $\ln N$. The graph of $y = \ln N$ is shown in Fig. 10.2.

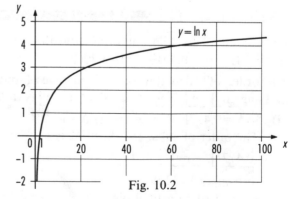

Fig. 10.2

Notice the graph crosses the horizontal axis at 1, i.e. $\ln 1 = 0$. Also $\ln e = 1$. The characteristics are much the same as for the common logarithmic graph. As N increases the gradient becomes less steep, tending to flatten a little. As N decreases, approaching 0, the graph tends to $-\infty$. The way we write this mathematically is $N \to 0$ then $\ln N \to -\infty$. The vertical axis acts as an asymptote.

For comparison the graphs of these different logarithmic functions are plotted together in Fig. 10.3. They do change at slightly different rates, but have similar tendencies.

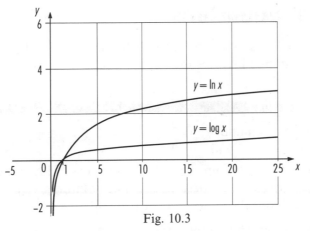

Fig. 10.3

Laws of logarithms

These laws apply to any logarithmic base. We can quote them generally.

1      $\log_b(MN) = \log_b M + \log_b N$

    i.e. multiplying the numbers, M and N, is related to adding their logs. This is similar to the law of indices

         $b^m \times b^n = b^{m+n}$

2      $\log_b\left(\dfrac{M}{N}\right) = \log_b M - \log_b N$

    i.e. dividing the numbers, M and N, is related to subtracting their logs.
    This is similar to the law of indices

         $b^m \div b^n = b^{m-n}$

3      $\log_b(N^a) = a \log_b N$

    i.e. raising a number, N, to a power is related to multiplying its log, $\log_b N$, by the power.
    This law looks at repeated multiplication and how it affects the logs. It is similar to our third law of indices. That law is not quoted again to avoid confusion with the letters we are using here.

The laws may be re-written simply in both our bases.

Common logs, i.e. base 10	*Natural logs, i.e. base e*
$\log MN = \log M + \log N$	$\ln MN = \ln M + \ln N$
$\log \dfrac{M}{N} = \log M - \log N$	$\ln \dfrac{M}{N} = \ln M - \ln N$
$\log N^p = p \log N$	$\ln N^p = p \ln N$

Indicial equations

We can use the laws of logarithms to help solve some indicial equations. An indicial equation involves the unknown (e.g. x) in the index position.

■■■■■ **Examples 10.5** ■■■■■

Solve the indicial equations   i) $3^x = 7$
and   ii) $3^x = 6$.

The hint before these examples was to use the laws of logarithms. We can choose logarithms to either base 10 or e. For both these examples it is the third law that looks the closest one of them all.

i)   In this case we choose the base e.

For $3^x = 7$

take natural logarithms of both sides,

> To maintain the balance.

i.e. $\ln 3^x = \ln 7$

$x \ln 3 = \ln 7$

> 3rd law.

$x(1.0986) = 1.9459$

$$x = \frac{1.9459}{1.0986}$$

$x = 1.77$ (3 sf).

ii)   In this case we choose the base 10.

For $3^x = 6$

take common logarithms of both sides,

> To maintain the balance.

i.e. $\log 3^x = \log 6$

$x \log 3 = \log 6$

> 3rd law.

$x(0.4771) = 0.7782$

$$x = \frac{0.7782}{0.4771}$$

$x = 1.63$ (3 sf).

As an exercise you should repeat these 2 examples. For $3^x = 7$ use common logarithms and for $3^x = 6$ use natural logarithms. In each case you will find the different bases do not affect the final answers.

████████ **Examples 10.6** ████████████████████████████████

Solve the indicial equations i) $3^x = 15.4^{x-2}$,

ii) $3^x = 15 \times 4^{x-2}$.

Like the previous examples we can use either common or natural logarithms. We choose to use common logarithms.

i) Starting with $3^x = 15.4^{x-2}$

take common logarithms of both sides

i.e. $\log 3^x = \log 15.4^{x-2}$

$x \log 3 = (x - 2) \log 15.4$ | 3rd law.

$x(0.4771) = (x - 2)1.1875$

$x(0.4771 - 1.1875) = -2 \times 1.1875$ | 2nd law.

i.e. $x(-0.7104) = -2.3750$ | 3rd law.

$x = \dfrac{-2.3750}{-0.7104}$

$x = 3.34$ (3 sf).

ii) Starting with $3^x = 15 \times 4^{x-2}$

take common logarithms of both sides,

i.e. $\log 3^x = \log(15 \times 4^{x-2})$

$= \log 15 + \log 4^{x-2}$

i.e. $x \log 3 = \log 15 + (x - 2) \log 4$

$x(0.4771) = 1.1761 + (x - 2)0.6021$

$x(0.4771 - 0.6021) = 1.1761 - (2 \times 0.6021)$

i.e. $x(-0.1250) = 1.1761 - 1.2042$

$x = \dfrac{-0.0281}{-0.1250}$

$x = 0.225$ (3 sf).

████████ **Example 10.7** ████████████████████████████████

Solve the indicial equations i) $6^{2x} = 5 \times 6^x$,

ii) $8^{2x} - 9 \times 8^x + 20 = 0$.

One of the laws we remembered at the beginning of the chapter is particularly useful:

$$(b^m)^n = b^{mn} = b^{nm} = (b^n)^m$$

i) In $6^{2x} = 5 \times 6^x$

we may think of 6^{2x} as $(6^x)^2$

Now let $q = 6^x$ so the original equation becomes

$$q^2 = 5q$$

i.e. $q^2 - 5q = 0$

$$q(q - 5) = 0$$

$\therefore$ $q = 0$ or $q - 5 = 0$

$$q = 0, 5.$$

When $q = 0$ we have $6^x = 0$. This equation has no solution because logarithms are defined only for positive numbers. Of course 0 is **not** a positive number. Also when $q = 5$ we have $6^x = 5$.

Take common logarithms of both sides

i.e. $\log 6^x = \log 5$

$$x \log 6 = \log 5$$

$$x\,(0.7782) = 0.6990$$

$$x = \frac{0.6990}{0.7782}$$

$$x = 0.898 \qquad (3\text{ sf}).$$

ii) In $8^{2x} - 9 \times 8^x + 20 = 0$

think of 8^{2x} as $(8^x)^2$

Now let $q = 8^x$ so the original equation becomes

$$q^2 - 9q + 20 = 0$$

This quadratic equation will factorise,

i.e. $(q - 4)(q - 5) = 0$

$\therefore$ $q - 4 = 0$ or $q - 5 = 0$

$$q = 4, 5.$$

When $q = 4$ we have $8^x = 4$.

Take common logarithms of both sides,

i.e. $\log 8^x = \log 4$

$$x \log 8 = \log 4$$

$$x\,(0.9031) = 0.6021$$

$$x = \frac{0.6021}{0.9031}$$

$$x = 0.667 \qquad (3\text{ sf}).$$

Also when $q = 5$ we have $8^x = 5$.

Using common logarithms in much the same way gives

$$x = \frac{0.6990}{0.9031}$$

$$x = 0.774 \qquad (3\text{ sf}).$$

■■■■■ EXERCISE 10.3 ■■■■■■

Solve the indicial equations for x in each case.

1 $2^x = 8$
(You should be able to spot this solution.)

2 $5^x - 14 = 0$

3 $2 \times 5^x - 14 = 0$

4 $4^x = 6^{x-1}$

5 $3^x = 7.5^{x-2}$

6 $2^x \times 3^x = 4^{x+1}$

7 $3 \times 4^x = 6^{x-1}$

8 $2^x \times 5^{x+2} = 6^{x-1}$

9 $4^{2x} = 5 \times 4^x$

10 $3^{2x} - 10 \times 3^x + 21 = 0$

Change of base

There is a link between common and natural logarithms. We need the option of being able to convert between them.

Consider $\log N = x$.

We can re-write this in its alternative form,

i.e. $N = 10^x$.

> 3rd law.

Now take natural logarithms of both sides,

i.e. $\ln N = \ln 10^x$

$\ln N = x \ln 10$

i.e. $\dfrac{\ln N}{\ln 10} = x$

i.e. $\dfrac{\ln N}{\ln 10} = \log N.$

We may re-write this as

$$\log N = \frac{\ln N}{2.3026}$$

or $\log N = 0.4343 \ln N.$

We can use this formula in the next example.

■■■■■ Example 10.8 ■■■■■

If we have $\ln N = 1.6094$ what is the value of $\log N$?

Using $\log N = \dfrac{\ln N}{\ln 10}$

we get $\log N = \dfrac{1.6094}{2.3026}$

i.e. $\log N = 0.699$ (3 sf).

A quick calculator check reveals the value of N to be 5. You can see from this example that not knowing it to be 5 was no handicap.

We can look again to create the opposite conversion.

Consider   ln $N = x$.

We can re-write this in its alternative form,

i.e. $N = e^x$.

Now take common logarithms of both sides,

i.e. $\log N = \log e^x$

$\log N = x \log e$ | 3rd law.

i.e. $\dfrac{\log N}{\log e} = x$

i.e. $\dfrac{\log N}{\log e} = \ln N.$

We may re-write this as

$$\ln N = \frac{\log N}{0.4343}$$

or   ln $N = 2.3026 \log N$

We can use this formula in the next example.

▰▰ Example 10.9 ▰▰

If we have $\log N = 0.7910$ what is the value of ln N?

Using $\ln N = \dfrac{\log N}{\log e}$

we get $\ln N = \dfrac{0.7910}{0.4343}$

i.e.   ln $N = 1.821$ (4 sf).

A quick calculator check reveals the value of N to be 6.18. Again you can see that not knowing it to be 6.18 was no handicap.

▰▰ EXERCISE 10.4 ▰▰

Use the conversion formula to find the value of $\log N$ in each case.

1   ln $N = 2.5$ 4   ln $N = 4.60$

2   ln $N = 5.175$ 5   ln $N = 1.123$

3   ln $N = -0.35$

Use the other conversion formula to find the value of ln N in each case.

6    $\log N = 2.5$ 9 $\log N = 7.38$

7    $\log N = -0.65$ 10 $\log N = 3.0$

8    $\log N = 5.0$

Now let us look at some examples applying these techniques to practical problems.

━━━━━ **Example 10.10** ━━━━━

In a cooling system every second $V\,\mathrm{m}^3$ of cooling fluid flows a distance $x\,\mathrm{m}$ according to $V = 3.95x^{2.55}$. In the design prototype x may be adjusted to test various specifications. Using this formula estimate the volume of cooling fluid when $x = 0.95$ m. Also find the value of x needed for a flow of $7.50\,\mathrm{m}^3$.

There are several ways to solve this problem. This method uses techniques we have learned in this chapter.

Given $x = 0.95$ we substitute into our formula

$$V = 3.95x^{2.55}$$

to get $\qquad V = 3.95\,(0.95)^{2.55}.$

We can find the value of this using the calculator.

Input 0.95, press the $\underline{x^y}$ button and then 2.55 followed by $\underline{=}$ to display 0.877.

Then $\qquad V = 3.95 \times 0.877\ldots$

$$= 3.47\ \mathrm{m}^3 \qquad (3\ \mathrm{sf})$$

i.e. a volume of $3.47\,\mathrm{m}^3$ of cooling fluid flows each second for this design specification.

The second part of the problem uses the substitution of $V = 7.50$ into our formula.

$$V = 3.95x^{2.55}$$

becomes $\qquad 7.50 = 3.95x^{2.55}$

i.e. $\qquad \dfrac{7.50}{3.95} = x^{2.55}$

$$1.899 = x^{2.55}.$$

Take common logarithms of both sides to give

$$\log 1.899 = \log x^{2.55}$$

i.e. $\qquad \log 1.899 = 2.55 \log x$

i.e. $\qquad \dfrac{0.278}{2.55} = \log x$

$$0.109 = \log x$$

i.e. $\qquad x = 10^{0.109}$

$$x = 1.29\ \mathrm{m}.$$

■■■■■■■ **Example 10.11** ■■■■■■■

In a submarine cable the speed of the signal, v, is related to the radius of the cable's covering, R, by $v = \dfrac{25}{R^2} \ln\left(\dfrac{R}{5}\right)$.

Plot a graph of v against R using values of R from $R = 2\,\text{mm}$ to $10\,\text{mm}$ at intervals of $1\,\text{mm}$. With the aid of the graph find the maximum signal speed. For what value of R does this occur?

Our first step is to construct a table of values. This has quite a few extra rows of working. Each row is labelled to aid the explanation.

R	2	3	4	5	6	7	8	9	10	①
$\dfrac{R}{5}$	0.400	0.600	0.800	1.000	1.200	1.400	1.600	1.800	2.000	②
$\ln\left(\dfrac{R}{5}\right)$	−0.916	−0.511	−0.223	0.000	0.182	0.336	0.470	0.588	0.693	③
R^2	4	9	16	25	36	49	64	81	100	④
$\dfrac{25}{R^2}$	6.250	2.778	1.563	1.000	0.694	0.510	0.391	0.309	0.250	⑤
v	−5.73	−1.42	−0.35	0.00	0.13	0.17	0.18	0.18	0.17	⑥

We can emphasise the meaning of each row.

Row ① is the specimen values of R.

Row ② shows each value of R divided by 5.

Row ③ takes the natural logarithms of Row ②.

Row ④ shows each specimen value of R squared.

Row ⑤ uses Row ④, dividing 25 by each value.

Row ⑥ is the product of Rows ⑤ and ③, the values of v.

Fig. 10.4 shows the plot of v against R. You can see that above the horizontal axis the curve bends only slightly. If you plot this for yourself take care.

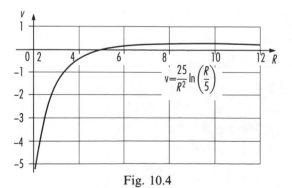

Fig. 10.4

Because of the slight bend Fig. 10.5 looks in more detail at this section of the graph.

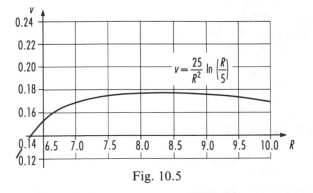

Fig. 10.5

The maximum signal speed occurs at the peak of the curve. Here a tangent to the curve is horizontal. From our graph we see the maximum signal speed is approximately $v = 0.18$ occurring at $R = 8.2$ mm, each correct to 2 significant figures.

EXERCISE 10.5

1 The power gain of an amplifier relates input power, P_1, to output power, P_0, by

$$\text{Power gain} = 10 \log \frac{P_0}{P_1}$$

The units for power gain are decibels, dB.
Calculate the power gain if $P_0 = 18$ W and $P_1 = 300$ mW. For an input power of 300 mW and a power gain of 20 dB what must be the output power?

2 The cutting speed, S m/min, and tool life, T min, of a machine tool for roughing cuts in steel are related by $ST^n = c$. c is a constant. Take common logarithms of both sides to make n the subject of this formula.

3 In a transmitting antenna the attenuation, A, is given by

$A = 10 \log \left(1 + \dfrac{R_L}{R_A}\right)$. If $R_L = 2.75\,\Omega$ is the loss resistance and $R_A = 90\,\Omega$ is the antenna loss calculate A.

4 Atmospheric pressure, p cm of mercury, is related to height, h m, by $pe^{ah} = c$. a and c are constants. Make h the subject of this formula using natural logarithms.

Given $a = \dfrac{1}{15000}$, $c = 76.2$ and $p = 50$ cm find h.

5   The diagram shows a belt in contact with a pulley. The length of belt in contact with the pulley subtends an angle θ radians with the centre. The coefficient of friction is $\mu = 0.32$. The tension on the taut side is $T_1 = 34.75\,\text{N}$ and on the slack side is $T_0 = 22.50\,\text{N}$.

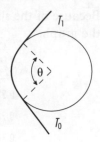

If $\theta = \dfrac{1}{\mu}\ln\left(\dfrac{T_1}{T_0}\right)$ calculate θ for these values.

Alternatively if θ is to be $60°$ what must be the new coefficient of friction if T_1 and T_0 remain unchanged?

6   $A = P\left(1 + \dfrac{R}{100}\right)^n$ is the formula for compound interest.

n is the number of years for the investment,
R is the interest rate,
P is the principal sum invested,
A is the total amount of the principal sum and interest.

The accountant of a local engineering firm earns a profit-related bonus of £10 000. He invests it at an interest rate of 8.5%. How many years will it be before his original bonus is doubled (i.e. $A = £20\,000$)?

If there is an alternative investment with an interest rate of 10.5% re-calculate the new time to double his money.

What would be the interest rate if his original investment doubled in 5 years?

7   The diagram shows a submarine signalling cable. The radius of the core is r and the radius of the covering is R. The ratio of these radii is given by $x = \dfrac{r}{R}$. The speed of the signal, v, is given

by $v = x^2 \ln\dfrac{1}{x}$. Use a law of logarithms to re-arrange

this formula into $v = -x^2 \ln x$.
For values of x from 0.1 to 0.8 plot a graph of v against x. When $v = 0.125$ what is the value of x?
Given the value of r as 5.5 mm what is the associated radius of the covering for this signal speed?

8   A local engineering firm uses the formula $V = P\left(1 - \dfrac{R}{100}\right)^T$ to

calculate depreciation costs for its equipment.
T is the working life of the equipment (years),
R is the rate of depreciation,
P is the original equipment cost,
V is the remaining second-hand value of the equipment.
The firm buys some capital equipment for £17 500 with an expected working life of 4 years. The second-hand value when the firm sells it

is approximately £5500. What rate of depreciation has the firm used (correct to 2 significant figures)?

9   The electrical current, C, in a wire is related to time, t, by
$$kt = -\ln(1 + C).$$
Show that this formula may be re-written as
$$t = \frac{1}{k}\ln\left(\frac{1}{1+C}\right).$$

10   The diagram shows an electrical circuit with an inductor, L, and resistor, R, connected in series to an emf, E. There is a time constant, T, given by $T=\dfrac{L}{R}$. Calculate the value of T.

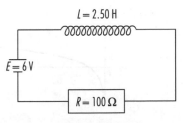

The formula for the current, i, flowing in this circuit during time, t, is given by $i=\dfrac{E}{R}(1 - e^{-t/T})$.

Make t the subject of this formula using natural logarithms. If $i = 51$ mA $(= 0.051$ A) calculate the value of time t (seconds).

■ MULTI-CHOICE TEST 10 ■

1   A natural logarithm, x of a number, N, can be represented by
A) $N=e^x$ and $\ln x = N$
B) $N=e^{-x}$ and $\ln x = N$
C) $N=e^x$ and $\ln N = x$
D) $N=x^e$ and $\ln N = x$

2   The natural logarithm of 42.09 is
A) 0.8654
B) 1.4372
C) 2.4372
D) 3.7398

3   If -1.0099 is the natural logarithm of a number then that number is
A) 0.0977
B) 0.3643
C) 2.7453
D) 10.2306

4   The value of $\log -0.85$
A) is -0.071
B) is -0.163
C) is 7.079
D) cannot be found

5 Log $N = -1.235$. The value of N
   A)   is 0.058
   B)   is 0.291
   C)   is 17.179
   D)   cannot be found

6 The graph of $y = \log x$ is

A)   B)

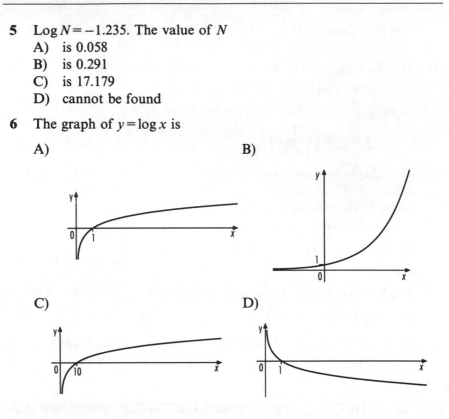

C)   D)

7 The value of ln 1.936 is
   A)   0.287
   B)   0.661
   C)   6.931
   D)   86.298

8 Ln $N = 0.747$. The value of $N =$
   A)   −0.292
   B)   −0.127
   C)   2.111
   D)   5.585

9 The graph of $y = \ln x$ is

A)   B)

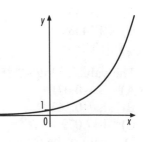

C)   D)

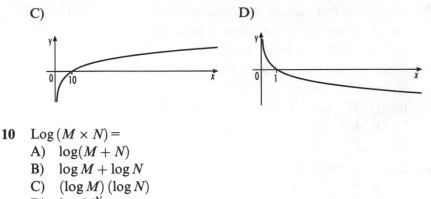

10 $\text{Log}(M \times N) =$
A) $\log(M + N)$
B) $\log M + \log N$
C) $(\log M)(\log N)$
D) $\log M^N$

11 $\text{Log}(M \div N) =$
A) $\log(M - N)$
B) $\log M - \log N$
C) $(\log M) \div (\log N)$
D) $\log M^{1/N}$

12 $\text{Log } N^p =$
A) $\log pN$
B) $p \log N$
C) $\ln N^p$
D) $p \ln N$

13 Written in terms of natural logarithms $\log N =$
A) $\dfrac{\ln N}{\ln 10}$
B) $\dfrac{\ln 10}{\ln N}$
C) $\ln N \ln 10$
D) $\ln N \ln e$

14 Decide whether each of these statements is True (T) or False (F).

   i) Naperian logarithms are to any base, b.
  ii) Common logarithms are to the base 10.

Which option best describes the two statements?

A) i) T   ii)   T
B) i) T   ii)   F
C) i) F   ii)   T
D) i) F   ii)   F

15 Decide whether each of these statements is True (T) or False (F).

   i) As x tends to 0 the graph of $\log x$ tends to $-\infty$

  ii) As x tends to ∞ the graph of $\log x$ tends to 1.

Which option best describes the two statements?

  A) i) T   ii) T
  B) i) T   ii) F
  C) i) F   ii) T
  D) i) F   ii) F

16 $\log_e y = x$ means that

  A) $x = e^y$
  B) $y = e^x$
  C) $e = y^x$
  D) $y = x^e$

17 If $2^x = 12$ then $x =$

  A) $\ln 6$
  B) 6
  C) $\ln 10$
  D) $\dfrac{\ln 12}{\ln 2}$

18 $4^x \times 5^x$ simplifies to

  A) 9^x
  B) 9^{2x}
  C) 20^x
  D) 20^{2x}

19 If $5 \times 3^x - 20 = 0$ then $x =$

  A) $\dfrac{\ln 4}{\ln 3}$
  B) $\ln 4 - \ln 3$
  C) $\dfrac{\ln 20}{\ln 15}$
  D) $\dfrac{\ln 20}{\ln 15} - \ln 3$

20 An electrical circuit is shorted. The intial current is i_0. After a time, t (s), the current, i (A), is given $i = i_0 e^{-4t}$. The time for i to decay to $0.2i_0$ is

  A) $0.18\,\text{s}$
  B) $0.25\,\text{s}$
  C) $0.40\,\text{s}$
  D) $4.00\,\text{s}$

11 Using Logarithms and Exponential Graphs

Element: Use graphs to solve engineering problems.

PERFORMANCE CRITERIA
- Coordinates and scales appropriate to the engineering problem are selected.
- Data are accurately plotted.
- The relationship between variables is identified.
- Values are accurately determined from graphs, and engineering problems are solved.

RANGE
Engineering problems: electrical, electronic; mechanical.
Coordinates: Cartesian.
Scales: linear, logarithmic.
Relationship: linear, exponential, logarithmic.
Values: two variables, gradient, intercept.

Introduction

In this chapter we are going to link together the natural logarithm, $\ln x$, and the exponential function, e^x. We will apply them in various forms and graphical sketches. By changing the scales on the axes we will present some logarithmic graphs as straight lines instead of curves.

ASSIGNMENT

The Assignment for this chapter involves a belt around part of a pulley. They are in contact along the arc AB. θ is the angle subtended at the centre of the pulley, i.e. $\angle AOB = \theta$. There is friction between the belt and the pulley. The

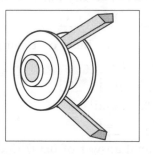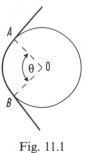

Fig. 11.1

321

coefficient of friction is μ. This means the tensions on either side of the pulley are different.

The exponential function

In Chapter 10 we looked at natural (or Naperian) logarithms. They are logarithms to the base e,

i.e. $\log_e N = x$ or $\ln N = x$ and $N = e^x$.

The value of e is 2.71828..., often shortened to 2.718. When no index is written we understand the index is 1. This means we understand e to be e^1. Using a calculator we can find the value of e for ourselves. Often e^x is shown above the $\underline{\ln}$ button. The calculator order is

$$\boxed{1}\quad \boxed{\text{inv}}\quad \boxed{\ln}$$

to display 2.71828...

The $\underline{\text{inv}}$ button together with the $\underline{\ln}$ button shows that e^x is the inverse function of ln. This is why it is often found above the $\underline{\ln}$ button on a calculator.

▬▬▬▬ **Examples 11.1** ▬▬▬▬

Find the values of   i) $e^{2.5}$,   ii) $e^{-2.5}$,   iii) $3.74e^{-2.5}$.

i) For $e^{2.5}$ the calculator order is

$$\boxed{2.5}\quad \boxed{\text{inv}}\quad \boxed{\ln}$$

to display 12.18 (4 sf).

ii) Comparing $e^{-2.5}$ with Example 11.1i) what changes is the sign of the index. Notice how we make that change in the calculator order

$$\boxed{2.5}\quad \boxed{^{+/-}}\quad \boxed{\text{inv}}\quad \boxed{\ln}$$

to display 0.0821 (4 dp).

Also remember the laws of indices. $e^{-2.5}$ is $\dfrac{1}{e^{2.5}}$. This means we can use our result from $e^{2.5}$ and then use the $\underline{1/x}$ button to reach our display of 0.0821. The alternative complete order for $e^{-2.5}$ is

$$\boxed{2.5}\quad \boxed{\text{inv}}\quad \boxed{\ln}\quad \boxed{1/x}$$

iii) For $3.74e^{-2.5}$ again we build on Example 11.1ii), multiplying by 3.74. The calculator order is

$$\boxed{2.5}\quad \boxed{^{+/-}}\quad \boxed{\text{inv}}\quad \boxed{\ln}$$

to display 0.08... and

$$\boxed{\times}\quad \boxed{3.74}\quad \boxed{=}$$

to display a final answer of 0.307 (3 dp).

Remember that the laws of indices we recalled in Chapter 10 continue to apply for similar bases.

▰▰▰▰▰ Examples 11.2 ▰▰▰▰▰▰▰▰▰▰▰▰▰▰▰▰▰▰▰▰▰▰▰▰▰▰▰

Find the values of   i) $e^{2.5} \times e^{1.7}$,   ii) $\dfrac{e^{2.7}}{e^{1.7}}$,   iii) $(e^{2.5})^4$.

We apply the appropriate law of indices **before** using the calculator. This saves time and effort.

i)   For $e^{2.5} \times e^{1.7}$ the bases are the same so we can add the indices, i.e. $e^{2.5} \times e^{1.7} = e^{2.5+1.7} = e^{4.2}$.
Now the calculator order is

$$\underline{4.2|} \quad \underline{inv|} \quad \underline{ln\,|}$$

to display 66.69 (4 sf).

ii)   Again, using another law of indices $\dfrac{e^{2.5}}{e^{1.7}} = e^{2.5-1.7} = e^{0.8}$.
The calculator order is

$$\underline{0.8|} \quad \underline{inv|} \quad \underline{ln\,|}$$

to display 2.226 (4 sf).

iii)   Yet another law of indices gives $(e^{2.5})^4 = e^{2.5 \times 4} = e^{10}$.
The calculator order is

$$\underline{10|} \quad \underline{inv|} \quad \underline{ln\,|}$$

to display 22026 (5 sf).

We can look again at this example using the power button $\underline{x^y|}$. This time we find the value of $e^{2.5}$ and then raise that value to the power 4.
The alternative order is

$$\underline{2.5|} \quad \underline{inv|} \quad \underline{ln\,|} \quad \underline{x^y|} \quad \underline{4|}$$

▰▰▰▰▰▰ ASSIGNMENT ▰▰▰▰▰▰▰▰▰▰▰▰▰▰▰▰▰▰▰▰▰▰▰▰▰▰

Let us take a first look at our pulley problem. The friction between the belt and the pulley is important. The tension of the belt before it passes around the pulley differs from the tension afterwards. On either side of the pulley they are related by the formula $T_1 = T_0 e^{\mu\theta}$. T_1 is the tension on the taut side and T_0 is the tension on the slack side. Suppose the slack side tension is 22.50 N, the coefficient of friction is 0.35 and the angle subtended is 80°, i.e. $T_0 = 22.50$, $\mu = 0.35$ and $\theta = 80°$. For the formula to work we need θ to be in radians. We saw how to convert degrees to radians in Chapter 4, i.e.

$$\theta = 80° = \frac{80 \times \pi}{180} = 1.396 \text{ radians.}$$

We can substitue our values into the formula so that

$$T_1 = 22.50 \, e^{0.35 \times 1.396}$$

$$= 22.50 \, e^{0.489}$$

$$= 22.50 \times 1.630$$

i.e. $T_1 = 36.7 \, \text{N}$.

This means the tension on the taut side of the pulley is 36.7 N.

■ EXERCISE 11.1 ■

Find the values of

1 $e^{1.7}$

2 $e^{7.4}$

3 e^{-1}

4 e^{0}

5 $2e^{7.4}$

6 $\dfrac{2e^{7.4}}{e^{1.7}}$

7 $2e^{1.7} \times 3e^{7.4}$

8 $7.4e^{2}$

9 $\dfrac{7.4}{e^{2}}$

10 $7.4 \, e^{-2}$

11 $\dfrac{7.4}{e^{-2}}$

12 $\dfrac{e^{2.35}}{e^{1.2}}$

13 $\dfrac{e^{2.35}}{e^{-1.2}}$

14 $\left(e^{2.35}\right)^{2}$

15 $\left(e^{2}\right)^{2.35}$

16 $\left(5e^{2.35}\right)^{2}$

17 $5\left(e^{2.35}\right)^{2}$

18 $\dfrac{e^{-1.2}}{e^{-2.35}}$

19 $\dfrac{1}{4e^{1.75}}$

20 $\dfrac{1}{4e^{-1.75}}$

Exponential graphs

Exponential graphs involve either e^{x} or e^{-x}. e^{x} represents exponential **growth** because the index is **positive**. e^{-x} represents exponential **decay** because the index is **negative**. It is interesting to see a diagram of a function. As we have done so many times we can construct a table and plot a graph of each function. The table is built up using specimen values of x and the inv| ln | calculator buttons.

$y = e^{x}$

x	−2.5	−2.0	−1.0	0	1.0	2.0	2.5
e^{x}	0.082	0.135	0.368	1	2.718	7.389	12.182

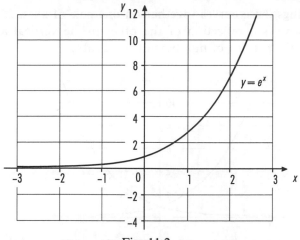

Fig. 11.2

The graph is shown in Fig. 11.2. Notice how it approaches the *x*-axis but never quite touches it. This means the horizontal axis is an asymptote to the curve. An asymptote is a straight line that a graph gets closer and closer to without actually touching it. Also notice that the curve crosses the vertical axis at $y=1$. The gradient of the curve is always positive.

Remember that gradient is $\text{gradient} = \dfrac{\text{vertical change}}{\text{horizontal change}}$

$y=e^{-x}$

x	-2.5	-2.0	-1.0	0	1.0	2.0	2.5
e^{-x}	12.182	7.389	2.718	1	0.368	0.135	0.082

You can see a pattern when comparing this table of values with the previous table. The graph is shown in Fig. 11.3. Again, notice how the *x*-axis acts as an asymptote. Again we see it crosses the vertical axis at $y=1$. This time the gradient of the curve is always negative.

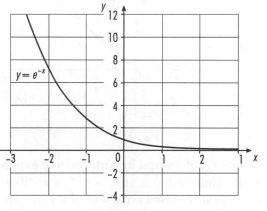

Fig. 11.3

Fig. 11.4 brings these curves together on one pair of axes. It emphasises that each curve is a reflection of the other in the vertical axis. This is hinted at by the pattern of numbers in the 2 tables.

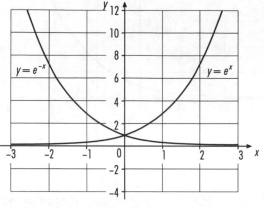

Fig. 11.4

We can look more closely at the gradients of these curves. Fig. 11.5 shows some specimen tangents. With care you find the gradients are related to the curve. (The quality and accuracy of your own graphs are important here.) The gradient of an exponential function is equal to an exponential function. At this stage we will not deepen the discussion with a precise rule. As an exercise for yourself you should accurately plot the graphs of $y = e^x$ and $y = e^{-x}$. Then you should choose a few points on each curve and find the gradients at those points. Each gradient should have the same numerical value as y at that point.

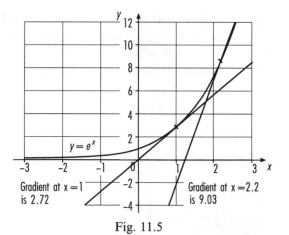

Fig. 11.5

Let us look at the graphs of natural logarithmic and exponential functions together. We know that $\ln N = x$ and $N = e^x$ are different forms of a relationship. The first one is based on a natural logarithm. The second one is based on an exponential. Fig. 11.6 shows them together with the line

$y = x$ on one pair of axes. In fact each graph is a reflection of the other in the line $y = x$. Such a reflective property occurs for a function and its inverse function. For yourself look at the asymptotes: the x-axis for one function, the y-axis for the other function. One of the graphs crosses an axis at $(0, 1)$, the other at $(1, 0)$.

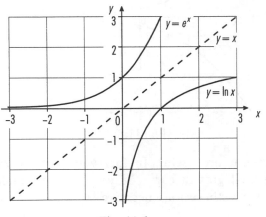

Fig. 11.6

The same situation applies to the common logarithmic function. As an exercise for yourself plot on one pair of axes the graphs of $y = \log x$ and $y = 10^x$. If you then plot the graph of $y = x$ on those axes you will notice the reflections in that line.

Fig. 11.4 shows the graphs of $y = e^x$ and $y = e^{-x}$ being reflections of each other in the vertical axis. Notice the negative sign in the index associated with this reflection. Also it is possible to have a negative sign immediately before the e. This causes a reflection in the horizontal axis. Fig. 11.7 shows the graphs of $y = e^x$ and $y = -e^x$ as reflections of each other in the horizontal axis. Also Fig. 11.8 shows the graphs of $y = e^{-x}$ and $y = -e^{-x}$ as reflections of each other, again in the horizontal axis.

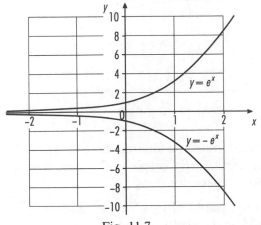

Fig. 11.7

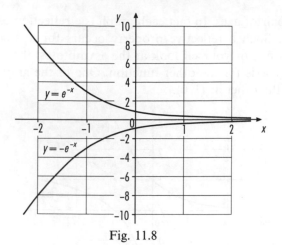

Fig. 11.8

All these reflections are useful. In fact you need remember only the original exponential growth curve of $y = e^x$. Simply understanding the reflections created with various negative signs leads to so many other curves.

We can apply a similar building technique to sketch the graph of $y = 1 - e^x$. We do it in stages. Firstly sketch the graph of $y = e^x$. Reflect that curve in the horizontal axis to get $y = -e^x$. If we re-write $y = 1 - e^x$ as $y = 1 + (-e^x)$ the final stage is reached by an upward shift of 1. Fig. 11.9 shows these graphs of $y = 1 - e^x$.

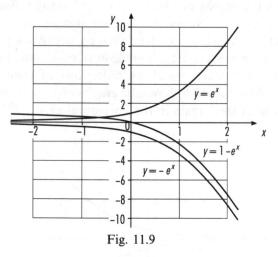

Fig. 11.9

So far we have concentrated on altering the $+/-$ signs. These have been in the index and/or immediately before e. Our next step is to change some numbers in either of these positions.

Fig. 11.10 shows the graphs of $y = e^x$, $y = e^{2x}$ and $y = e^{0.75x}$, i.e. our indices have 1, 2 and 0.75 as the coefficients of x.

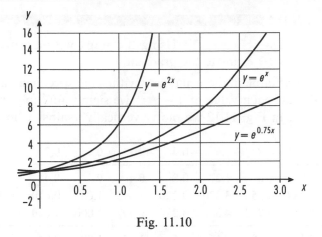

Fig. 11.10

Each graph crosses the vertical axis at $y=1$. The change in the gradient is evident for each curve. The greater the coefficient of x the steeper the curve.

What happens if we have a multiplying factor immediately before the exponential? Fig. 11.11 shows the graphs of $y=e^x$, $y=2e^x$ and $y=0.75e^x$. This time where each graph crosses the vertical axis is different. In each case $x=0$, but in turn

$$y = e^0 \quad\quad = 1$$
$$y = 2e^0 \quad = 2 \times 1 \quad = 2$$
$$y = 0.75e^0 \ = 0.75 \times 1 = 0.75.$$

There are changes in the gradient. The changes are not as great as those caused by numerical changes to the index.

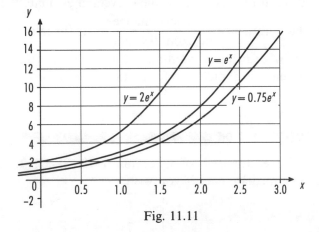

Fig. 11.11

We have looked at changes related to the curve for **exponential growth**, $y=e^x$. The same types of change apply to **exponential decay**, $y=e^{-x}$. Remember that each exponential growth curve has a decay curve as a reflection in the vertical axis.

Example 11.3

Plot the graph of $y=\frac{1}{2}(1+e^{-2x})$. Using the graph with a tangent at the point where $x=-0.3$ estimate the gradient.

We construct our table in the usual way with some specimen values of x. Once we calculate each value of $1+e^{-2x}$ we simply multiply by $\frac{1}{2}$ to get each value of y. In Fig. 11.12 we plot y vertically against x horizontally.

x	-1.0	-0.8	-0.6	-0.4	-0.2	0	0.2	0.4	0.6
$-2x$	2.0	1.6	1.2	0.8	0.4	0	-0.4	-0.8	-1.2
e^{-2x}	7.389	4.953	3.320	2.226	1.492	1.000	0.670	0.449	0.301
$1+e^{-2x}$	8.389	5.953	4.320	3.226	2.492	2.000	1.670	1.449	1.301
y	4.19	2.98	2.16	1.61	1.25	1.00	0.84	0.72	0.65

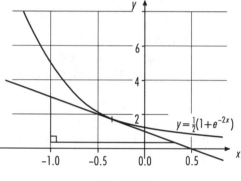

$y=\frac{1}{2}(1+e^{-2x})$

Fig. 11.12

At $x=-0.3$ we draw a tangent to the curve and estimate the gradient. Remember that $\text{gradient}=\dfrac{\text{vertical change}}{\text{horizontal change}}$. As usual, we get these change values by drawing a right-angled triangle. If you do this for yourself accurately you should aim for a gradient of approximately -1.8.

EXERCISE 11.2

Use one set of axes for each of Questions **1–10**. Sketch the following curves, with intermediate ones if you wish.

1   $y=e^{-2x}$

2   $y=-e^{2x}$

3   $y=-2e^{-2x}$

4   $y=3(1-e^x)$

5   $y=3(1+e^{-x})$

6   $y=1+e^{2x}$

7   $y=1-e^{-x}$

8   $y=3(1+e^{2x})$

9   $y=2(1-3e^x)$

10   $y=2(1+3e^{-x})$

Use one set of axes for each of Questions **11–16**, and values of x from $x=-2.5$ to $x=2.5$ at intervals of 0.5. Plot the following curves.

11 $y=e^{0.5x}$

12 $y=2e^{0.5x}$

13 $y=1+e^{2x}$

14 $y=1+2e^{3x}$

15 $y=2(1+e^{3x})$

16 $y=3(1+2e^{-0.5x})$

This final selection of 4 questions is based on some of the curves you have plotted accurately in Questions **11–16**.

17 By drawing a tangent to the curve $y=e^{0.5x}$ at the point where $x=1.5$ calculate this gradient of the curve.

18 For the curve $y=2e^{0.5x}$ at the point where $x=1.5$ draw a tangent and find its gradient.

19 With the aid of a tangent at $x=1.8$ find the gradient of the curve $y=1+2e^{3x}$ at that point.

20 With the aid of a tangent at $x=1.8$ find the gradient of the curve $y=2(1+e^{3x})$ at that point.

Your answers to Question **20** and to Question **19** should be identical. The closeness of your 2 answers depends on the quality and accuracy of your own graphs. How close are your answers?

■■■■■■■■ **Example 11.4** ■■■■■■■■

Newton's law of cooling relates the temperature of a body to its surroundings over a period of time, t seconds. If the body is hotter than its surrounding environment then its temperature will decrease over time. Let $T°C$ be the difference in temperature between a body and its environment. Suppose that the relation is $T=400e^{-t/1500}$.

Plot a graph of $T°C$ against t seconds at intervals of 60 seconds (i.e. every minute) for the first 10 minutes.

i) What is the initial temperature difference?

ii) How long will it be before the value of T has fallen by 25%?

iii) What is the temperature gradient at 8 minutes?

As usual our first step is to construct a table of values.

t	0	60	120	180	240	300	360 $\longleftrightarrow$
$e^{-t/1500}$	1.000	0.961	0.923	0.887	0.852	0.819	0.787
T	400	384	369	355	341	327	315 $\updownarrow$

	420	480	540	600 $\longleftrightarrow$
	0.756	0.726	0.698	0.670
	302	290	279	268 $\updownarrow$

The middle line of working, $e^{-t/1500}$, is quoted to 3 decimal places. In truth all the decimal places have been retained in the table calculations; only for the temperature, T, have the answers been approximated.

The graph is plotted in Fig. 11.13 from where we have deduced our answers.

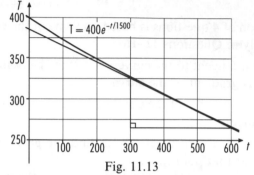

Fig. 11.13

i) The initial temperature is the temperature at $t=0$, i.e. 400°C.

ii) If the temperature has been reduced by 25% this means we are interested in 75% of the original temperature,

i.e. $400 \times \dfrac{75}{100} = 300°C.$

From the graph, when $T=300°C$, $t=432$ seconds, i.e. 7 minutes 12 seconds. You can check this result for yourself. Use $T=300$ in the equation

$$T = 400e^{-t/1500}.$$

Take natural logarithms of both sides to find $t=431.5$.

iii) We find the gradient by drawing a tangent to the curve at $t=480$ seconds (i.e. 8 minutes). We complete a right-angled triangle and use

$$\text{gradient} = \frac{\text{vertical change}}{\text{horizontal change}}.$$

This gives a value of about $-0.2°C/second$. The minus sign indicates a cooling effect. The body's temperature tends towards the temperature of its surrounding environment. If left alone for long enough the body would cool to that temperature.

◼◼◼ ASSIGNMENT ◼◼◼

Let us take another look at our pulley problem. This time we are going to plot a graph of results. The following table relates the angle, θ (radians), with the belt tension on the taut side, T_1.

θ	1.00	1.10	1.20	1.25	1.30	1.40	1.50
T_1	32.0	33.3	33.9	34.9	35.5	36.9	38.2

We know from our previous look at the Assignment that $T_1 = T_0 e^{\mu\theta}$. T_0 is the tension on the slack side and μ is the coefficient of friction. Using our graph we are going to see how closely the table values compare with $T_1 = T_0 e^{\mu\theta}$. Remember $T_0 = 22.50$ N and $\mu = 0.35$.

Fig. 11.14 shows the graph. We have attempted to draw a smooth curve through the plotted points. However you can see that some of our results do *not* lie on the curve. Our table of results and graph compare with our relationship, $T_1 = 22.50 e^{0.35\theta}$. It is not a totally accurate comparison because not all points lie on the curve. At this stage we have no way of finding the true values for T_0 to replace 22.50 N and μ to replace 0.35.

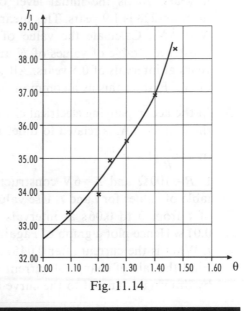

Fig. 11.14

1   $p = 76 e^{-h/15000}$ relates the pressure, p (cm of mercury), to the height, h (m), above sea-level. Construct a table of values for h and p. Use values of h from sea-level up to and including 6000 m at intervals of 500 m. Plot a graph of p against h. Drawing a tangent at $h = 2500$ m find the gradient of the curve at this point.

2   The voltage, v, across the inductor, L, in the electrical circuit drops exponentially over time, t. The relation is

$v = E e^{-t/T}$ where $T = \dfrac{L}{R}$. If L is 2.50 H

and R is 125 Ω calculate T. The emf is given by $E = 6.2$ V. For values of t from 0 to 0.05 s at intervals of 0.01 s construct a table of values.

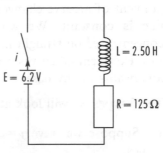

Plot the exponential curve of v against t. Using your graph
i)   draw a tangent at $t = 0.025$ and so find the gradient,
ii) find the time taken to halve the initial voltage.

3   A bacterial growth, B, is related exponentially to time, t hours, by $B = B_0 e^{ct}$. $B_0 = 1.75 \times 10^3$ and $c = 0.8$. Using this relation and these values construct a table and plot an exponential curve. Use values of t from $\frac{1}{2}$ hour to 5 hours at intervals of $\frac{1}{2}$ hour.

From your graph read off the number of bacteria at $t=\frac{1}{2}$ hour. By what time has this number trebled?

4   The relationship of $N = N_0 e^{-ct}$ relates radioactivity, N, to time, t, years. N_0 is the initial level of radioactivity. The half-life of thorium-228 is 1.9 years. This means that after 1.9 years the value of N is $\frac{1}{2}N_0$. Calculate the value of c. Use this value of c to help construct a table of values of N and t. Use values of t from 0 to 5 years at intervals of 0.5 years. All your values of N will be multiples of N_0. Extract this as a common factor when you plot N against t.

5   In the accompanying electrical circuit the current, i A, is related to time, t s,

by $i = \dfrac{E}{R}(1 - e^{-40t})$.

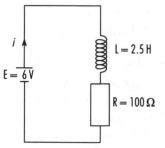

If $R = 100\,\Omega$ and $E = 6$ V construct a table of values for i and t. Use values of t from 0 to 0.06 s at intervals of 0.01 s. Hence plot a graph of i against t. What is the current after 0.045 s?
Find the rate of change of current with respect to time at $t = 0.025$ s by drawing a tangent to the curve at this point.

Logarithmic graphs

We know the basic shapes of curves involving logarithms and exponentials. A curve is not as simple to draw as a straight line. The gradient of a curve changes continually whilst the gradient of a straight line is constant. We would save ourselves time and effort if we concentrated on straight lines. In fact we can do this for various curves. For exponential curves we can do it using logarithms. We can use logarithms to any base.

The 3 types we will look at involve natural logarithms.

1.   Suppose we have $y = ae^{bx}$ where a and b are constant, numerical, values.

Take natural logarithms of both sides,

i.e. $\ln y = \ln(ae^{bx})$

$\qquad\quad = \ln a + \ln e^{bx}$ | First log law.

$\qquad\quad = \ln a + bx \ln e$ | Third log law.

$\qquad\quad = \ln a + bx$ | $\ln e = 1$

i.e. $\ln y = bx + \ln a$.

Let us compare this relation with the usual one for a straight line,

i.e. $\quad Y = MX + C.$

The comparison is based on pattern and position. On the left sides we have $\ln y$ and Y. Hence $\ln y$ is plotted vertically like Y in Figs. 11.15.

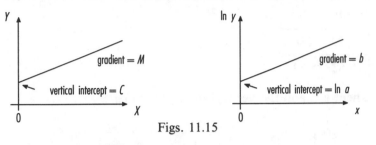

Figs. 11.15

On the right sides the other variables are x and X. Hence as usual we plot x horizontally. Also MX indicates that M is the gradient, so bx indicates that b is the corresponding gradient. Remember this may be negative just as easily as positive. To complete the comparison we have vertical intercepts of C and $\ln a$. These may be negative or zero or positive. This comparison shows we have reduced $y = ae^{bx}$ to a straight line in the form $\ln y = bx + \ln a$.

■ Example 11.5 ■

Suppose we have the following table of values

t	0	60	120	180	240	300	360	420	480	540	600
T	400	384	369	355	341	327	315	302	290	279	268

It is thought that t and T are related by $T = ae^{bt}$. Check that this is true and find the values of a and b.

You will remember part of this table from a few pages ago. We know that T and t are related according to an exponential graph. We know as well that accurate drawings are difficult. Thus we will look to attempt a straight line based on our new theory.

Using $\quad T = ae^{bt}$

take natural logarithms of both sides

i.e. $\quad \ln T = \ln(ae^{bt})$

$\qquad = \ln a + \ln e^{bt}$ | First log law.

$\qquad = \ln a + bt \ln e$ | Third log law.

$\qquad = \ln a + bt$ | $\ln e = 1$

i.e. $\quad \ln T = bt + \ln a.$

We plot $\ln T$ vertically and t horizontally. Thus we need another row in our table of values.

t	0	60	120	180	240	300	360	420	480	⟷
T	400	384	369	355	341	327	315	302	290	
$\ln T$	5.99	5.95	5.91	5.87	5.83	5.79	5.75	5.71	5.67	↕

	540	600	⟷
	279	268	
	5.63	5.59	↕

The gradient is b and the vertical intercept is $\ln a$.

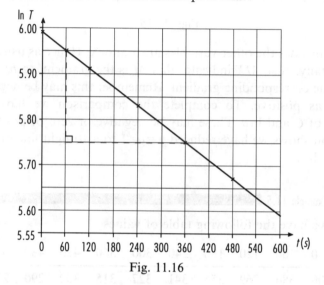

Fig. 11.16

Fig. 11.16 shows the graph is indeed a straight line. You might have expected this from the pattern of values of $\ln T$. From our graph we can read off the vertical intercept as 5.99,

i.e. $\qquad \ln a = 5.99$

i.e. $\qquad a = e^{5.99} = 399.$

Also $\quad$ gradient $= \dfrac{\text{vertical change}}{\text{horizontal change}}$

i.e. $\qquad b = \dfrac{5.95 - 5.75}{60 - 360}$

$\qquad\qquad = \dfrac{0.20}{-300}$

i.e. $\qquad b = -\dfrac{1}{1500}.$

We may link these values together to give

$$\ln T = -\frac{1}{1500}t + 5.99.$$

A preferred version uses our original relation of
$$T = ae^{bt}$$
i.e. $\qquad\qquad T = 399e^{-t/1500}$

You can see this is only slightly different from our original relation with 399 in place of 400. Such a minor change is acceptable.

---

2. Suppose we have $y = ax^b$ where a and b are constant, numerical, values. Take natural logarithms of both sides,

i.e. $\qquad \ln y = \ln (ax^b)$

$\qquad\qquad = \ln a + \ln x^b$ | First log law.

$\qquad\qquad = \ln a + b \ln x.$ | Third log law.

Writing this as
$$\ln y = b \ln x + \ln a$$
again we can compare it with
$$Y = MX + C.$$

The comparison is based on pattern and position. On the left sides we have $\ln y$ and Y. Hence $\ln y$ is plotted vertically like Y in Figs. 11.17.

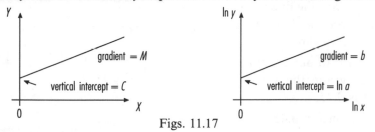

Figs. 11.17

On the right sides the other variables are $\ln x$ and X. Hence we plot $\ln x$ horizontally. Also MX indicates that M is the gradient, so $b \ln x$ indicates that b is the corresponding gradient. Remember this may be negative just as easily as positive. To complete the comparison we have vertical intercepts of C and $\ln a$. These may be negative or zero or positive. This comparison shows we have reduced $y = ax^b$ to a straight line in the form $\ln y = b \ln x + \ln a$.

■■■■ Example 11.6 ■■■■

Suppose we have the following table of values

x	1	2	3	4	5
y	1.50	8.50	23.00	48.00	84.10

It is thought that x and y are related by $y = ax^b$. Check that this is true and find the values of a and b.

We look to plot a straight line based on this second new piece of theory.

Using　　$y = ax^b$

we take natural logarithms of both sides and eventually get

　　$\ln y = b \ln x + \ln a.$

Hence we must use our table of values to calculate $\ln x$ and $\ln y$.

$\ln x$	0	0.69	1.10	1.39	1.61
$\ln y$	0.41	2.14	3.14	3.87	4.43

We plot $\ln y$ vertically and $\ln x$ horizontally in Fig. 11.18. The points do not quite all lie on one straight line. We need to draw a line of best fit through them. If they do not all lie on the line ensure there are some points above and some below the line.

Suppose all the spare points are above your line. This means the line is too low and needs a slight shift vertically upwards. Suppose all the spare points are below your line. This means the line is too high and needs a slight shift vertically downwards.

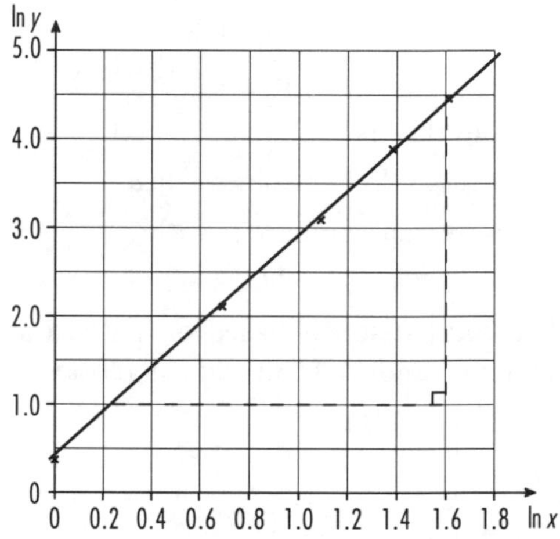

Fig. 11.18

Let us return to our example. From our graph the vertical intercept is 0.45, i.e.

　　　　$\ln a = 0.45$

　　　　$a = e^{0.45} = 1.57.$

Also　　gradient $= \dfrac{\text{vertical change}}{\text{horizontal change}}$

i.e.　　　　$b = \dfrac{4.40 - 1.00}{1.60 - 0.22}$

　　　　　　$= \dfrac{3.40}{1.38}$

　　　　　　$= 2.46.$

These values give the relationship $y = 1.57x^{2.46}$.

3.　Suppose we have $y = ab^x$ where a and b are constant, numerical, values.
Take natural logarithms of both sides

i.e.　　　$\ln y = \ln(ab^x)$

　　　　　　$= \ln a + \ln b^x$

　　　　　　$= \ln a + x \ln b.$

> First log law.
> Third log law.

Writing this as
$$\ln y = (\ln b)\,x + \ln a$$
again we can compare it with
$$Y = MX + C.$$
The comparison is based on pattern and position. On the left sides we have $\ln y$ and Y. Hence $\ln y$ is plotted vertically like Y in Figs. 11.19.

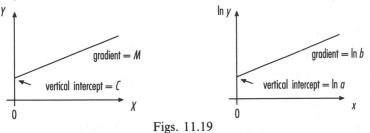

Figs. 11.19

On the right sides the other variables are x and X. Hence we continue to plot them both horizontally. Also MX indicates that M is the gradient, so $(\ln b)\,x$ indicates that $\ln b$ is the corresponding gradient. Remember this may be negative just as easily as positive. To complete the comparison we have vertical intercepts of C and $\ln a$. These may be negative or zero or positive. This comparison shows we have reduced $y = ab^x$ to a straight line in the form $\ln y = (\ln b)\,x + \ln a$.

Example 11.7

Suppose we have the following table of values

x	0	2	4	5	6	8	10
y	3.00	2.40	1.97	1.77	1.59	1.10	1.05

It is thought that x and y are related by $y = ab^x$. Check that this is true and find the values of a and b.
One of the table values is probably inaccurate. Decide which one it is from your graph.

We will look to plot a straight line based on this third new piece of theory.
Using $\qquad y = ab^x$
we take natural logarithms of both sides and eventually get
$$\ln y = (\ln b)\,x + \ln a.$$
Hence we must use our table of values to calculate $\ln y$.

x	0	2	4	5	6	8	10	$\longleftrightarrow$
y	3.00	2.40	1.97	1.77	1.59	1.10	1.05	
$\ln y$	1.099	0.875	0.678	0.571	0.464	0.095	0.049	$\updownarrow$

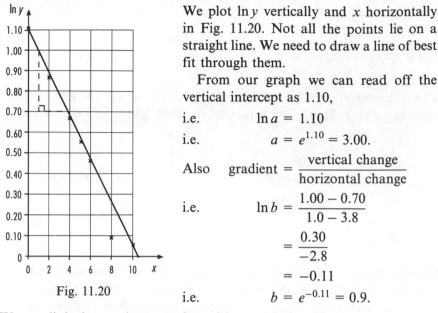

Fig. 11.20

We plot $\ln y$ vertically and x horizontally in Fig. 11.20. Not all the points lie on a straight line. We need to draw a line of best fit through them.

From our graph we can read off the vertical intercept as 1.10,

i.e. $\qquad \ln a = 1.10$

i.e. $\qquad a = e^{1.10} = 3.00.$

Also $\quad$ gradient $= \dfrac{\text{vertical change}}{\text{horizontal change}}$

i.e. $\qquad \ln b = \dfrac{1.00 - 0.70}{1.0 - 3.8}$

$\qquad\qquad\quad = \dfrac{0.30}{-2.8}$

$\qquad\qquad\quad = -0.11$

i.e. $\qquad b = e^{-0.11} = 0.9.$

We can link these values together with our relationship to get

$$y = 3.00 \times 0.9^x.$$

In addition, from our graph we see that the point where $x = 8$ is probably inaccurate. This is because it lies so far from the straight line.

███ ASSIGNMENT ███

Let us take a third look at our pulley problem. Again we are going to plot a graph of results, but this time as a straight line. The table relates the angle, θ (radians), with the belt tension on the taut side, T_1.

θ	1.00	1.10	1.20	1.25	1.30	1.40	1.50
T_1	32.0	33.3	33.9	34.9	35.5	36.9	38.2

We know that $T_1 = T_0 e^{\mu\theta}$. T_0 is the tension on the slack side and μ is the coefficient of friction. Previously we thought $T_0 = 22.50\,\text{N}$ and $\mu = 0.35$. Using our graph we are going to see how closely the table values compare with $T_1 = 22.50 e^{0.35\theta}$. Now we are going to use a straight line to check these values.

Using $\qquad T_1 = T_0 e^{\mu\theta}$

we take natural logarithms of both sides

i.e. $\qquad \ln T_1 = \ln(T_0 e^{\mu\theta})$

$\qquad\qquad\quad = \ln T_0 + \ln(e^{\mu\theta})$

$\qquad\qquad\quad = \ln T_0 + \mu\theta \ln e$

$\qquad\qquad\quad = \ln T_0 + \mu\theta.$

> First log law.
> Third log law.
> $\ln e = 1.$

We can re-write this as
$$\ln T_1 = \mu\theta + \ln T_0.$$
Comparing it with the standard equation for a straight line
$$Y = MX + C$$
we plot $\ln T_1$ vertically against θ horizontally. This means we must extend our table of values to

θ	1.00	1.10	1.20	1.25	1.30	1.40	1.50	$\longleftrightarrow$
T_1	32.0	33.3	33.9	34.9	35.5	36.9	38.2	
$\ln T_1$	3.47	3.51	3.52	3.55	3.57	3.61	3.64	$\updownarrow$

Fig. 11.21 shows the graph. We have attempted to draw a straight line of best fit. Not all our results lie on the straight line.

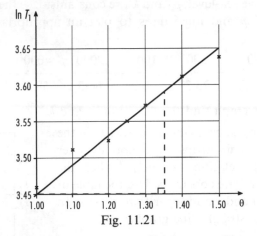

Fig. 11.21

Immediately the straight line with points lying on either side casts doubt on the accuracy of our original $T_1 = 22.50e^{0.35\theta}$. It is not a totally accurate relation. However it is of more use than our previous exponential curve. Let us see if we can improve on our original values for T_0 and μ.

We cannot read off the vertical intercept because we have not included where $\theta = 0$. Instead let us start with the gradient.
$$\text{Gradient} = \frac{\text{vertical change}}{\text{horizontal change}}$$
i.e.
$$\mu = \frac{3.59 - 3.45}{1.35 - 1.00}$$
$$= \frac{0.14}{0.35}$$
$$= 0.40.$$
We can substitute this value of $\mu = 0.40$ into
$$\ln T_1 = \mu\theta + \ln T_0$$
to get
$$\ln T_1 = 0.40\theta + \ln T_0.$$

To find the value of $\ln T_0$ we choose any point on the line and substitute for those values,

i.e.
$$3.45 = 0.40 \times 1.00 + \ln T_0$$
$$3.45 = 0.40 + \ln T_0$$
$$3.45 - 0.40 = \ln T_0$$
$$3.05 = \ln T_0$$

i.e.
$$T_0 = e^{3.05} = 21.12\,\text{N}.$$

Hence our new complete relationship is $T_1 = 21.12 e^{0.400}$.

EXERCISE 11.4

1 $p = p_0 e^{-kH}$ relates the pressure, p (cm of mercury), to the height, H (m), above sea-level. p_0 and k are constants. Use the table of values for H and p and logarithms to plot an appropriate straight line graph.

H	0	500	1000	2000	4000	8000	10000
p	75.8	73.3	70.9	66.3	58.1	44.5	38.9

From your graph find the values of p_0 and k.

2 The voltage, v, across the inductor, L, in the electrical circuit drops exponentially over time, t s. The relation is $v = E e^{-t/T}$ where the emf, E, and T are constants. Use the table of values for t and v and logarithms to plot an appropriate straight line graph.

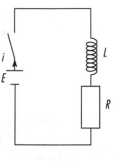

t	0	0.010	0.020	0.025	0.030	0.040	0.050
v	5.50	2.95	1.60	1.15	0.85	0.45	0.25

Use your graph to find the values of E and T. How long did it take to halve the initial voltage?

3 Gas pressure, p bars, and temperature, $T°K$, are thought to be related by $T = \alpha p^\gamma$. α and γ are constants to be found from your graph. Attempt to plot a straight line graph using logarithms and this table.

p	1	1.5	2.0	2.5	5.0	7.5
T	510	570	610	650	780	860

4 b, the intensity of light varies with V, the voltage according to $b = V^c$. c is a constant. Take natural logarithms of both sides of this equation. Hence extend the table so that you can plot a straight line graph.

V	14	16	18	20	22	24
$b\,(\times 10^{-3})$	5.82	4.49	3.57	2.90	2.41	2.04

From your graph find the value of c.

5 $V=a^R$ relates a variable resistance, R, with an output voltage, V, for an amplifier. a is a constant. Using logarithms, the table of values and a straight line graph show that this is true.

R	35	40	45	50	55	60	65
V	7.70	10.30	13.80	18.40	24.70	33.00	44.10

From your graph find the value of a.

6 It is thought that a bacterial growth, B, is related exponentially to time, t hours, by $B=B_0 e^{ct}$. B_0 and c are constants. Use the table of values and an appropriate straight line graph to find B_0 and c.

t	0.25	0.50	1.00	1.50	2.00	2.50	4.00	5.00
$B\,(\times 10^3)$	1.93	2.33	3.39	4.93	7.17	10.43	32.14	68.00

7 A microwave oven includes an element with a non-linear resistance, $R\,\Omega$. The current, $I\,A$, and the resistance, R, are thought to be related by $I=VR^k$. k is a constant. Take natural logarithms of both sides of this equation. Hence extend your table so that you can plot a straight line graph. From your graph find the value of k.

R	2	4	6	8	10	12
I	5.70	16.00	29.40	45.25	63.00	83.00

8 The relationship of $N=N_0 e^{-ct}$ relates radioactivity, N, to time t years. N_0 is the initial level of radioactivity. The half-life of thorium-228 is 1.9 years. This means that after 1.9 years the value of N is $\frac{1}{2}N_0$. Calculate the value of c. Use this value of c to help construct a table of values of N and t. Use values of t from 0 to 5 years at intervals of 0.5 years. All your values of N will be multiples of N_0. Extract this as a common factor when you plot your graph. This was Question 3 in Exercise 11.3. That plot of N against t produced an exponential curve. This time attempt to plot $\ln N$ against t and so create a straight line graph.

9 In the accompanying electrical circuit the current, $i\,A$, is related to time, t s, by

$$i=\frac{E}{R}(1-e^{-40t}).$$

If $R=100\,\Omega$ and $E=5\,V$ construct a table of values for i and t. Use values of t from 0 to 0.06 s at intervals of 0.01 s.

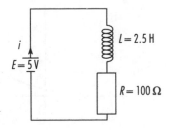

Now re-arrange your original equation into the form $1 - 20i = e^{-40t}$. Hence include another line in your table for $1 - 20i$.

Take natural logarithms of both sides of this re-arranged equation. Hence include a further line in your table for $\ln(1 - 20i)$. Now plot a graph of $\ln(1 - 20i)$ against t expecting to get a straight line. From your graph what is the current after 0.045 s?

10 A regional engineering company increases its profits each year. During a recent period each year's profits have been 20% higher than those for the previous year. At the end of Year 1 the profits were £150 000. Complete the following table using 3 significant figures.

Year (y)	1	2	3	4	5	6
Profit ($£ \times 10^3$) (P)	150					373

It is thought that P and y are related by $P = P_1 r^y$ where P_1 and r are constants. Does a plot of $\ln P$ against y give a straight line? What are the values of P_1 and r?

◼◼◼ MULTI-CHOICE TEST 11 ◼◼◼

1 The value of $-2.75e^{-1.25}$, correct to 2 decimal places, is:
   A) −9.60
   B) −2.46
   C) −0.79
   D) 0.08

2 The value of $(3e^{0.5})^2$, correct to 2 decimal places, is:
   A) 3.85
   B) 8.15
   C) 11.56
   D) 24.46

Questions **3** and **4** refer to the following information.
Pressure, p, and height, h, above sea-level are connected by $p = p_0 e^{-kh}$. p_0 and k are positive constants.

3 A correct sketch of this equation is:
   A)   B)

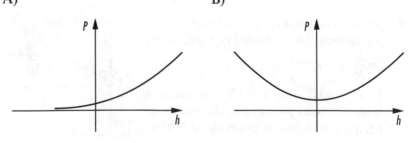

C)   D)

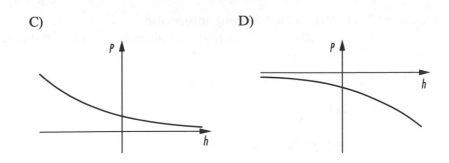

4 The sign on the gradient of this curve:
   A) is always positive
   B) may be found only by numerical substitution
   C) is always negative
   D) is sometimes positive and sometimes negative

5 The diagram shows the graphs of $y=e^x$, $y=e^{0.5x}$ and $y=e^{1.5x}$.

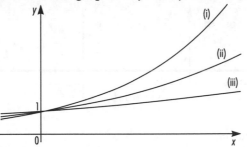

   In order they are:
   A) (i), (ii) and (iii)
   B) (ii), (iii) and (i)
   C) (iii), (i) and (ii)
   D) (i), (iii) and (ii)

6 The diagram shows the graph of $y=e^{-2x}$, $y=2e^{-2x}$ and $y=0.5e^{-2x}$.

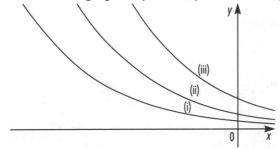

   In order they are:
   A) (i), (ii) and (iii)
   B) (ii), (iii) and (i)
   C) (iii), (i) and (ii)
   D) (i), (iii) and (ii)

Questions **7–11** refer to the following information.
A capacitor is being charged. Its current, i, is given by $i = 10e^{-t/CR}$ during time, t. $C = 5 \times 10^{-6}$ F and $R = 0.25 \times 10^{6} \Omega$.

7   The logarithmic form of this equation is:

A) $\ln i = \ln 10 - \dfrac{t}{CR}$

B) $\log_{10} i = \log_{10} 10 - \log_{10} \left(\dfrac{t}{CR} \right)$

C) $\ln i = \ln 10 + \dfrac{CR}{t}$

D) $\ln i = \ln 10 - \ln \left(\dfrac{t}{CR} \right)$

8   To obtain a straight line graph you use the axes:

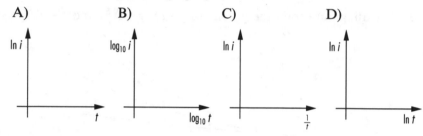

9   The gradient is:

A) $-\dfrac{1}{CR}$

B) $\log_{10} 10$

C) $\ln 10$

D) $-\ln \left(\dfrac{1}{CR} \right)$

10   The intercept is:

A)   10

B) $\ln 10$

C) $-\ln \left(\dfrac{1}{CR} \right)$

D) $-\dfrac{1}{CR}$

11   Current halves in a time of:

A)   0.693 s

B)   0.866 s

C)   2.019 s

D)   6.703 s

Questions **12–16** refer to the following information.

A cooling system is purchased by an engineering company. Its depreciated value, £D, is based on its initial value, £C, time, t years, and rate of depreciation, r. You are given $D = Cr^t$

12 The logarithmic form of this equation is:
A) $\log D = \log C + t \log r$
B) $\log D = \log C + r \log t$
C) $\log D = t + \log Cr$
D) $\log D = t \log r + C$

13 You need to obtain a straight line graph from this equation.
Decide whether each of the statements is True (T) or False (F).
  i) The vertical axis should be $\log D$.
  ii) The horizontal axis should be $\log t$.
Which option best describes the two statements?
A) i) T   ii) T
B) i) T   ii) F
C) i) F   ii) T
D) i) F   ii) F

14 The gradient is:
A) r
B) $\log C$
C) t
D) $\log r$

15 The intercept is:
A) r
B) $\log t$
C) $\log C$
D) C

16 The initial cost of the equipment is £400 000. If $r = 0.60$, the value of the cooling system after $1\frac{1}{2}$ years is approximately:
A) £110 100
B) £117 600
C) £185 900
D) £360 000

Questions **17–20** refer to the following information.
The luminosity, I, of a lamp is connected to a varying voltage, V.

17 If the logarithmic form of the relationship is $\ln I = \ln a + n \ln V$ then the original equation is:
A) $I = aV^n$
B) $I = V^{a+n}$
C) $I = an^V$
D) $I = a + n^V$

18 A straight line graph is obtained using the axes:

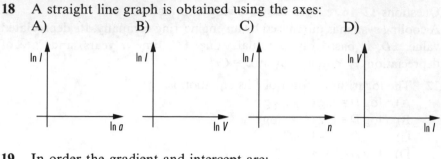

A)   B)   C)   D)

19 In order the gradient and intercept are:
  A) n, a
  B) $\ln a, n$
  C) $\ln V, \ln a$
  D) $n, \ln a$

20 Given $n=4$, $V=120$ and $I=180$ the value of a is:
  A) -13.96
  B) 1.10×10^{-14}
  C) 8.68×10^{-7}
  D) 3.73×10^{10}

12 Using Graphs: II

Introduction

This chapter may be split into 3 sections:

  i)   some standard curves; the parabola and the circle;
 ii)   polar graphs;
iii)   reduction to a straight line graph.

In Chapter 8 we solved quadratic equations graphically. There we first met the parabola. Following on we look at the circle using Cartesian axes and polar axes. We also investigate some more simple polar graphs. Finally we alter the scales on the horizontal Cartesian axis to create simple line graphs. We have done this before with logarithmic graphs reduced to straight lines in Chapter 11.

■■■■ ASSIGNMENT ■■■■

The Assignment for this chapter involves a collection of numbers. You can see they are displayed as pairs of coordinates. The aim is to find how these pairs of coordinates are **related**. During the chapter we will attempt to do this using the graphs we introduce.

The pairs of coordinates are:

(1.35, −1.11),	(0.40, −0.03),	(3.06, −1.54),
(1.74, −1.26),	(0.10, 2.34),	(2.48, −1.45),
(3.88, −1.64),	(0.90, −0.82),	(0.15, 1.47).

The first task is to put them in some order. Perhaps a display in a table is the best method of presentation.

x	0.10	0.15	0.40	0.90	1.35	1.74	2.48	3.06	3.88
y	2.34	1.47	−0.03	−0.82	−1.11	−1.26	−1.45	−1.54	−1.64

The plot is shown in Fig. 12.1. It does look as though there is some relationship. At present it is not clear what it might be. However, our work so far does tell us something. Obviously these points do not lie on a straight line and the graph does not pass through the origin. Hence our knowledge of proportion tells us that x and y are not directly proportional to each other. We may find that there is some other proportional relationship, but at the moment nothing is obvious.

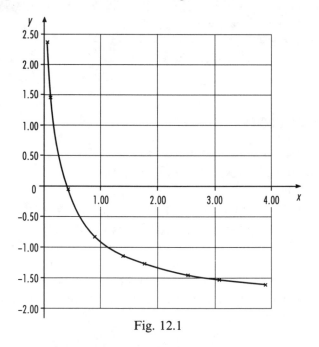

Fig. 12.1

Some standard curves

There are many interesting curves in Mathematics. Here are just two standard ones.

1. The parabola, $y^2 = 4ax$

Usually when we want to plot a graph we have only y on the left-hand side of the relationship. To change our relation into this form we take the square root of both sides, i.e. $y = \pm\sqrt{4ax}$.

a can have a range of values. The form $4a$ is included to make the algebra simpler in Pure Mathematics.

Various parabolas are described in a variety of ways. We saw an alternative version of a parabola in Chapter 8 on Quadratic Equations.

Example 12.1

On one set of axes plot the graphs of

i) $y = \pm\sqrt{x}$ i.e. $y^2 = x$,

ii) $y = \pm\sqrt{3x}$ i.e. $y^2 = 3x$,

iii) $y = \pm\sqrt{\frac{1}{2}x}$ i.e. $y^2 = \frac{1}{2}x$.

We choose some specimen values of x and construct our tables as usual. Because we are using $\sqrt{\ }$ we know that none of the x values can be negative.

x	0	1	2	3	4	5	6	$\leftrightarrow$
$\pm\sqrt{x}$	0	±1.000	±1.414	±1.732	±2.000	±2.236	±2.449	$\updownarrow$
$3x$	0	3	6	9	12	15	18	
$\pm\sqrt{3x}$	0	±1.732	±2.449	±3.000	±3.464	±3.873	±4.243	$\updownarrow$
$\frac{1}{2}x$	0	0.5	1.0	1.5	2.0	2.5	3.0	
$\pm\sqrt{\frac{1}{2}x}$	0	±0.707	±1.000	±1.225	±1.414	±1.581	±1.732	$\updownarrow$

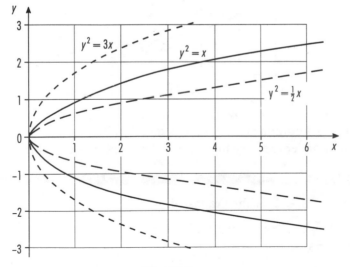

Fig. 12.2

Fig. 12.2 shows the different curvatures of the parabolas. The smaller the value in place of $4a$ the closer the parabola stays to the horizontal axis. The greater the value in place of $4a$ the more the parabola curves away from the horizontal axis.

Example 12.2

On one set of axes plot the graphs of

  i) $y = \pm\sqrt{x}$ i.e. $y^2 = x$,

  ii) $y = \pm\sqrt{x+1}$ i.e. $y^2 = x + 1$,

 iii) $y = \pm\sqrt{x-2}$ i.e. $y^2 = x - 2$.

We choose some specimen values of x and construct our tables as usual. Because we are using $\sqrt{}$ we know that none of the x values can be negative.

x	0	1	2	3	4	5	6	$\leftrightarrow$
$\pm\sqrt{x}$	0	± 1.000	± 1.414	± 1.732	± 2.000	± 2.236	± 2.449	$\updownarrow$
$x + 1$	1	2	3	4	5	6	7	
$\pm\sqrt{x+1}$	± 1.000	± 1.414	± 1.732	± 2.000	± 2.236	± 2.449	± 2.646	$\updownarrow$
$x - 2$	-2	-1	0	1	2	3	4	
$\pm\sqrt{x-2}$	not defined		0	± 1.000	± 1.414	± 1.732	± 2.000	$\updownarrow$

Fig. 12.4

In Fig. 12.3 we see how the basic parabola, $y = \pm\sqrt{x}$ is shifted horizontally. Addition under the $\sqrt{}$ symbol shifts the graph to the left. Subtraction under the $\sqrt{}$ symbol shifts the graph to the right.

Example 12.3

On one set of axes plot the graphs of

  i) $y = \pm\sqrt{x}$ i.e. $y^2 = x$,

  ii) $y + 1 = \pm\sqrt{x}$ i.e. $(y + 1)^2 = x$,

 iii) $y - 2 = \pm\sqrt{x}$ i.e. $(y - 2)^2 = x$.

We choose some specimen values of x in the usual way. The $\pm$ option needs a little more care in the second and third graphs.

i) The table for $y = \pm\sqrt{x}$ is the usual one seen in previous examples.

x	0	1	2	3	4	5	6
$\pm\sqrt{x}$	0	±1.000	±1.414	±1.732	±2.000	±2.236	±2.449

ii) For $y + 1 = \pm\sqrt{x}$

$$y = -1 \pm \sqrt{x}$$

i.e. $y = -1 + \sqrt{x}, \qquad y = -1 - \sqrt{x}.$

This means there are 2 options for y which we use. The table shows these options.

x	0	1	2	3	4	5	6
$\sqrt{x}$	0	1.000	1.414	1.732	2.000	2.236	2.449
$-1+\sqrt{x}$	-1.000	0	0.414	0.732	1.000	1.236	1.449
$-1-\sqrt{x}$	-1.000	-2.000	-2.414	-2.732	-3.000	-3.236	-3.449

iii) For $y - 2 = \pm\sqrt{x}$

$$y = 2 \pm \sqrt{x}$$

i.e. $y = 2 + \sqrt{x}, \qquad y = 2 - \sqrt{x}.$

Again there are 2 options for y. The table shows these options.

x	0	1	2	3	4	5	6
$\sqrt{x}$	0	1.000	1.414	1.732	2.000	2.236	2.449
$2+\sqrt{x}$	2.000	3.000	3.414	3.732	4.000	4.236	4.449
$2-\sqrt{x}$	2.000	1.000	0.586	0.268	0	-0.236	-0.449

In Fig. 12.4 we see how the basic parabola, $y^2 = x$, is shifted. $y = -1 \pm \sqrt{x}$ is a vertical shift of -1. $y = 2 \pm \sqrt{x}$ is a vertical shift of $+2$.

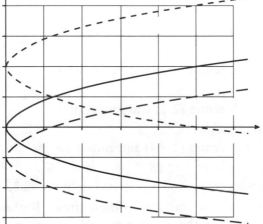

We can continue to investigate and experiment with other parabolas. These examples are just a selection of possibilities.

■■■■■ EXERCISE 12.1 ■■■■■■■

For each question first attempt to sketch the graph. Follow up the sketch with an accurate plot of the graph. In each case the sketch should influence your choice of specimen table values for x.

1 $y = \pm\sqrt{2x}$ or $y^2 = 2x$.

2 $y = \pm\sqrt{x+3}$ or $y^2 = x + 3$.

3 $y^2 = x$ and $y^2 = -x$

4 $x^2 = y$ and $x^2 = -y$.

5 $(y - 2)^2 = x + 3$.

2. The circle, $x^2 + y^2 = r^2$

r is the radius of the circle.

$x^2 + y^2 = r^2$ describes a circle with centre $(0,0)$ and radius r.

More generally $(x - \alpha)^2 + (y - \beta)^2 = r^2$ describes a circle with centre (α, β) and radius r.

When using $x^2 + y^2 = r^2$ we can re-arrange it to get

$$y^2 = r^2 - x^2$$

i.e. $$y = \pm\sqrt{r^2 - x^2}.$$

You will find this form more useful when constructing a table of values.

▰▰▰▰ Examples 12.4 ▰▰▰▰▰▰▰▰

In this set of examples we describe a series of circles.

i) $x^2 + y^2 = 9$, i.e. $x^2 + y^2 = 3^2$
   is a circle with centre $(0,0)$ and radius of 3.

ii) $(x - 2)^2 + y^2 = 3^2$
   is a circle with centre $(2,0)$ and radius of 3.

iii) $x^2 + (y - 1)^2 = 3^2$
   is a circle with centre $(0,1)$ and radius of 3.

iv) $(x - 2)^2 + (y - 1)^2 = 25$
   is a circle with centre $(2,1)$ and radius of 5.

v) $(x - 2)^2 + (y + 4)^2 = 1^2$
   is a circle with centre $(2,-4)$ and radius 1.

vi) $(x + 0.5)^2 + (y + 3)^2 = 42.25$
   is a circle with centre $(-0.5, -3)$ and radius $\sqrt{42.25} = 6.5$.

It is left as an exercise for you to draw these circles. Rather than plot them you may prefer to draw them with a pair of compasses.

Example 12.5

Plot the circle $(x-2)^2 + y^2 = 3^2$.

We know from the previous set of examples this circle has centre $(2,0)$ and a radius of 3. This means the horizontal diameter stretches 3 units either side of 2, i.e. from $x = -1$ to $x = 5$. The vertical diameter stretches 3 units either side of 0, i.e. from $y = -3$ to $y = 3$. The values of x we have just deduced will influence our specimen table values.

Also we can re-arrange our equation,

i.e. $(x-2)^2 + y^2 = 3^2$

becomes $y^2 = 3^2 - (x-2)^2$

$\therefore \qquad y = \pm\sqrt{9 - (x-2)^2}$.

The table is constructed below.

x	-1.00	-0.50	0	0.50	1.00	1.50	2.00	$\longleftrightarrow$
$x-2$	-3.00	-2.50	-2.00	-1.50	-1.00	-0.50	0	
$(x-2)^2$	9.00	6.25	4.00	2.25	1.00	0.25	0	
$9-(x-2)^2$	0	2.75	5.00	6.75	8.00	8.75	9.00	
y	0	±1.66	±2.24	±2.60	±2.83	±2.96	±3.00	$\updownarrow$

	2.50	3.00	3.50	4.00	4.50	5.00		$\longleftrightarrow$
	0.50	1.00	1.50	2.00	2.50	3.00		
	0.25	1.00	2.25	4.00	6.25	9.00		
	8.75	8.00	6.75	5.00	2.75	0		
	±2.96	±2.83	±2.60	±2.24	±1.66	0		$\updownarrow$

Fig. 12.5 shows the circle. It confirms the centre as $(2,0)$ and the radius as 3.

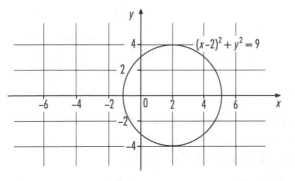

Fig. 12.5

Also you can use the diagram to check the points where it passes through the coordinate axes. These are consistent with the initial circle equation for this example.

It crosses the x-axis when $y = 0$,

i.e. $(x - 2)^2 + y^2 = 9$

becomes $(x - 2)^2 = 9$

∴ $x - 2 = \pm 3$

i.e. $x = 2 - 3, \quad 2 + 3$

i.e. $x = -1, \quad 5.$

This is confirmed by Fig. 12.5.

Also it crosses the y-axis when $x = 0$,

i.e. $(x - 2)^2 + y^2 = 9$

becomes $(0 - 2)^2 + y^2 = 9$

i.e. $4 + y^2 = 9$

$y^2 = 9 - 4 = 5$

∴ $y = \pm\sqrt{5} = \pm 2.24.$

Again this is confirmed by Fig. 12.5.

▰▰▰ EXERCISE 12.2 ▰▰▰

For Questions 1–10 describe the circles by giving their centres and radii.

1   $x^2 + y^2 = 36$

2   $x^2 + y^2 = 18$

3   $\left(x + \dfrac{1}{2}\right)^2 + y^2 = 36$

4   $\left(x - \dfrac{1}{2}\right)^2 + y^2 = 25$

5   $x^2 + (y - 3)^2 = 49$

6   $x^2 + \left(y + \dfrac{1}{4}\right)^2 = 6.25$

7   $\left(x + \dfrac{1}{4}\right)^2 + \left(y - \dfrac{1}{3}\right)^2 = 9$

8   $(x - 2)^2 + (y - 2.5)^2 = 1$

9   $(x + 2)^2 + (y + 1.7)^2 = 4$

10 $\left(x - \dfrac{2}{5}\right)^2 + \left(y + \dfrac{3}{4}\right)^2 = 1$

11 The circle $(x - 2)^2 + (y - 2.5)^2 = 25$ cuts the x-axis when $y = 0$. Find the coordinates of these 2 points. Similarly find the coordinates where the circle crosses the y-axis.

12 Why does the circle $(x - 2)^2 + (y - 2.5)^2 = 1$ not intersect either axis? Check your answer with an accurate plot of this circle. Use specimen values of x from $x = -3$ to $x = 3$ at intervals of 0.5.

Polar coordinates

This is an alternative format to the Cartesian coordinates we have used in earlier chapters. Remember that for Cartesian coordinates generally we use (x, y). We plot x horizontally and y vertically.

Polar coordinates use r and θ, written as (r, θ). r is the radial distance from the origin to the point (r, θ). The positive x-axis is replaced with the axis $\theta = 0°$ (or 0 radians). We measure positive θ from this axis in an anticlockwise direction. This implies that when θ is negative we move in a clockwise direction. We plot some specimen coordinates.

Examples 12.6

Fig. 12.6 shows some polar coordinates using degrees:
i) $(2, 30°)$,   ii) $(3.5, 115°)$,   iii) $(1.25, 240°)$,   iv) $(2, 390°)$.

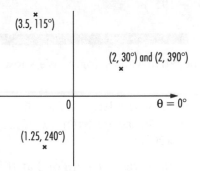

Fig. 12.6

Notice that $(2, 30°)$ and $(2, 390°)$ have the same position. $390°$ equals a complete revolution and $30°$.

Examples 12.7

Fig. 12.7 shows some polar coordinates using radians:

i) $\left(2, \dfrac{\pi}{4}\right)$,   ii) $\left(5, \dfrac{3\pi}{5}\right)$,   iii) $(2.5, 5.5)$,   iv) $(1, 3.6)$.

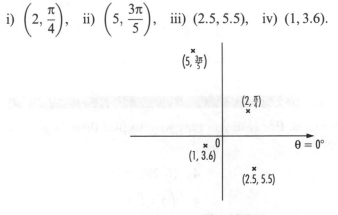

Fig. 12.7

■■■■■■ **Examples 12.8** ■■■■■■

In Fig. 12.8 we use negative values for θ in degrees:

i) (2, −60°),   ii) (3.5, −165°)

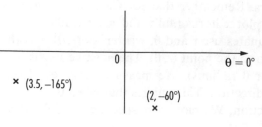

Fig. 12.8

Notice that (2, −60°) has the same position as (2, 300°). Also (3.5, −165°) has the same position as (3.5, 195°). These are because −60° = −(360° − 300°) and −165° = −(360° − 195°).

■■■■■■ **Examples 12.9** ■■■■■■

We can also have negative values for *r*. We show some coordinates in Fig. 12.9.

i)   For (−5, 60°) we start with a length of 5 at 60° anticlockwise from the axis θ = 0°. We interpret the −5 from the origin to be in the opposite direction, i.e. (5, −120°).

ii) For (−3, 105°) we start with a length of 3 at 105° anticlockwise from the axis θ = 0°. We interpret the −3 from the origin to be in the opposite direction, i.e. (3, −75°).

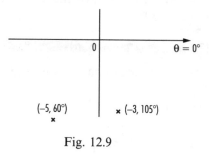

Fig. 12.9

■■■■■■ **EXERCISE 12.3** ■■■■■■

Plot the polar coordinates. For ease you may wish to plot them in groups of five to each set of axes.

1   (10, 45°)

2   (5, 150°)

3   (3, 200°)

4   (6, 270°)

5   $\left(7.5, \dfrac{\pi}{4}\right)$

6 $\left(1, \dfrac{2\pi}{3}\right)$

7 $\left(10, \dfrac{7\pi}{4}\right)$

8 $\left(2, \dfrac{11\pi}{6}\right)$

9 $\left(2, -\dfrac{\pi}{4}\right)$

10 $(1, -100°)$

11 $(5, -45°)$

12 $(4, -180°)$

13 $(3, -20°)$

14 $(6, -90°)$

15 $\left(7.5, -\dfrac{\pi}{4}\right)$

16 $\left(-4, \dfrac{2\pi}{3}\right)$

17 $\left(-12, \dfrac{7\pi}{6}\right)$

18 $\left(-2, \dfrac{11\pi}{4}\right)$

19 $(2, 450°)$

20 $(-5, -100°)$

Conversions between Cartesian and polar coordinates

In Fig. 12.10 we see that r and θ are related to x and y. These are based upon simple trigonometry and Pythagoras' theorem. We have drawn a right-angled triangle in the first (positive) quadrant.

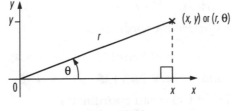

Fig. 12.10

We have some general rules for the conversions. However we must emphasise that the techniques are easier with the aid of a diagram.
When converting from **from polar to Cartesian** coordinates

$$x = r \cos \theta$$

and $\qquad y = r \sin \theta$

> Using trigonometry in the right-angled triangle.

When converting **from Cartesian to polar** coordinates

$$r = \sqrt{x^2 + y^2}$$

and $\quad \tan \theta = \dfrac{y}{x}$

> Pythagoras' theorem.
> Positive root is generally used.

We apply these conversion rules in the following examples. Afterwards we look at how simply the calculator does the conversions.

███████ **Examples 12.10** ████████████████████████████████████

With the aid of a diagram convert the following polar coordinates to Cartesian coordinates

i) $(6, 30°)$,   ii) $(9.5, 125°)$ iii) $(2.75, -50°)$,   iv) $(11, -149°)$.

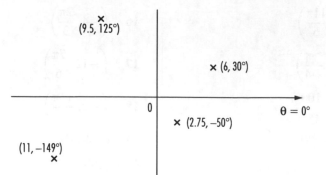

Fig. 12.11

The diagram shows the position of the 4 coordinates, one in each quadrant. For the conversions we apply the general rules. Remember that $x = r \cos \theta$ and $y = r \sin \theta$.

i) For $(6, 30°)$ $x = r \sin \theta$

        is $x = 6 \cos 30°$ $= 5.20$

            $y = r \sin \theta$

        is $y = 6 \sin 30°$ $= 3.00$

    i.e. $(5.20, 3)$ in Cartesian coordinates.

ii) For $(9.5, 125°)$ $x = 9.5 \cos 125°$ $= -5.45$

           and $y = 9.5 \sin 125°$ $= 7.78$

    i.e. $(-5.45, 7.78)$ in Cartesian coordinates.

iii) For $(2.75, -50°)$ $x = 2.75 \cos(-50°)$ $= 1.77$

           and $y = 2.75 \sin(-50°)$ $= -2.11$

    i.e. $(1.77, -2.11)$ in Cartesian coordinates.

iv) For $(11, -149°)$ $x = 11 \cos(-149°)$ $= -9.43$

           and $y = 11 \sin(-149°)$ $= -5.67$

    i.e. $(-9.43, -5.67)$ in Cartesian coordinates.

███████ **Examples 12.11** ████████████████████████████████████

With the aid of diagrams convert the following Cartesian coordinates to polar coordinates

i) $(3, 4)$,   ii) $(-5, 12)$,   iii) $(2, -1)$,   iv) $(-8, -6)$.

i) For (3, 4),

$$r = \sqrt{x^2 + y^2} \qquad \tan \theta = \frac{y}{x}$$

is $\quad r = \sqrt{3^2 + 4^2} \qquad$ is $\quad \tan \theta = \frac{4}{3}$

$\qquad = \sqrt{9 + 16} \qquad$ i.e. $\quad \theta = 53.13°$

$\qquad = \sqrt{25}$

i.e. $r = 5$

i.e. (5, 53.13°) in polar coordinates.

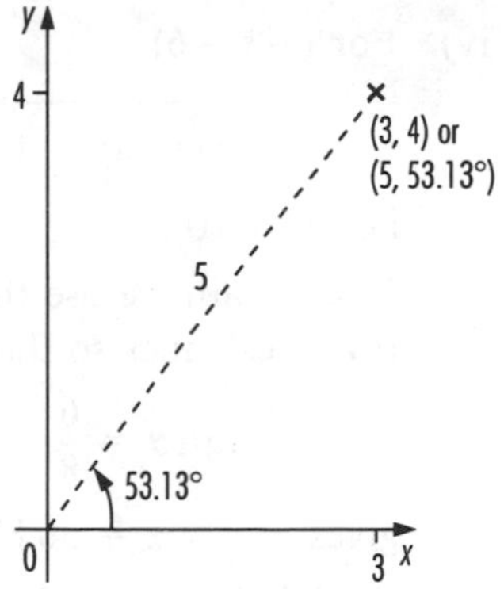

Fig. 12.12

ii) For (−5, 12),

$$r = \sqrt{(-5)^2 + 12^2}$$

i.e. $r = 13$

When we attempt to find θ we need to be careful. The calculator does *not* give an immediate answer. We use an acute angle, α, with the *sizes* of the opposite and adjacent sides,

i.e. $\qquad \tan \alpha = \dfrac{12}{5}$

gives $\qquad \alpha = 67.38°$.

This means that

$$\theta = 180° - \alpha$$

$$= 180° - 67.38°$$

$$= 112.62°$$

i.e. (13, 112.62°) in polar coordinates.

Fig. 12.13

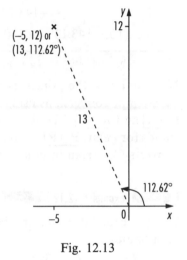

iii) For (2, −1),

$$r = \sqrt{2^2 + (-1)^2}$$

i.e. $r = 2.24$.

Again we use the *sizes* of the triangle's sides so that

$$\tan \alpha = \frac{1}{2}$$

gives $\qquad \alpha = 26.57°$.

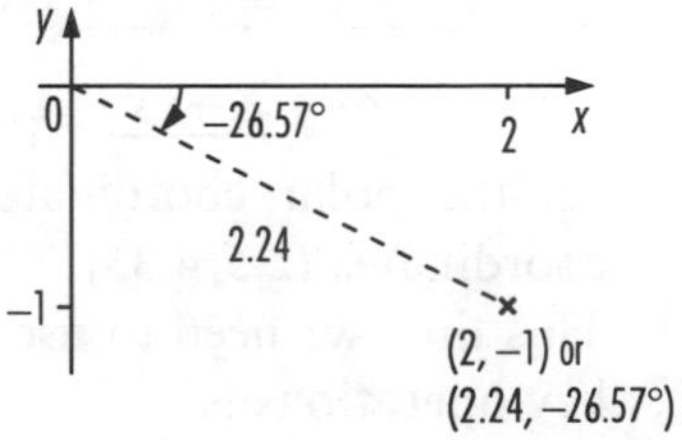

Fig. 12.14

We know this clockwise direction is negative, i.e. $\theta = -\alpha = -26.57°$, i.e. (2.24, −26.57°) or (2.24, 333.43°) in polar coordinates.

iv) For $(-8, -6)$

$$r = \sqrt{(-8)^2 + (-6)^2}$$

i.e. $r = 10$.

Once again we use the *sizes* of the triangles' sides so that

$$\tan \alpha = \frac{6}{8}$$

gives $\alpha = 36.87°$.

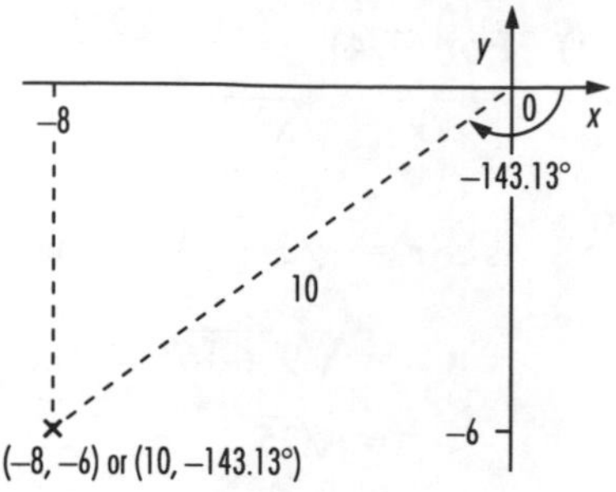

Fig. 12.15

We label θ in our diagram, remembering it is in a negative direction,

i.e. $\theta = -(180° - \alpha)$

$$= -(180° - 36.87°)$$

$$= -143.13°,$$

i.e. $(10, -143.13°)$ or $(10, 216.87°)$ in polar coordinates.

Now we know the principles behind the conversions. In this next section we attempt conversions using a calculator. The function keys we need are P→R| and R→P|. The **R** stands for **rectangular** (i.e. **Cartesian**) and the **P** stands for **polar**. P→R| converts polar to Cartesian coordinates and R→P| converts Cartesian to polar coordinates.

████████ **Examples 12.12** ████████████████████████████████

Using a calculator convert the following polar coordinates to Cartesian coordinates

i) $(5, 60°)$,   ii) $\left(7.4, \frac{3\pi}{4}\right)$ or $(7.4, 2.356 \ldots)$.

i)   First we must check that the calculator is in the degree mode. For $(5, 60°)$ our order of key operations is

5| P→R| 60| =| to display 2.5

followed by   X↔Y| to display 4.33;

i.e. the polar coordinates $(5, 60°)$ are the same as the Cartesian coordinates $(2.5, 4.33)$.

ii) This time we need to use radian mode. For $(7.4, 2.356)$ our order of key operations is

7.4| P→R| 2.356| =| to display -5.23

followed by   X↔Y| to display 5.23;

i.e. the polar coordinates $(7.4, 2.356)$ are the same as the Cartesian coordinates $(-5.23, 5.23)$.

████████ **Examples 12.13** ████████████████████

Using a calculator convert the following Cartesian coordinates to polar coordinates   i) $(3, 4)$,   ii) $(-5, -12)$.

In these examples we may use either degree or radian mode.

i) For $(3, 4)$ our order of key operations is

$\underline{3|}$ $\underline{R{\rightarrow}P|}$ $\underline{4|}$ $\underline{=|}$ to display 5

followed by $\underline{X{\leftrightarrow}Y|}$ to display 53.13°.

53.13° is 0.927 in radian mode,
i.e. the Cartesian coordinates $(3, 4)$ are the same as the polar coordinates $(5, 53.13°)$.

ii) For $(-5, -12)$ our order of key operations is

$\underline{5|}$ $\underline{+/-|}$ $\underline{R{\rightarrow}P|}$ $\underline{12|}$ $\underline{+/-|}$ $\underline{=|}$ to display 13

followed by $\underline{X{\leftrightarrow}Y|}$ to display 112.62°.

−112.62° is −1.966 in radian mode,
i.e. the Cartesian coordinates $(-5, -12)$ are the same as the polar coordinates $(13, -112.62°)$.

████████ **EXERCISE 12.4** ████████████████████

Convert the following polar coordinates to Cartesian coordinates.

1 $(1.8, 13°)$ **4** $(11.4, -135°)$

2 $(4.5, 117°)$ **5** $(2.7, -90°)$

3 $(2.5, -76°)$

Convert the following Cartesian coordinates to polar coordinates.

6 $(5, 2)$ **9** $(-3, -4)$

7 $(6, -7)$ **10** $(-15, -15)$

8 $(-1, 3)$

Polar graphs

Our Cartesian format of x and y generally is replaced by r and θ. θ is the independent variable and r is the dependent variable. We plot r against θ. Remember the x axis is replaced by the axis $\theta = 0$. The values of θ, like the spokes in a wheel, are referred to this axis. Along each 'spoke' we measure the length, r.

▬▬▬▬▬ **Examples 12.14** ▬▬▬▬▬

Between $\theta = 0°$ and $360°$ (i.e. 0 and 2π radians) we plot the polar graphs
i) $r = 2\sin\theta$,   ii) $r = 3\sin\theta$.

We show the table of values built up in the usual way. Here are some specimen calculations.

Say $\theta = 60°$ for the curve $r = 2\sin\theta$. Our order of key operations is

60 | sin | to display 0.866 . . .

×| 2| =| to display 1.732.

These coordinates for (r, θ) in the table are $(1.732, 60°)$.
Say $\theta = 225°$ for the curve $r = 3\sin\theta$. Our order of key operations is

225| sin| to display -0.707 . . .

×| 3| =| to display -2.121.

These coordinates for (r, θ) would be $(-2.121, 225°)$.

$\theta°$	0	30	60	90	120	150	180	210
$\sin\theta$	0	0.500	0.866	1.000	0.866	0.500	0	-0.500
$r = 2\sin\theta$	0	1.000	1.732	2.000	1.732	1.000	0	-1.000
$r = 3\sin\theta$	0	1.500	2.598	3.000	2.598	1.500	0	-1.500

	240	270	300	330	360
	-0.866	-1.000	-0.866	-0.500	0
	-1.732	-2.000	-1.732	-1.000	0
	-2.598	-3.000	-2.598	-1.500	0

If we continue our table with another revolution to 720° we trace over the graph exactly, i.e. another cycle. In Fig. 12.16 we show both graphs.

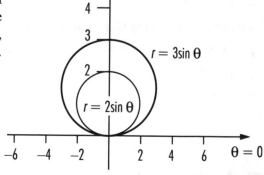

▬▬▬▬▬ **Examples 12.15** ▬▬▬▬▬

Between $\theta = 0°$ and $360°$ (i.e. 0 and 2π radians) we plot the polar graphs
i) $r = \cos 3\theta$,   ii) $r = \cos 4\theta$

We omit the table of values, leaving it as an exercise to be built up in the usual way. However here are some specimen calculations.

Say $\theta = 15°$ for the curve $r = \cos 3\theta$. Our order of key operations is

$\underline{15}$ $\underline{\times}$ $\underline{3}$ $\underline{=}$ to display 45

$\underline{\cos}$ to display 0.707...

i.e. these coordinates are $(0.707, 15°)$.

Say $\theta = 210°$ for the curve $r = \cos 4\theta$. Our order of key operations is

$\underline{210}$ $\underline{\times}$ $\underline{4}$ $\underline{=}$ to display 840

$\underline{\cos}$ to display -0.5.

i.e. these coordinates are $(-0.5, 210°)$.

We show both graphs in Fig. 12.17.

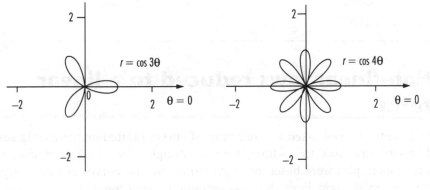

Fig. 12.17

Example 12.16

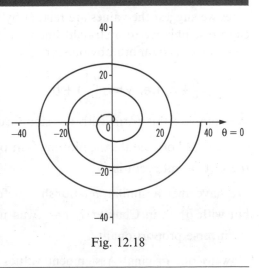

Between $\theta = 0$ and 2π radians we plot the polar graph $r = 2\theta$. Notice we use radians. A length for r in degrees has no meaning. According to the equation $r = 2\theta$ we simply multiply each value of θ by 2 to obtain the value for r. Again we omit the table. In this example the graph is a never ending spiral. We show the first 3 revolutions for the graph in Fig. 12.18.

Fig. 12.18

In Exercise 12.5 we have a selection of polar graphs for you to try.

▬▬▬ EXERCISE 12.5 ▬▬▬▬▬▬▬▬

For each question construct a table of values from $0°$ to $360°$ (or 0 to 2π radians) and then plot your graph. You need to obtain each graph as a smooth curve. In some cases you may wish to plot at frequent intervals rather than the $30°$ we demonstrated.

1 $r = 2\cos\theta$

2 $r = \cos 2\theta$

3 $r = 5\sin\theta$

4 $r = 4\cos\theta$

5 $r = 3\cos 2\theta$

6 $r = 5\sin 2\theta$

7 $r = 2\sin^2\theta$

8 $r = \cos^2\theta$

9 $r = 4\cos^2\theta$

10 $r = 2\sin^2 2\theta$

11 $r = 5(1 + \cos\theta)$

12 $r = 3(1 + 2\cos\theta)$

13 $r = 1 + 2\cos\theta$

14 $r = 2 + \sin\theta$

15 $r = 2 + 4\sin\theta$

Non-linear laws reduced to a linear form

In Chapter 11 we looked at a selection of curves plotted on the usual axes. Then we saw how to re-draw them as straight lines on amended axes. Those examples were based on logarithms. Not all curves can be simply reduced to straight lines. Not all reductions need logarithms. For each trial we are going to test our assignment table of values against a possible straight line type relationship. The values may all lie on a straight line. Alternatively we may be able to draw a line of best fit. In either of these cases we suggest the values are related by the particular law. If there is no close resemblance to a straight line then that particular law fails. Let us test a few of the more obvious types.

1. $y = \dfrac{a}{x} + b$, i.e. $y = a\left(\dfrac{1}{x}\right) + b$

We compare this form with the standard straight line $Y = MX + C$. We change the horizontal axis, plotting $\dfrac{1}{x}$ in place of X. The gradient is a and the vertical intercept is b.

We have met a similar relationship before, but with $b = 0$, in Chapter 2. $y = \dfrac{a}{x}$ was used for inverse proportionality.

Now to our original Assignment values we add another row in the table for $\dfrac{1}{x}$.

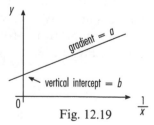

Fig. 12.19

x	0.10	0.15	0.40	0.90	1.35	1.74	2.48	3.06	3.88	
$\dfrac{1}{x}$	10.00	6.67	2.50	1.11	0.74	0.57	0.40	0.33	0.26	↔
y	2.34	1.47	−0.03	−0.82	−1.11	−1.26	−1.45	−1.54	−1.64	↕

Fig. 12.20 shows our plot of y against $\dfrac{1}{x}$ with no obvious straight line. This means that our values are *not* related by $y = \dfrac{a}{x} + b.$

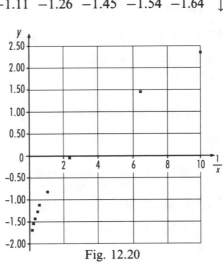

Fig. 12.20

2. $y = ax^2 + b$

We compare this form with the standard straight line $Y = MX + C$. We change the horizontal axis, plotting x^2 in place of X. The gradient is a and the vertical intercept is b (Fig. 12.21).

To our original Assignment values we add another row in the table for x^2.

Fig. 12.21

x	0.10	0.15	0.40	0.90	1.35	1.74	2.48	3.06	3.88	
x^2	0.01	0.02	0.16	0.81	1.82	3.03	6.15	9.36	15.05	↔
y	2.34	1.47	−0.03	−0.82	−1.11	−1.26	−1.45	−1.54	−1.64	↕

Fig. 12.22 shows our plot of y against x^2 with no obvious straight line. This means that our values are *not* related by $y = ax^2 + b.$

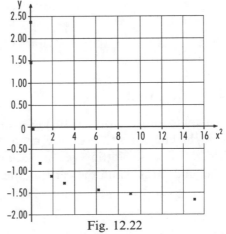

Fig. 12.22

3. $y = \dfrac{a}{x^2} + b$, i.e. $y = a\left(\dfrac{1}{x^2}\right) + b$

We compare this form with the standard straight line $Y = MX + C$. We change the horizontal axis, plotting $\dfrac{1}{x^2}$ in place of X. The gradient is a and the vertical intercept is b (Fig 12.23).

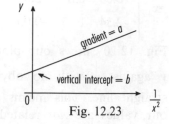

Fig. 12.23

To our original Assignment values we add another row in the table for $\dfrac{1}{x^2}$.

x	0.10	0.15	0.40	0.90	1.35	1.74	2.48	3.06	3.88	
$\dfrac{1}{x^2}$	100.00	44.44	6.25	1.23	0.55	0.33	0.16	0.11	0.07	↔
y		2.34	1.47	−0.03	−0.82	−1.11	−1.26	−1.45	−1.54	−1.64 ↕

Fig. 12.24 shows our plot of y against $\dfrac{1}{x^2}$ with no obvious straight line. This means that our values are *not* related by $y = \dfrac{a}{x^2} + b$.

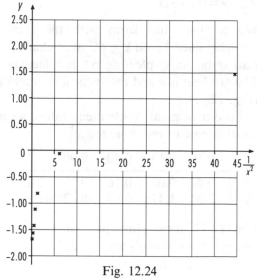

Fig. 12.24

4. $y = a\sqrt{x} + b$

We compare this form with the standard straight line $Y = MX + C$. We change the horizontal axis, plotting $\sqrt{x}$ in place of X. The gradient is a and the vertical intercept is b (Fig 12.25).

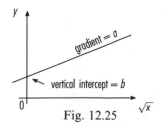

Fig. 12.25

To our original Assignment values we add another row in the table for $\sqrt{x}$.

x	0.10	0.15	0.40	0.90	1.35	1.74	2.48	3.06	3.88	
$\sqrt{x}$	0.32	0.39	0.63	0.95	1.16	1.32	1.57	1.75	1.97	$\leftrightarrow$
y	2.34	1.47	−0.03	−0.82	−1.11	−1.26	−1.45	−1.54	−1.64	$\updownarrow$

Fig. 12.26 shows our plot of y against $\sqrt{x}$ with no obvious straight line. This means that our values are *not* related by $y = a\sqrt{x} + b$.

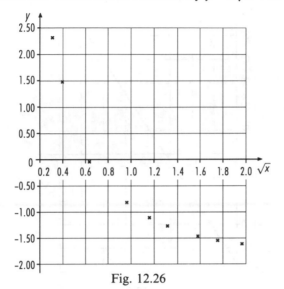

Fig. 12.26

5. $y = \dfrac{a}{\sqrt{x}} + b$, i.e. $y = a\left(\dfrac{1}{\sqrt{x}}\right) + b$

We compare this form with the standard straight line $Y = MX + C$. We change the horizontal axis, plotting $\dfrac{1}{\sqrt{x}}$ in place of X. The gradient is a and the vertical intercept is b (Fig. 12.27).

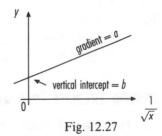

Fig. 12.27

To our original Assignment values we add another row in the table for $\dfrac{1}{\sqrt{x}}$.

x	0.10	0.15	0.40	0.90	1.35	1.74	2.48	3.06	3.88	
$\dfrac{1}{\sqrt{x}}$	3.16	2.58	1.58	1.05	0.86	0.76	0.64	0.57	0.51	$\leftrightarrow$
y	2.34	1.47	−0.03	−0.82	−1.11	−1.26	−1.45	−1.54	−1.64	$\updownarrow$

Fig. 12.28 shows our plot of y against $\dfrac{1}{\sqrt{x}}$. Immediately we see we can draw a straight line of best fit. All the values are very close to our line. This indicates that our values *are* related by $y = \dfrac{a}{\sqrt{x}} + b$. We can read off the vertical intercept, b, as -2.40. In the usual way we can draw a right-angled triangle and calculate the gradient, a, to be 1.5. These values mean we can write our complete relation as $y = \dfrac{1.5}{\sqrt{x}} - 2.4$.

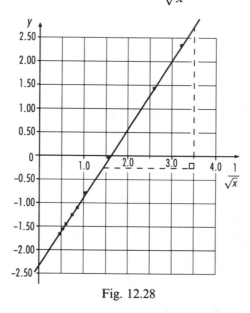

Fig. 12.28

Now the final exercise will give you some practice at finding various relationships. In each case the relation is suggested. Firstly all you need to do is calculate the extra table row of values. Then you need to plot the graph to confirm the relation. Finally you should be able to determine the values for gradient and vertical intercept from your graph.

■■■■ EXERCISE 12.6 ■■■■■■■■■■■■■■■■■■■

1   Performance trials for a medium size 5 door saloon car recorded the following results. They were from a standing start, relating distance travelled, s m, to time, t s.

t (s)	1	2	4	6	8	10
s (m)	1.45	5.95	24.00	54.05	96.25	150.00

It is thought that these values are related by $s = mt^2$. Using a straight line graph check that this is true. From your graph find the value of m.

2 The time period, Ts, for a simple pendulum is related to the length of the pendulum, lm. It is thought that $T=k\sqrt{l}+c$. Using the test results draw a straight line graph to show that this is true for $c=0$.

l (m)	0.950	0.960	0.980	0.990	0.995	1.000
T (s)	1.949	1.962	1.977	1.989	1.995	2.002

From your graph find the value of k correct to 2 decimal places.

3 Tests have been carried out in the treatment of a bacterial infection. The number of bacteria recorded is B ($\times10^3$) during time t hours. They are probably related by $B=\dfrac{a}{\sqrt{t}}+b$ for the new ointment. Using the table of data test whether this relationship is true within experimental limits.

t	12	18	24	60	120	180
B ($\times10^3$)	500	380	310	140	50	10

What are the values of a and b?

4 The rate of heat energy transfer for a piece of brass is Q. It is thought to be related to the thickness of the brass, x, by $Q=\dfrac{m}{x}+c$. Using the following results and a straight line graph test whether this relationship is true. From your graph find values for m and c.

x (m)	0.050	0.075	0.100	0.150	0.200	0.250
Q (W)	25 000	17 000	12 500	8 500	6 500	5 000

5 One particular machine in an engineering workshop costs £C to lease each week. The costs are related to the number of hours/week, t, that the machine is worked. It is thought that $C=at^3+b$. Using the data from the first six weeks plot a straight line graph. From your graph find the values of a and b.

t (hour)	35.0	37.5	40.0	42.0	46.0	50.0
C (£)	2600	3100	3700	4200	5400	6800

6 We know the power, P, across a resistor, R, is related to the current, I, by $P=RI^2$. Use the following test results to draw a straight line graph.

I (A)	1.20	1.25	1.40	1.50	1.60	1.75
P (W)	4.30	4.70	5.90	6.75	7.70	9.20

From your graph check that this resistor is approximately $3\,\Omega$.

7 For a gas, pressure, p, is thought to be inversely proportional to volume, v. We may write this as $p = \dfrac{k}{v}$. Using the table of results draw a straight line graph.

v	10.0	12.5	15.0	18.0	20.0	25.0
p	200	160	130	110	100	80

From your graph find the value of k.

8 μ is the coefficient of friction between a pulley and a belt. $v\,\text{ms}^{-1}$ is the velocity of the pulley. It is thought that they are related according to $\mu = a\sqrt{v} + b$. Use the table of results and an appropriate straight line graph to check this.

μ	0.223	0.249	0.258	0.273	0.286	0.301
$v\ (\text{ms}^{-1})$	2.5	5.0	6.0	8.0	10.0	12.5

From your graph what are the values of a and b?

9 For a van the distance, $s\,\text{m}$, it travels is related to its velocity, $v\,\text{ms}^{-1}$, by $s = mv^2 + c$. Using the available data and a straight line graph check that this is true.

$v\ (\text{ms}^{-1})$	4	6	8	10	15	20
$s\ (\text{m})$	5.45	15.00	28.50	45.75	106.00	189.75

Find values for m and c from your graph.

10 The second moment of area, I, about xy is related to the outer diameter, D, of the ring by $I = aD^4 + b$. Using a straight line graph based on the table of results find values for a and b.

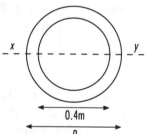

D	0.50	0.55	0.60	0.65	0.70	0.75	0.80
$I\ (\times 10^{-3})$	1.81	3.24	5.11	7.51	10.53	14.27	18.85

▄▄▄▄ MULTI-CHOICE TEST 12 ▄▄▄▄

1 The number of straight lines that may be drawn through 2 given points is:
A) none
B) one
C) two
D) three or more

2   An example of a zero gradient and a negative gradient is:

A)   B)   C)   D)

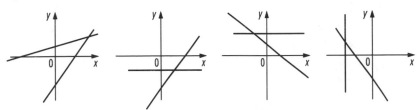

Questions **3** and **4** refer to the following information.
Experimental values of x and y are thought to satisfy the equation $y = ax^2 + b$. a and b are constants.

3   To obtain a straight line graph you use the axes:

A)   B)   C)   D)

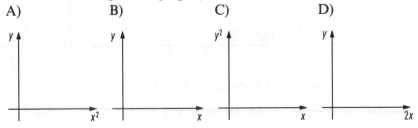

4   Decide whether each of the statements is True (T) or False (F).
   i)   The gradient is a^2
   ii) The intercept on the vertical axis is b.
   A) i) T   ii)   T
   B) i) T   ii)   F
   C) i) F   ii)   T
   D) i) F   ii)   F

Questions **5, 6** and **7** refer to the following information.
μ is the coefficient of friction between a pulley and a belt. v (ms^{-1}) is the velocity of the pulley. It is believed μ and v are connected by the law $\mu = a\sqrt{v} + b$.

5   To obtain a straight line graph you need to plot:
   A) μ against $\sqrt{v}$
   B) $\sqrt{v}$ against μ
   C) μ^2 against v
   D) $a\sqrt{v}$ against μ

6   Given $a = 0.04$, $b = 0.15$ and $\mu = 0.25$ the value of v, correct to 2 decimal places, is:
   A) 0.06
   B) 1.58
   C) 6.25
   D) 100

7 If the graph passes through the origin $b=0$ and the law may be re-written:

A) $\mu = a\sqrt{v}$

B) $\mu^2 = av$

C) $\mu = a^2 v$

D) $\sqrt{\dfrac{\mu}{a}} = v$

8 $y = 3 - \dfrac{2}{\sqrt{x}}$ is plotted as a straight line on the axes:

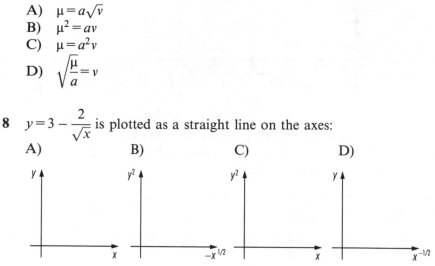

A)

B)

C)

D)

9 $B \ (\times 10^3)$ represents bacteria and t (hours) represents time in the relation $B = \dfrac{2500}{t^{1/2}} - 169$. In order the gradient and intercept on the vertical axis are:

A)   50 and 13

B)   2500 and -169

C)   2500 and 169

D)   6 250 000 and 28 561

10   The graph is the result of plotting y against $\sqrt{x}$. The law connecting x and y is:

A) $y = 3\sqrt{x}$

B) $y = \dfrac{1}{3}x$

C) $y = \sqrt{x} + 1$

D) $y^2 = 3x + 1$

11   The graph of $y^2 = x$ is shown by:

A)

B)

C)

D)

12 $y^2 = x$ is a standard parabola. From this graph we obtain the graph of
$(y-2)^2 = x + 1$.
Decide whether each of the statements is True (T) or False (F).
i) We shift 2 units vertically upwards.
ii) We shift 1 unit horizontally to the left.
A) i) T   ii)   T
B) i) T   ii)   F
C) i) F   ii)   T
D) i) F   ii)   F

13 An example of an equation of a circle is:
A) $x^2 = 9 + y^2$
B) $2x^2 - 2y^2 = 15$
C) $(x+2)^2 + (y-1)^2 = 7$
D) $x^2 + y^2 + 2 = 0$

14 The standard labelling of axes for polar coordinates is:
A)   B)   C)   D)

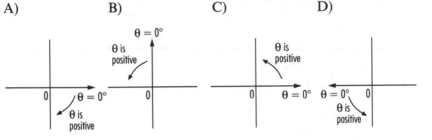

15 Polar coordinates are (r, θ).
Decide whether each of the statements is True (T) or False (F).
i) r is the independent variable.
ii) θ is measured from the horizontal.
A) i) T   ii)   T
B) i) T   ii)   F
C) i) F   ii)   T
D) i) F   ii)   F

16 In polar coordinates, (r, θ), r is found using:
A) $\pm\sqrt{x^2 - y^2}$
B) $\tan^{-1}\dfrac{y}{x}$
C) the distance OP
D) the angle PON

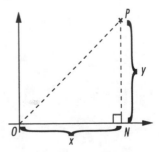

17 When converting polar coordinates to Cartesian coordinates, in order (x, y) is:
   A) $(r \cos \theta, r \sin \theta)$
   B) $\left(\sqrt{r^2 + \theta^2},\ \tan^{-1} \dfrac{\theta}{r} \right)$
   C) $(r \sin \theta, r \cos \theta)$
   D) $\left(\tan^{-1} \dfrac{r}{\theta},\ \sqrt{r^2 - \theta^2} \right)$

18 The Cartesian coordinates $(-4, 0)$ converted to polar coordinates are:
   A) $(-4, 90°)$
   B) $(-4, 180°)$
   C) $(4, 0°)$
   D) $(4, 180°)$

19 The polar coordinates $(1, -120°)$ converted to Cartesian coordinates are:
   A) $(-0.866, -0.5)$
   B) $(-0.866, 0.5)$
   C) $(-0.5, -0.866)$
   D) $(-0.5, 0.866)$

20 The polar equation of the curve is:
   A) $r = 2 \sin \theta$
   B) $r = 4 \cos \theta$
   C) $4r = \sin \theta$
   D) $r = 4 \sin \theta$

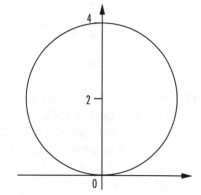

Element: Use calculus to solve engineering problems.

PERFORMANCE CRITERIA
- Functions appropriate to the engineering problem are identified.
- Functions are differentiated.
- Numerical values are substituted and specified values relevant to the engineering problem are obtained.

RANGE
Engineering problems: electrical, electronic; mechanical.
Functions: algebraic, trigonometrical, exponential, logarithmic.
Specified values: rates of change, gradient, maximum value, minimum value.

Element: Use graphs to solve engineering problems.

PERFORMANCE CRITERIA
- Coordinates and scales appropriate to the engineering problem are selected.
- Data are accurately plotted.
- The relationship between variables is identified.
- Values are accurately determined from graphs, and engineering problems are solved.

RANGE
Engineering problems: electrical, electronic; mechanical.
Coordinates: Cartesian.
Scale: linear, logarithmic.
Relationship: linear, parabolic, exponential, logarithmic, sinusoidal.
Values: two variables, gradient.

Introduction

In this chapter we are going to look at change. Many things change continually, though some do remain constant. On page 45 in Chapter 2 we looked at change when we calculated the gradients of straight line graphs. You will remember the gradient is the same all along a straight line. Later in this volume we looked at various curves where the gradient changed along the curve. Mathematics has a set of rules for these changes. We will briefly introduce these rules now, and extend them in Volume 2.

![ASSIGNMENT]

The Assignment for this chapter links together those for Chapters 1 and 2. We have a vehicle travelling at a speed of $30 \, \text{ms}^{-1}$ when the brakes are applied to produce a deceleration of $4 \, \text{ms}^{-2}$.

Allowing for a minor measurement error we have the distance travelled given by

> Deceleration and retardation, being negative acceleration, mean a slowing down.

$$s = 1 + 30t - 2t^2$$

where s is the distance travelled

and t is the time from when the brakes are applied.

Also the speed, v, is linked to time, t, by the formula $v = 30 - 4t$.

The gradient of a graph

When we looked at a gradient before we considered the change in y due to a change in x. Remember x is the independent variable and y is the dependent variable. This means the change in x automatically causes a change in y. For many things this change takes place continually, though obviously a constant **never changes**. We may wish to look at an average change or an instantaneous change, as in this first example.

![Example 13.1]

Let us look at the graph of $y = 3x^2$.

First we can look at the change over a portion of the curve. We may start anywhere on the graph and either increase or decrease the x value. Suppose we start at the point $(1, 3)$ and move along the curve increasing the x values. Automatically we change the y values, which in this case also increase. We draw this in Fig. 13.1, stopping at the point $(4, 48)$. From our start to finish the curve is getting steeper. We can look at the average

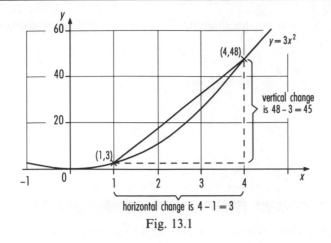

Fig. 13.1

change by drawing a straight line from $(1, 3)$ to $(4, 48)$. This is a **chord**. We know from Chapter 2 that

$$\text{gradient} = \frac{\text{vertical change}}{\text{horizontal change}}$$

so that gradient $= \dfrac{48 - 3}{4 - 1}$

$$= \frac{45}{3}$$

$$= 15.$$

This gradient calculation only tells us about the **average value**. It tells us nothing about what happens to the gradient of this portion of the curve at particular points.

Alternatively we may find the change at a specific, instantaneous, point on the curve. In this case we draw a tangent to the curve at that point. In Fig. 13.2 we are interested in the gradient of the curve $y = 3x^2$ at the points where $x = 2.5$ and $x = -2$.

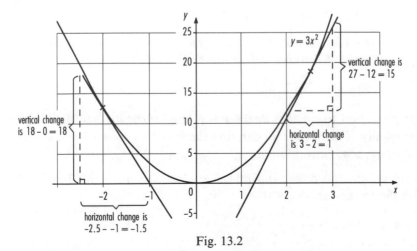

Fig. 13.2

Figure 13.2 shows the tangents together with their associated right-angled triangles.

Using gradient $= \dfrac{\text{vertical change}}{\text{horizontal change}}$

at $x = -2$ at $x = 2.5$

gradient $= \dfrac{18}{-1.5}$ gradient $= \dfrac{15}{1}$

$= -12.$ $= 15.$

ASSIGNMENT

We have used the techniques we learned in Chapter 1 to construct a table of $s = 1 + 30t - 2t^2$ and plot its graph. The table from that chapter is not repeated but the graph is given in Fig. 13.3.

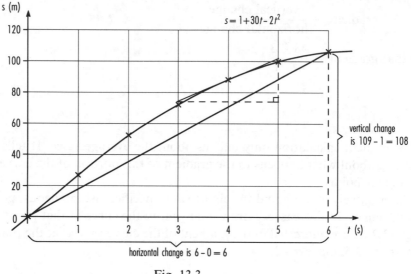

Fig. 13.3

Distance travelled, s, and time, t, are associated with speed, v. Speed is the rate of change of distance with respect to time,

i.e.   Speed $= \dfrac{\text{Distance}}{\text{Time}}.$

It is important to see that **distance is plotted vertically and time is plotted horizontally**. This means we can find the speed by looking at the gradient of the curve we have plotted.

By drawing a chord we can find the average speed for the first 6 seconds after the brakes are applied.

Using gradient $= \dfrac{\text{vertical change}}{\text{horizontal change}}$

i.e. $\qquad$ speed $= \dfrac{\text{distance travelled}}{\text{time taken}}$

we get $\qquad$ speed $= \dfrac{108 \text{ m}}{6 \text{ s}}$

$\qquad\qquad\qquad = 18 \text{ ms}^{-1}$.

Also we can find the speed at a specific time, say 4 seconds after the brakes are applied. We draw a tangent at the point on the curve where $t = 4$. Using the usual gradient formula we get

$$\text{speed} = \frac{103 - 75}{5 - 3}$$

$$= \frac{28}{2}$$

$$= 14 \text{ ms}^{-1}.$$

This speed is correct only after 4 seconds because we drew the tangent only at this point.

■ EXERCISE 13.1 ■

1 Plot the graph of $y = x^2 - 2x$. In your table use values of x from $x = -1$ to $x = 5$ at intervals of 1.

From your graph find the average change of y over this range. Also draw a tangent to your graph at the point where $x = 3.25$. What is the gradient at this point?

2 From $x = -2.0$ to $x = 2.0$ at intervals of 0.5 construct a table of values for $\quad$ i) $y = 2e^x$

and $\quad$ ii) $y = 2e^{-x}$.

On one set of axes plot these graphs. For each graph find the average change in y between $x = -0.5$ and $x = 0.5$.

Also find the gradient at the point where $x = 0.2$ in each case.

Do you notice any similarities/differences between your answers for each pair of calculations?

3 For values of x from $x = 0$ to $x = \pi$ radians plot the graph of $y = \sin x$. At $x = \frac{1}{2}\pi$ what is the gradient of this curve?

Find the average gradients from $x =$ $\quad$ i) 0 to $\frac{1}{2}\pi$,

$\qquad\qquad\qquad\qquad\qquad$ ii) $\frac{1}{2}\pi$ to π,

$\qquad\qquad\qquad$ and $\quad$ iii) 0 to π.

4 A rigid beam of length 10 m carries a uniformly distributed total load of 10 000 N. The bending moment, M, is related to its distance from one end of the beam, x, according to

$$M = -10\,000 + 5000x - 500x^2.$$

What is the bending moment at either end of the beam? What is the difference between the values?

5   Newton's law of cooling relates temperature of a body to its surroundings over a period of time, t seconds. If the body is hotter than its surrounding environment then its temperature will decrease over time. Let $T°C$ be the difference in temperature between a body and its environment. Suppose that the relation is $T = 400e^{-t/1500}$. Find the average temperature gradient over the first 10 minutes without plotting the graph.

The graphical methods for finding gradients take too much time. Drawing an accurate tangent can prove to be difficult. To overcome this problem we derive a method that avoids the use of graphs. It is **differentiation from first principles**.

Differentiation from first principles

In this section we establish the idea behind differentiation. Once you have seen what is involved all you need to do is obey the rules.

Fig. 13.4 shows a curve, any curve. Think of it as a magnification under a microscope with A and B *very* close together. Let A be a general point (x, y). Now consider a small change in x. This small change is δx where δ is the Greek letter delta. δx is one complete symbol. The original x together with the small change, δx, means we have moved to a new point with the horizontal value $x + \delta x$.

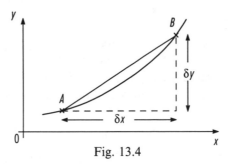

Fig. 13.4

Because y depends on x this causes a small change in y. This change is δy where δy is again one complete symbol. The original y together with the small change, δy, means the new point we mentioned before has a vertical value of $y + \delta y$. Let B be the point $(x + \delta x, y + \delta y)$. In Fig. 13.4 we have labelled A and B and the horizontal and vertical changes.

Using   gradient $= \dfrac{\text{vertical change}}{\text{horizontal change}}$

we have the gradient of the chord AB to be $\dfrac{\delta y}{\delta x}$.

If A is fixed and we move B closer and closer to A the chord gets nearer to the curve. Eventually, as B reaches A, the chord resembles a tangent to the curve at A. We show this in Figs. 13.5. As B approaches A the changes, δx and δy, get smaller and smaller. We write that δx tends to zero, written as $\delta x \to 0$. We are interested in the limit of δy as $\dfrac{\delta y}{\delta x} \to 0$.

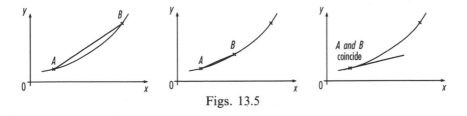

Figs. 13.5

$\dfrac{dy}{dx}$ (spoken as "*dy* by *dx*") represents the gradient of the curve at a general point. This gives us $\dfrac{dy}{dx} = \underset{\delta x \to 0}{\text{Lim}} \dfrac{\delta y}{\delta x}$.

It means that the gradient at a point is the limiting value (Lim) as δx tends to 0 ($\delta x \to 0$) of $\dfrac{\delta y}{\delta x}$.

$\dfrac{dy}{dx}$ has various interpretations:

i)   the change in y due to the change in x,
ii) the differentiation of y with respect to x,
iii) the first derivative,
iv) the gradient,
v) $\tan \alpha$ where α is the angle of inclination of the tangent to the horizontal.

In the next example we apply this new technique of differentiation from first principles.

▬▬▬ **Examples 13.2** ▬▬▬

Differentiate from first principles   i) $y = 3x^2$

and   ii) $y = \dfrac{5}{4x}$.

i)   Let us look at a general point (x, y) on the curve. Now consider a small change in x, δx, to a new point on the curve $x + \delta x$. In turn this causes a small change in y, δy, to $y + \delta y$. Because the original point (x, y) and the new point $(x + \delta x, y + \delta y)$ lie on the curve they both satisfy the equation of that curve.

We write $y + \delta y = 3(x + \delta x)^2$

and $y\ \ \ \ = 3x^2.$ | The original equation. |

We have 4 unknowns in this pair of equations; x, y, δx and δy. By subtraction we can attempt to eliminate some of them,

i.e. $\delta y = 3(x + \delta x)^2 - 3x^2$

$= 3((x + \delta x)^2 - x^2)$

$= 3(x^2 + 2x(\delta x) + (\delta x)^2 - x^2)$ | Expanding $(x + \delta x)^2$. |

$= 3(2x(\delta x) + (\delta x)^2)$

i.e. $\delta y = 3(2x + \delta x)(\delta x)$ | Removing a common factor of δx. |

so that $\dfrac{\delta y}{\delta x} = 3(2x + \delta x).$ | Dividing by δx. |

Using $\dfrac{dy}{dx} = \underset{\delta x \to 0}{\mathrm{Lim}} \dfrac{\delta y}{\delta x}$

we see that $2x + \delta x \to 2x$ | Because $\delta x \to 0$. |

to give $\dfrac{dy}{dx} = 3\,(2x) = 6x$

i.e. the gradient of the curve $y = 3x^2$ is always $6x$.

ii)   Let us look at a general point (x, y) on the curve. Now consider a small change in x, δx, to a new point on the curve $x + \delta x$. In turn this causes a small change in y, δy, to $y + \delta y$. Because the original point (x, y) and the new point $(x + \delta x, y + \delta y)$ lie on the curve they both satisfy the equation of that curve.

We write $y + \delta y = \dfrac{5}{4(x + \delta x)}$

and $y\ \ \ \ = \dfrac{5}{4x}.$ | The original equation. |

We attempt to eliminate some of the variables by subtraction

i.e. $\delta y = \dfrac{5}{4(x + \delta x)} - \dfrac{5}{4x}$

$= \dfrac{5}{4}\left(\dfrac{1}{x + \delta x} - \dfrac{1}{x}\right)$

$= \dfrac{5}{4}\left(\dfrac{x - (x + \delta x)}{(x + \delta x)x}\right)$

i.e. $\delta y = \dfrac{-5\delta x}{4(x + \delta x)x}$

so that $\dfrac{\delta y}{\delta x} = \dfrac{-5}{4(x + \delta x)\,x}.$ | Dividing by δx. |

Using $\qquad \dfrac{dy}{dx} = \underset{\delta x \to 0}{\text{Lim}} \dfrac{\delta y}{\delta x}$ $\qquad$ Because $\delta x \to 0$.

we see that $x + \delta x \to x$

to give $\qquad \dfrac{dy}{dx} = \dfrac{-5}{4\,(x)\,x} = \dfrac{-5}{4x^2}$

i.e. the gradient of the curve $y = \dfrac{5}{4x}$ is always $\dfrac{-5}{4x^2}$.

We can add/subtract gradients according to the usual rules of algebra. Also we can multiply them by a scalar.

▰▰▰▰ **Examples 13.3** ▰▰▰▰

We use our two previous results to demonstrate some algebraic variations.

For $y = 3x^2$, $\dfrac{dy}{dx} = 6x$ and for $y = \dfrac{5}{4x}$, $\dfrac{dy}{dx} = \dfrac{-5}{4x^2}$.

i) If $y = \dfrac{5}{4x} + 3x^2$ then $\dfrac{dy}{dx} = \dfrac{-5}{4x^2} + 6x$.

ii) If $y = 3x^2 - \dfrac{5}{4x}$ then $\dfrac{dy}{dx} = 6x + \dfrac{5}{4x^2}$. $\qquad$ $(-)(-) = +.$

iii) If $y = 21x^2$, i.e. $7\,(3x^2)$, then $\dfrac{dy}{dx} = 42x$, i.e. $7\,(6x)$.

▰▰▰▰ **ASSIGNMENT** ▰▰▰▰

We can move on from our graphical attempts to differentiate $s = 1 + 30t - 2t^2$ from first principles.

Let us look at a general point (t, s) on the curve. Now consider a small change in t, δt, to a new point on the curve $t + \delta t$. In turn this causes a small change in s, δs, to $s + \delta s$. Because the original point (t, s) and the new point $(t + \delta t, s + \delta s)$ lie on the curve they both satisfy the equation of that curve.

We write $\qquad s + \delta s = 1 + 30(t + \delta t) - 2(t + \delta t)^2$

and $\qquad\quad s \qquad = 1 + 30t \qquad\quad - 2t^2.$

We subtract these equations to eliminate some of these variables,

i.e. $\qquad\qquad \delta s = 30\,(t + \delta t - t) - 2(t^2 + 2t\,(\delta t) + (\delta t)^2 - t^2)$

$\qquad\qquad\qquad = 30\delta t - 2(2t\,(\delta t) + (\delta t)^2)$

$\qquad\qquad\qquad = 30\delta t - 4t\,(\delta t) - 2(\delta t)^2$

i.e. $\qquad\qquad \delta s = (30 - 4t - 2\,(\delta t))\,(\delta t)$

so that $\qquad\quad \dfrac{\delta s}{\delta t} = 30 - 4t - 2\delta t.$ $\qquad$ Dividing by δt.

Using $\qquad \dfrac{ds}{dt} = \underset{\delta t \to 0}{\text{Lim}} \dfrac{\delta s}{\delta t}$

we get $\qquad \dfrac{ds}{dt} = 30 - 4t$

i.e. the rate of change with respect to time for this distance equation is the speed, $30 - 4t$.

Generally we write $v = \dfrac{ds}{dt}$ because speed is the rate of change of distance with respect to time.

◼◼◼ EXERCISE 13.2 ◼◼◼

For the following curves differentiate from first principles to find general expressions for their gradients.

1 $y = x$

2 $y = 3x$

3 $y = x^2$

4 $s = 2t^2$

5 $s = t^2 + t$

6 $y = 2x^2 - 3x + 1$

7 $y = x^3$

8 $y = \dfrac{2}{x}$

9 $s = \dfrac{4}{t^2}$

10 $y = \dfrac{2}{x} - 3x$

Differentiating algebraic functions, $y = ax^n$

In the previous exercise you should have obtained some general results including

For $y = x$, $\dfrac{dy}{dx} = 1$ or $1x^0$.

$\boxed{x^0 = 1 \text{ and } 1 \times 1 = 1.}$

For $y = x^2$, $\dfrac{dy}{dx} = 2x$ or $2x^1$.

For $y = x^3$, $\dfrac{dy}{dx} = 3x^2$.

Following a similar pattern you would expect that

for $y = x^4$, $\dfrac{dy}{dx} = 4x^3$.

Generally for $y = x^n$ we have $\dfrac{dy}{dx} = nx^{n-1}$ where n is a constant.

Also we can compare two of our earlier results.

For $y = x^2$ we have $\dfrac{dy}{dx} = 2x$ and

for $y = 3x^2$ we have $\dfrac{dy}{dx} = 6x$, i.e. $3(2x)$.

This is a particular case of a more general result:

For $y = ax^n$ we have $\dfrac{dy}{dx} = anx^{n-1}$ where n and a are constants.

There is an alternative notation of

$$\frac{d}{dx}(ax^n) = anx^{n-1}.$$

It is important to follow the pattern of differentiation. The old power n comes forward to multiply and the power is reduced by 1.

Earlier in this chapter we mentioned that not all things change, i.e. some things are constant. This leads us to the general result that if $y =$ constant then $\dfrac{dy}{dx} = 0$.

■■■■■■■ **Examples 13.4** ■■■■■■■

In this set of examples we start with y in terms of x and find a general expression for $\dfrac{dy}{dx}$.

i) $y = 18.5$, which is a constant, $\dfrac{dy}{dx} = 0$.

ii) $y = x^7$, $\dfrac{dy}{dx} = 7x^{7-1}$ $= 7x^6$.

iii) $y = 4x^7$, $\dfrac{dy}{dx} = 4 \times 7x^{7-1}$ $= 28x^6$.

iv) $y = \dfrac{1}{2}x^{10}$, $\dfrac{dy}{dx} = \dfrac{1}{2} \times 10x^{10-1}$ $= 5x^9$.

v) $y = \dfrac{1}{2x} = \dfrac{1}{2}x^{-1}$, $\dfrac{dy}{dx} = \dfrac{1}{2} \times (-1)x^{-1-1} = -\dfrac{1}{2}x^{-2}$ or $\dfrac{-1}{2x^2}$.

vi) $y = \dfrac{7}{2x^4} = \dfrac{7x^{-4}}{2}$, $\dfrac{dy}{dx} = \dfrac{7}{2} \times (-4)x^{-4-1} = -14x^{-5}$ or $\dfrac{-14}{x^5}$.

Notice how we move the x term to the numerator before we differentiate it. In Volume 2 we will look at alternative methods of differentiation.

Not all expressions that we differentiate have only one term. In this next example we simplify the algebra first. The next step is to differentiate each term according to our rule.

██████ **Example 13.5** ██████

For $y = 4x^3 + \dfrac{(2x^2 - x)}{3x}$ first we simplify the algebra. We put each term in the bracket separately upon the common denominator. This gives

$$y = 4x^3 + \frac{2x^2}{3x} - \frac{x}{3x}$$

i.e. $y = 4x^3 + \dfrac{2x}{3} - \dfrac{1}{3}.$

Differentiating term by term gives

$$\frac{dy}{dx} = 12x^2 + \frac{2}{3}.$$

██████ **ASSIGNMENT** ██████

The last time we looked at the Assignment we differentiated from first principles. Now our improved technique using the differentiation rules allows us to write

$$s = 1 + 30t - 2t^2$$

and $\dfrac{ds}{dt} = \quad 30 - 4t$ immediately.

This is a simpler method of finding the general expression for speed, v, which in this case is $v = 30 - 4t$.

██████ **EXERCISE 13.3** ██████

Differentiate the following expressions by rule.

1 $y = 9$

2 $y = 3x$

3 $y = 3x - 9$

4 $y = 4x^2$

5 $s = 4t^2 - 3t + 9$

6 $y = 9 - 3x - 4x^2$

7 $y = x^2 + x$

8 $y = 2x^2 - 3x + 1$

9 $s = 6t^{-1} + 2t^{-2}$

10 $s = \dfrac{4}{t^2}$

11 $y = \dfrac{6}{x^2} + \dfrac{2}{x}$

12 $y = 3x - \dfrac{2}{x}$

13 $y = \dfrac{2}{x^6} - 3x^2$

14 $y = \dfrac{3x^3 + 2x^2}{x}$

15 $y = \dfrac{x - x^2}{x^4}$

So far we have concentrated on powers that are whole numbers (integers). Exactly the same ideas apply to fractional or decimal powers.

Examples 13.6

Again in this set of examples we start with y in terms of x and find a general expression for $\dfrac{dy}{dx}$ by rule.

i) $y = \sqrt{x} = x^{1/2}$, $\dfrac{dy}{dx} = \dfrac{1}{2} x^{1/2-1}$ $= \dfrac{1}{2} x^{-1/2}$

or $\dfrac{1}{2x^{1/2}}$ or $\dfrac{1}{2\sqrt{x}}$.

ii) $y = 3\sqrt{x} = 3x^{1/2}$, $\dfrac{dy}{dx} = 3 \times \dfrac{1}{2} x^{1/2-1}$ $= \dfrac{3}{2} x^{-1/2}$

or $\dfrac{3}{2x^{1/2}}$ or $\dfrac{3}{2\sqrt{x}}$.

iii) $y = 8\sqrt[4]{x^3} = 8x^{3/4}$, $\dfrac{dy}{dx} = 8 \times \dfrac{3}{4} x^{3/4-1}$ $= 6x^{-1/4}$

or $6x^{-0.25}$.

iv) $y = \dfrac{5}{x^{1/2}} = 5x^{-1/2}$, $\dfrac{dy}{dx} = 5 \times \left(-\dfrac{1}{2}\right) x^{-1/2-1}$ $= -\dfrac{5}{2} x^{-3/2}$

or $-2.5x^{-1.5}$.

v) $y = \dfrac{2}{3\sqrt[3]{x}} = \dfrac{2}{3x^{1/3}}$ $\dfrac{dy}{dx} = \dfrac{2}{3} \times \left(-\dfrac{1}{3}\right) x^{-1/3-1}$ $= \dfrac{-2}{9} x^{-4/3}$.

$= \dfrac{2}{3} x^{-1/3}$,

We leave the answer for the final part as a fraction. If you convert it to a decimal make sure you are accurate with the decimal places.

The next set of exercises continues with differentiation by rule for these types of powers.

EXERCISE 13.4

Differentiate the following expressions by rule.

1 $y = 4\sqrt{x}$

2 $y = 3.5x^{0.5}$

3 $y = 8x^{1/3}$

4 $y = \dfrac{1}{3} x^{1/8}$

5 $s = 4t^{1/2}$

6 $y = \sqrt{x^3} - 4$

7 $y = \sqrt[3]{x^2} + 3$

8 $s = 4t^{0.2} + 2t^{0.4}$

9 $y = 7x^{0.5} - 2x^{-0.5}$

10 $y = \dfrac{2}{3x^{0.5}}$

11 $y = \dfrac{2}{3} x^{-0.5} + \dfrac{1}{3} x^{-1.5}$

12 $y = \dfrac{3}{2\sqrt[3]{x}} + \dfrac{1}{\sqrt{x}}$

13 $s = \sqrt{t} - \dfrac{1}{\sqrt{t}}$

14 $y = 4\sqrt{x} - \dfrac{1}{2\sqrt{x}}$

15 $y = \dfrac{\sqrt{x}+1}{x^2}$

Differentiating sine and cosine

We know the shape of the graphs of $y = \sin\theta$ and $y = \cos\theta$ from Chapter 4. From our early work in this chapter we can find the gradient of a curve graphically. For each of these curves we can find the instantaneous gradients by drawing a series of tangents. In Example 13.7 we start off the technique and leave you to complete the task.

Example 13.7

For the curve $y = \sin\theta$ we plot the graph over one cycle (i.e. from 0 to 2π radians) at intervals of $\dfrac{\pi}{6}$ radians. At the same intervals draw tangents to the curve and find their gradients.

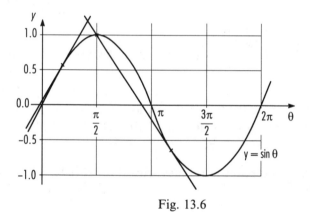

Fig. 13.6

In Fig. 13.6 we have drawn 3 specimen gradients to the sine curve to show the idea. You need to do this for yourself and check your gradient values against those in our table.

θ	0	$\dfrac{\pi}{6}$	$\dfrac{\pi}{3}$	$\dfrac{\pi}{2}$	$\dfrac{2\pi}{3}$	$\dfrac{5\pi}{6}$
$\sin\theta$	0	0.500	0.866	1	0.866	0.500
gradient	1	0.866	0.500	0	−0.500	−0.866

	π	$\dfrac{7\pi}{6}$	$\dfrac{4\pi}{3}$	$\dfrac{3\pi}{2}$	$\dfrac{5\pi}{3}$	$\dfrac{11\pi}{6}$	2π
	0	−0.500	−0.866	−1	−0.866	−0.500	0
	−1	−0.866	−0.500	0	0.500	0.866	1

You ought to be able to spot a pattern. Each gradient value is the **value of the cosine at that point**.

For yourself you should repeat this technique using the cosine curve, again over one cycle using radians. In this case the gradient at each point should be the negative of the sine value.

The patterns give us the differentiation rules

for $\qquad y = \sin\theta \qquad$ and for $\qquad y = \cos\theta$

$$\frac{dy}{d\theta} = \cos\theta \qquad\qquad\qquad \frac{dy}{d\theta} = -\sin\theta$$

or using the alternative notation

$$\frac{d}{d\theta}(\sin\theta) = \cos\theta \qquad \text{and} \qquad \frac{d}{d\theta}(\cos\theta) = -\sin\theta$$

More generally we quote the rules for

$$y = \sin(a\theta + b) \qquad \text{and} \qquad y = \cos(a\theta + b)$$

to give $\qquad \dfrac{dy}{d\theta} = a\cos(a\theta + b) \qquad$ and $\qquad \dfrac{dy}{d\theta} = -a\sin(a\theta + b)$

where a and b are constants.

When we differentiate all trigonometric functions we must use radians.

■■■■■■ **Examples 13.8** ■■■■■■

We differentiate the sine functions by rule.

i) For $\quad y = 3\sin\theta$

$$\frac{dy}{d\theta} = 3\cos\theta.$$

ii) For $\quad y = 3\sin 2\theta$

$$\frac{dy}{d\theta} = 3 \times 2\cos\theta = 6\cos 2\theta.$$

iii) For $\quad y = 3\sin(2\theta - 5)$

$$\frac{dy}{d\theta} = 3 \times 2\cos(2\theta - 5) = 6\cos(2\theta - 5).$$

■■■■■■ **Examples 13.9** ■■■■■■

Again we differentiate trigonometric functions by rule.

i) For $\quad y = 4\cos\theta$

$$\frac{dy}{d\theta} = -4\sin\theta.$$

ii) For $y = 4\cos\dfrac{3\theta}{2}$

$$\frac{dy}{d\theta} = -4 \times \frac{3}{2}\sin\frac{3\theta}{2} = -6\sin\frac{3\theta}{2}.$$

iii) For $y = 4\cos\left(\dfrac{3\theta}{2} + 7\right)$

$$\frac{dy}{d\theta} = -4 \times \frac{3}{2}\sin\left(\frac{3\theta}{2} + 7\right) = -6\sin\left(\frac{3\theta}{2} + 7\right).$$

iv) For $y = \cos(6 - 3\theta)$

$$\frac{dy}{d\theta} = -3 \times -\sin(6 - 3\theta) = 3\sin(6 - 3\theta).$$

v) We can add/subtract our functions and their derivatives. Using results from Examples 13.8 and 13.9 we have

$$y = 3\sin 2\theta - 4\cos\frac{3\theta}{2} + 3\sin(2\theta - 5)$$

$$\frac{dy}{d\theta} = 6\cos 2\theta + 6\sin\frac{3\theta}{2} + 6\cos(2\theta - 5)$$

$$= 6\left(\cos 2\theta + \sin\frac{3\theta}{2} + \cos(2\theta - 5)\right).$$

> 6 is a common factor.
> $(-)(-) = +.$

EXERCISE 13.5

Differentiate the following sines and cosines by rule.

1   $y = 2\sin\theta$

2   $y = 3\cos\theta$

3   $y = \sin 2\theta$

4   $y = 3\cos 2t$

5   $y = 4\sin(3t - 2)$

6   $y = 4\cos(\theta + 2)$

7   $y = 4\sin(3t - 2) + 3\cos 2t$

8   $y = 2\cos 3x - 3\sin(2x + 1)$

9   $y = 6\cos\dfrac{3x}{4} + 2\sin\dfrac{x}{2}$

10   $y = 9\sin\left(\dfrac{2\theta}{3} - 1\right) - 14\cos\left(\dfrac{3\theta}{2} + 1\right)$

Differentiating the exponential function

Again we can find a general rule for differentiation by looking at the tangents to the curve. This time the whole task is left as an exercise for you. You should construct a table and plot the graph of $y = e^x$ for values of x from $x = -2$ to $x = 2$. Within this range you should draw tangents to the curve at intervals of 0.25. Calculate the gradient of each tangent and include it in your table as we did for $y = \sin \theta$. With accuracy you should find each gradient value is the same as the y value at that point. This gives us the general rule for $y = e^x$

$$\frac{dy}{dx} = e^x$$

or using the alternative notation $\quad \dfrac{d}{dx}(e^x) = e^x.$

More generally for $y = e^{ax+b}$

$$\frac{dy}{dx} = ae^{ax+b}.$$

Examples 13.10

We differentiate exponential functions by rule.

i) For $\qquad y = 2.75e^x$

$$\frac{dy}{dx} = 2.75e^x.$$

ii) For $\qquad y = 2.75e^{2x}$

$$\frac{dy}{dx} = 2.75 \times 2e^{2x} = 5.50e^{2x}.$$

iii) For $\qquad y = 5e^{3x-4}$

$$\frac{dy}{dx} = 5 \times 3e^{3x-4} = 15e^{3x-4}.$$

iv) Any exponential function in a denominator position needs slight adjustment before we differentiate it.

For $\qquad y = \dfrac{1}{2e^{4x+1}} = \dfrac{1}{2}e^{-4x-1}$

$$\frac{dy}{dx} = \frac{1}{2} \times -4e^{-4x-1} = -2e^{-4x-1}.$$

EXERCISE 13.6

Differentiate the following expressions by rule.

1 $y = 4e^{3x}$

2 $y = 4e^{3x-2}$

3 $y = 7.5e^{-3t}$

4 $y = \dfrac{5}{e^{3x}}$

5 $y = \dfrac{5}{e^{3t+6}}$

6 $y = 0.25e^{7-4x}$

7 $y = 2.5t - 0.5e^{3+t}$

8 $y = 4e^{2\theta} + 2\cos 4\theta$

9 $y = 3\sin \theta + e^{4+2\theta}$

10 $y = 3\sin(2\theta - 3) + 2e^{3\theta-2}$

Differentiating the logarithmic function

Once again we can find the general rule for differentiation by looking at the tangents to the curve. The whole task is left as an exercise for you. You should construct a table and plot the graph of $y = \ln x$ for values of x from $x = 0.25$ to $x = 3$. Within this range you should draw tangents to the curve at intervals of 0.25. Calculate the gradient of each tangent and include it in your table as we did for the exponential function. With accuracy you should find each gradient value is the same as the $\dfrac{1}{x}$ value at that point. This gives us the general rule for $y = \ln x$

$$\frac{dy}{dx} = \frac{1}{x}$$

or using the alternative notation

$$\frac{d}{dx}(\ln x) = \frac{1}{x}.$$

More generally for $y = \ln(ax + b)$

$$\frac{dy}{dx} = \frac{a}{ax + b}.$$

Example 13.11

Differentiate the following logarithmic functions by rule

i) For $y = 3\ln x$

$$\frac{dy}{dx} = 3 \times \frac{1}{x} = \frac{3}{x}.$$

ii) For $y = \ln(3x)$

    i.e. $y = \ln 3 + \ln x$ | First log law.

    so that $\dfrac{dy}{dx} = 0 + \dfrac{1}{x} = \dfrac{1}{x}.$ | ln 3 is a constant.

iii) Again for $y = \ln 3x$

$$\frac{dy}{dx} = \frac{3}{3x} = \frac{1}{x}.$$

   $a = 3$ and $b = 0$ in $\dfrac{a}{ax + b}$.

iv) For $y = \ln\left(\dfrac{x}{4}\right)$

    i.e. $y = \ln x - \ln 4$ | Second log law.

    so that $\dfrac{dy}{dx} = \dfrac{1}{x} - 0 = \dfrac{1}{x}.$ | ln 4 is a constant.

v) Again for $y = \ln\left(\dfrac{x}{4}\right) = \ln\left(\dfrac{1}{4}x\right)$

$$\frac{dy}{dx} = \frac{1/4}{x/4} = \frac{1}{x}.$$

   $a = \dfrac{1}{4}$ and $b = 0$ in $\dfrac{a}{ax + b}$.

vi)   For $\quad y = \ln\left(\dfrac{1}{x}\right)$

i.e. $\qquad y = \ln 1 - \ln x$

> Second log law.
> $\ln 1 = 0$.

i.e. $\qquad y = -\ln x$

so that $\quad \dfrac{dy}{dx} = -\dfrac{1}{x}.$

vii)   For $\quad y = \ln(x^2)$

i.e. $\qquad y = 2\ln x$

> Third log law.

so that $\quad \dfrac{dy}{dx} = \dfrac{2}{x}.$

viii)   For $\quad y = \ln(3x - 5)$

$$\dfrac{dy}{dx} = \dfrac{3}{3x - 5}$$

The next set of exercises allows you to practise each of these differentiations by rule.

▰▰▰▰ EXERCISE 13.7 ▰▰▰▰▰▰

Differentiate the following expressions by rule.

1   $y = 2.5 \ln x$

2   $y = 4 \ln x + \ln(4x)$

3   $y = 2 \ln x + 3 \ln\left(\dfrac{1}{x}\right)$

4   $y = 2 \ln(4x) + 3 \ln\left(\dfrac{4}{x}\right)$

5   $y = 2 \ln(4x + 3)$

6   $y = 2 \ln(4x - 3)$

7   $y = 5 \ln(3 - 4x)$

8   $y = 3 \sin 2\theta + \ln(2\theta - 4)$

9   $y = 2e^{4t-3} + \ln(3t + 4)$

10   $y = 8 \ln\left(\dfrac{3x}{2} + 5\right) - \ln(2 - 4x)$

Gradient at a point

We know how to differentiate using a set of rules. Each time we have started with an equation and found a general expression for the gradient, $\dfrac{dy}{dx}$, or similar.

As we move along a curve the values of x and y change. This means the gradient changes. (Only along a straight line does the gradient remain the same.) Suppose we wish to find the gradient at a particular point. We find the general expression for the gradient *before* substituting any particular numbers.

Example 13.12

Find the gradient of the curve $y=3x^2$ at the points where
i) $x=-2$
and ii) $x=2.5$.

i) For this particular curve we know, differentiating by rule, that

$$\frac{dy}{dx} = 6x.$$

We substitute for $x=-2$ to get

$$\frac{dy}{dx} = 6(-2) = -12$$

i.e. the gradient of the curve where $x=-2$ is -12, or the gradient of the tangent to the curve where $x=-2$ is -12.

ii) Again using $\dfrac{dy}{dx} = 6x$

we substitute for $x=2.5$ to get

$$\frac{dy}{dx} = 6(2.5) = 15$$

i.e. the gradient of the curve where $x=2.5$ is 15.

Example 13.13

A bacterial growth, B, is related exponentially to time, t hours, by $B=B_0e^t$ where $B_0=1.75\times10^3$. Find the rate of growth of the bacteria after 2 hours.

We differentiate our equation by rule to get the general expression

$$\frac{dB}{dt} = 1.75\times10^3e^t.$$

Now we substitute for our particular value of $t=2$ so that

$$\frac{dB}{dt} = 1.75\times10^3e^2 = 12.9\times10^3,$$

i.e. the bacterial growth at this particular time is 12.9×10^3 per hour.

Example 13.14

For the curve $y=\frac{1}{2}(e^{2x}-e^{-2x})$ find the value of $\dfrac{dy}{dx}$ where $x=0.75$.

We differentiate $$y = \frac{1}{2}(e^{2x}-e^{-2x})$$

to get $$\frac{dy}{dx} = \frac{1}{2}(2e^{2x}-(-2)e^{-2x}).$$

This is our general expression for the first derivative.

Now we substitute for $x = 0.75$ so that

$$\frac{dy}{dx} = \frac{1}{2}(2e^{2(0.75)} + 2e^{-2(0.75)})$$

$$\frac{dy}{dx} = \frac{2}{2}(e^{1.5} + e^{-1.5})$$

We remove a common factor of 2 and tidy up the powers.

$$= 4.4817 + 0.2231$$

$$= 4.70.$$

ASSIGNMENT

Let us take a final look at our vehicle with

$$s = 1 + 30t - 2t^2$$

and $\quad v = \dfrac{ds}{dt} = 30 - 4t.$

s is distance (m).
t is time (s).
v is speed (ms^{-1}).

Suppose we wish to find the initial speed. Initial means "at the beginning", i.e. at $t = 0$ so that

$$v = 30 - 4(0) = 30\,\text{ms}^{-1} \quad \text{is the initial speed.}$$

To find the speed after 3.5 seconds we substitute for $t = 3.5$ in

$$v = 30 - 4t$$

to get $\quad v = 30 - 4(3.5) = 16\,\text{ms}^{-1} \quad$ as the speed.

EXERCISE 13.8

1   If $y = 2x + 3e^x$ find $\dfrac{dy}{dx}$ where $x = 1.5$.

2   $yx^{3.5} = 2$. Make y the subject of this formula. What is the value of $\dfrac{dy}{dx}$ where $x = 2.25$?

3   $y = 4\sin t - \cos t$. Find the rate of change of y with respect to t when $t = \dfrac{\pi}{3}$.

4   Given that $y = \dfrac{10 - \pi x^2}{\pi x}$ find a general expression for $\dfrac{dy}{dx}$. What is the value of this gradient where $x = 3.4$?

5   Find the value of $\dfrac{dy}{d\theta}$ where $\theta = \dfrac{\pi}{4}$ for the curve $y = 2(\theta - 3.2\sin\theta)$.

6   $s = 15.6t - 1.06t^2$ refers to the motion of a particular vehicle. s is the displacement in metres after t seconds. What is the initial velocity $\left(\text{i.e. } \dfrac{ds}{dt}\right)$ of the vehicle? After what time will the vehicle come to rest?

7   A rigid beam of length 10 m carries a uniformly distributed total load of 10 000 N. The bending moment, M, is related to its distance from one end of the beam, x, according to $M = -10\,000 + 5000x - 500x^2$.

What is the rate of change of bending moment $\left(\text{i.e. } \dfrac{dM}{dx}\right)$ at 3 m from either end of the beam?

8   The diagram shows a belt in contact with a pulley. The length of belt in contact with the pulley subtends an angle θ radians with the centre. The coefficient of friction is 0.4. The tension on the taut side is T and on the slack side is 25 N. If $\theta = 2.5 \ln \left(\dfrac{T}{25}\right)$ find $\dfrac{d\theta}{dT}$ when $T = 37.5$ N.

9   A disc is spun from rest. It spins through $y°$ in t seconds according to $y = 100t^2 - 5t^3$. What is the value of $\dfrac{dy}{dt}$ after 2.5 seconds?

10   One particular machine in an engineering workshop costs £C to lease each week according to the formula $C = 200 + \dfrac{t^3}{20}$. t is the number of hours/week worked by the machine. $\dfrac{dC}{dt}$ is the rate of increase of cost during the week. Find a general expression for $\dfrac{dC}{dt}$. When does this rate exceed £150 per hour?

The second derivative

For y as a function of x (i.e. $y = f(x)$) we know that $\dfrac{dy}{dx}$ (or $f'(x)$) is the change in y with respect to x. One interpretation of this is the gradient of the graph. By differentiating the expression for gradient we can discuss how that gradient in turn changes.

We start with y or $f(x)$

and differentiate to get $\dfrac{dy}{dx}$ or $f'(x)$.

> First derivative/differential.

We differentiate again and write $\dfrac{d}{dx}\left(\dfrac{dy}{dx}\right)$ or $f''(x)$.

> Second derivative/differential.

$f''(x)$ is said '*f* **double dashed**' or '*f* **double dashed of** *x*'. $\dfrac{d}{dx}\left(\dfrac{dy}{dx}\right)$ is a complete symbol stating that we intend to differentiate with respect to x whatever appears in the bracket. For simplicity we write $\dfrac{d}{dx}\left(\dfrac{dy}{dx}\right)$ as $\dfrac{d^2y}{dx^2}$ (said '*d* **two** *y* **by** *dx* **squared**'). Again this is a complete symbol. The positions of the 2s have *no* algebraic significance.

In the following examples we further apply the techniques of the previous chapter.

▬▬▬ **Example 13.15** ▬▬▬

If $y = 7x^2 + 3x - \dfrac{2}{x}$ find expressions for $\dfrac{dy}{dx}$ and $\dfrac{d^2y}{dx^2}$.

We rewrite the last term in our equation so that

$$y = 7x^2 + 3x - 2x^{-1}.$$

We differentiate once to get

$$\frac{dy}{dx} = 7\,(2x^{2-1}) + 3\,(1x^{1-1}) - 2\,(-1x^{-1-1})$$

$$= 14x + 3 + 2x^{-2}$$

$$\text{or} \quad 14x + 3 + \frac{2}{x^2}$$

Using the expression for $\dfrac{dy}{dx}$ we differentiate again to get

$$\frac{d^2y}{dx^2} = 14(1x^{1-1}) + 0 + 2(-2x^{-2-1})$$

$$= 14 - 4x^{-3}$$

$$\boxed{x^1 = x,\ x^0 = 1.}$$

$$\text{or} \quad 14 - \frac{4}{x^3}$$

Earlier in this chapter we found the gradient at a point. Remember that $\dfrac{d^2y}{dx^2}$ gives a general expression. Hence we can find its value for given values of x.

▬▬▬ **Examples 13.16** ▬▬▬

From the previous example we have $\dfrac{d^2y}{dx^2} = 14 - \dfrac{4}{x^3}$. We find its value where i) $x = 2$, ii) $x = -0.5$.

i) Using $\dfrac{d^2y}{dx^2} = 14 - \dfrac{4}{x^3}$

  we substitute for $x = 2$,

i.e. $\qquad \dfrac{d^2y}{dx^2} = 14 - \dfrac{4}{2^3}$

$\qquad\qquad\qquad = 14 - \dfrac{4}{8}$

$\qquad\qquad\qquad = 13.5.$

ii) Using $\qquad \dfrac{d^2y}{dx^2} = 14 - \dfrac{4}{x^3}$

we substitute for $x = -0.5$,

i.e. $\qquad \dfrac{d^2y}{dx^2} = 14 - \dfrac{4}{(-0.5)^3}$

$\qquad\qquad\qquad = 14 - \dfrac{4}{-0.125}$

$\qquad\qquad\qquad = 14 + 32$

$\qquad\qquad\qquad = 46.$

Velocity and acceleration

We know that $\dfrac{dy}{dx}$ is a change in y due to a change in x. Instead of y being a function of x, suppose that s is a function of t, i.e. $s = f(t)$. For a body moving s metres in a time t seconds we give the displacement as $s = f(t)$.

Now **velocity is defined as the change in displacement, s, due to a change in time, t, i.e. $v = \dfrac{ds}{dt}$ or $f'(t)$.**

For a displacement–time graph (i.e. s against t) v is the gradient at an instant. Early in this chapter we distinguished between average gradients (chord to the curve) and instantaneous gradients (tangent to the curve).

Acceleration is defined as the change in velocity, v, due to a change in time, t, i.e. $a = \dfrac{dv}{dt}$. Using $v = \dfrac{ds}{dt}$ we may write $a = \dfrac{d}{dt}\left(\dfrac{ds}{dt}\right) = \dfrac{d^2s}{dt^2}$ or $f''(t)$.

Let us look more closely at our definitions. In Chapter 7 we used **displacement** and **velocity** as examples of **vectors**. The corresponding **scalars** are **distance** and **speed**. If we differentiate the displacement to find an expression for velocity, then its size is speed, i.e. **speed** $= \left|\dfrac{ds}{dt}\right|$ or $|v|$.

Alternatively **we may define speed as the change in distance due to a change in time**.

Because vectors have both magnitude and direction we need to choose and label a positive direction. A negative displacement is a distance in the opposite direction. A negative velocity is a speed in the opposite direction. Rather differently we interpret a **negative acceleration (deceleration** or **retardation)** to be a slowing down.

Acceleration has no corresponding scalar term. We may treat acceleration as either a vector or a scalar. If an example uses displacement and/or velocity then acceleration is a vector. If it uses distance and/or speed then acceleration is a scalar.

Examples 13.17

The displacement of a body, s metres, in time, t seconds, is given by $s = 0.2(4 + 7t - 2t^2 - t^3)$. Find

i) the velocity after 3 seconds,
ii) when the velocity is zero,
iii) the change in speed during the 4th second,
iv) the acceleration when $t = 1.5$ seconds.

We differentiate our original expression for s to find an expression for velocity $\left(v \text{ or } \dfrac{ds}{dt} \right)$.

$$\therefore \quad \frac{ds}{dt} = 0.2(7 - 4t - 3t^2).$$

i) When $t = 3$, $v = 0.2(7 - 4(3) - 3(3)^2) = -6.4 \, \text{ms}^{-1}$, i.e. a velocity of $6.4 \, \text{ms}^{-1}$ in a direction opposite to the original direction.

ii) For $v = 0$, $\qquad 0 = 7 - 4t - 3t^2$ | Cancelling by 0.2. |

    i.e. $\qquad\qquad\quad 0 = (1 - t)(7 + 3t)$

    i.e. $\qquad\quad 1 - t = 0, \quad 7 + 3t = 0$

    i.e. $\qquad\qquad\quad t = 1, \ -\dfrac{7}{3}.$

Because time is positive (i.e. we *cannot* go back in time) the velocity is zero after 1 second.

iii) The 4th second starts at $t = 3$ and continues until $t = 4$. Already we have substituted for $t = 3$ in our expression for velocity. Also, when $t = 4$, $v = 0.2(7 - 4(4) - 3(4)^2) = -11.4 \, \text{ms}^{-1}$. In this time the velocity changes from $-6.4 \, \text{ms}^{-1}$ to $-11.4 \, \text{ms}^{-1}$, i.e. the size of the change is $5 \, \text{ms}^{-1}$, i.e. the change in speed is $5 \, \text{ms}^{-1}$.

iv) We differentiate our expression for velocity to find an expression for acceleration $\left(a \text{ or } \dfrac{dv}{dt} \text{ or } \dfrac{d^2s}{dt^2} \right),$

$$\frac{d^2s}{dt^2} = 0.2(-4 - 6t).$$

When $t = 1.5$, $\dfrac{d^2s}{dt^2} = 0.2(-4 - 6(1.5)) = -2.6 \text{ ms}^{-2}$,

i.e. the body is slowing down, the retardation (deceleration) indicated by the minus sign.

■■■ EXERCISE 13.9 ■■■

In Questions **1–10** find expressions for the first derivative $\left(\text{often } \dfrac{dy}{dx}\right)$ and the second derivative $\left(\text{often } \dfrac{d^2y}{dx^2}\right)$.

1 $y = 2x^3 + 5x^2 - 6$

2 $y = \dfrac{x + 2x^2}{x^4}$

3 $y = 2\sqrt{t} + \dfrac{4}{\sqrt{t}}$

4 $y = 3\cos 2x$

5 $y = 2\sin 3\theta - 4\cos 2\theta$

6 $y = 3e^{2x} + 4e^{-x}$

7 $y = 3x + 2e^{-2x}$

8 $y = 2\ln 4x$

9 $y = e^{-0.5x} + 2\ln x$

10 $y = 5e^x + 2\sin 2x$

11 Find the values of $\dfrac{dy}{dx}$ and $\dfrac{d^2y}{dx^2}$ where $x = 1.5$ given that $y = \dfrac{2x - 6}{x^2}$.

12 If $s = \sin(2t - 3)$ what is the value of $\dfrac{d^2s}{dt^2}$ when $t = 2.5$ radians?

13 Find general expressions for the first and second derivatives of $y = e^{-x} + 2\ln 3x$. Find the values of these expressions where i) $x = 1$, ii) $x = 2.5$. Why can you *not* do this where $x = -1$?

14 Given that $x = 2 - 5t + 2t^2 - 3t^4$ relates displacement, x, to time, t, find expressions for the i) velocity and ii) acceleration.

15 A body's displacement, x metres, over time, t seconds, is given by $x = 10\ln 3t$. Differentiate, and fully simplify your answers, to get expressions for velocity and acceleration.

16 Find expressions for the velocity and acceleration of a body where the displacement, x, is related to the time of motion, t, according to $x = 2 + t^2$.

17 A body moves with simple harmonic motion about an equilibrium point O. Its displacement from O, x metres, is related to time, t seconds, by $x = 2.5\sin 2t$. By differentiation find expressions for the velocity, $\dfrac{dx}{dt}$, and acceleration, $\dfrac{d^2x}{dt^2}$.

Displacement and acceleration are connected by $\dfrac{d^2x}{dt^2} = -kx$. What is the value of k?

18 A body moves according to the equation $x = 2t^2 + 16t + 4$. x metres is the displacement at time t seconds.
   i) How far does the body move in the 3rd second?
   ii) Calculate the speed at the beginning and end of that 3rd second. What is the change in speed during this time?

19 A particle oscillates about E, a point of equilibrium. From E in time t seconds its displacement x metres is given by $x = 4 \sin \dfrac{\pi t}{6} + 3 \cos \dfrac{\pi t}{6}$.
   What is the   i)   initial speed?
                 ii) velocity after 0.5 seconds?
                 iii) acceleration after 0.75 seconds?

20 Cargo is pushed from a military transport aircraft during flight. It falls subject to air resistance and the effects of its parachutes. During its descent the vertical displacement from the aircraft, y metres, is connected to time, t seconds, by $y = 49(t + 0.5e^{-2t})$. Find a general expression for the velocity.

Turning points

Fig. 13.7 shows y as some general function of x, i.e. $y = f(x)$. We are interested in points A and B. Here the tangents are parallel to the horizontal axis, i.e. the gradients are zero, i.e. $\dfrac{dy}{dx} = 0$. A and B are called **turning points**.

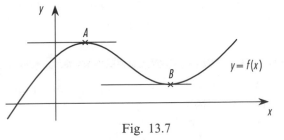

Fig. 13.7

████████ Example 13.18 ██

Has the graph of $y = 2x^2 + x + 1$ any turning points? If there are any, at what values of x do they occur?

   Fig. 13.8 shows there is a turning point where $x = -0.25$. Indeed the sketch includes a horizontal tangent at this point. We do not always need to draw a graph of a function just to find any turning points. Indeed, we can save ourselves time by not having to draw one. Our method uses differentiation.

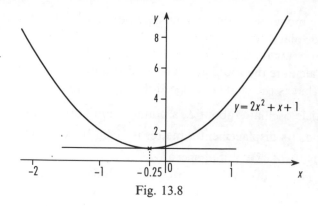

Fig. 13.8

Using $y = 2x^2 + x + 1$

we differentiate once,

i.e. $\dfrac{dy}{dx} = 4x + 1.$

We know that $\dfrac{dy}{dx} = 0$ at a turning point. This specific condition can be linked to our general expression, $4x + 1$, so that

$$4x + 1 = 0$$

i.e. $x = -0.25.$

This confirms what we discovered in Fig. 13.8, that there is a turning point where $x = -0.25$. Notice that we have only one turning point rather than several. We recall from Chapter 8, that quadratic functions only turn once. Also, when we differentiate a quadratic function (highest powered term being x^2) we get a linear function (highest powered term being x). The ensuing linear equation we create can have only one solution.

■ Example 13.19 ■

Has the graph of $y = x^3 - 3.5x^2 + 2x - 6$ any turning points? If there are any, at what values of x do they occur?

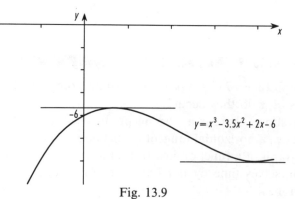

Fig. 13.9

Fig. 13.9 shows there to be two turning points. We have removed the scales from the axes on purpose. This is because the aim of this chapter is to find a turning point using differentiation.

Using $$y = x^3 - 3.5x^2 + 2x - 6$$

we differentiate once,

i.e. $$\frac{dy}{dx} = 3x^2 - 7x + 2.$$

Using our specific knowledge that $\frac{dy}{dx} = 0$ at a turning point we have

$$3x^2 - 7x + 2 = 0.$$

This quadratic equation factorises to give

$$(3x - 1)(x - 2) = 0$$

∴ $$3x - 1 = 0,\ x - 2 = 0$$

i.e. $$x = \frac{1}{3},\ 2.$$

This calculation confirms what we see in our sketch. We have two turning points, one where $x = \frac{1}{3}$ and the other where $x = 2$. Our particular quadratic equation factorised, but if necessary we can use the formula.

> For $ax^2 + bx + c = 0$
>
> $$x = \frac{-b \pm \sqrt{b^2 - 4ac}}{2a}.$$

Notice how we started with a cubic function (highest powered term being x^3) and differentiated to get a quadratic function. Generally, though *not* always, a quadratic equation has two solutions. These two solutions are consistent with a cubic function that has two turning points. Now this means we need to be able to distinguish between them. We look to do this after the next exercise.

EXERCISE 13.10

In each question find, by differentiation, the value(s) of x at the turning point(s).

1 $y = x^2 - 5x + 3.5$

2 $y = 2.25 + 3x - x^2$

3 $y = 2x^3 - 4.5x^2 + 3x + 8$

4 $y = 4x^3 - 3x^2 - 11x + 2.5$

5 $y = x^3 - 4.4x^2 + 5.6x - 1.6$

Maximum and minimum turning points

Let us redraw our first figure. In Fig. 13.10 we include the sign of the gradient, noticing how it changes.

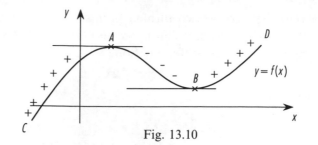

Fig. 13.10

We start from the left, moving to the right as x increases. The gradient changes from positive ($+$) through zero at A to negative ($-$). This type of change shows that A is a **maximum turning point**. It is also called a **local maximum point**. This distinguishes it from points like D which are absolutely bigger. The gradient changes from negative ($-$) through zero at B to positive ($+$). This type of change shows that B is a **minimum turning point**. It is also called a **local minimum point**. This distinguishes it from points like C which are absolutely smaller. These features mean we can test a turning point by examining the gradient closely on either side of it. This is called the **gradient test**.

Example 13.20

Has the graph of $y = 3 + 2x - 5x^2$ a turning point? If there is a turning point decide whether it is a maximum or minimum.

Using
$$y = 3 + 2x - 5x^2$$

we differentiate once,

i.e.
$$\frac{dy}{dx} = 2 - 10x.$$

At a turning point $\dfrac{dy}{dx} = 0$ so that

$$2 - 10x = 0$$

giving a solution $x = 0.2.$

So far we know there is a turning point at $x = 0.2$. Our next step is to check the gradients closely on either side of this turning point. We simply choose values of x slightly below and slightly above 0.2.

In $\dfrac{dy}{dx} = 2 - 10x$

we substitute for $x = 0.175$ to get

$$\frac{dy}{dx} = 2 - 10(0.175) = 0.25.$$

Also for $x = 0.225$ we get

$$\frac{dy}{dx} = 2 - 10(0.225) = -0.25.$$

Let us look at our numerical results in order.

x	0.175	0.2	0.225
$\dfrac{dy}{dx}$	0.25	0	−0.25
	(+)		(−)

As x increases the gradient changes from positive through zero to negative. This means we have a maximum turning point.

There is an alternative method to the gradient test that uses the second derivative.

We know the gradient changes from positive through zero to negative at a maximum turning point. This means the gradient is decreasing, i.e. the change is negative. Mathematically we write this gradient change as

$\dfrac{d}{dx}\left(\dfrac{dy}{dx}\right)$ is negative, i.e. $\dfrac{d^2y}{dx^2} < 0$.

This means our test for a **maximum turning point (local maximum)** is $\dfrac{dy}{dx} = 0$ **and** $\dfrac{d^2y}{dx^2} < 0$.

Also we know the gradient changes from negative through zero to positive at a minimum turning point. This means the gradient is increasing, i.e. the change is positive. Mathematically we write this gradient change as

$\dfrac{d}{dx}\left(\dfrac{dy}{dx}\right)$ is positive, i.e. $\dfrac{d^2y}{dx^2} > 0$.

This means our test for a **minimum turning point (local minimum)** is $\dfrac{dy}{dx} = 0$ **and** $\dfrac{d^2y}{dx^2} > 0$.

Example 13.21

Find the turning points of $y = 2x^3 - 9x^2 - 60x + 4$. Using the second derivative discover if these are maximum or minimum turning points.

Using $\qquad y = 2x^3 - 9x^2 - 60x + 4$

we differentiate once,

i.e. $\qquad \dfrac{dy}{dx} = 6x^2 - 18x - 60$.

At a turning point $\dfrac{dy}{dx} = 0$ so that

$\qquad 6x^2 - 18x - 60 = 0$

i.e. $6(x^2 - 3x - 10) = 0$

$(x + 2)(x - 5) = 0$

$\therefore$ $x + 2 = 0, x - 5 = 0$

i.e. $x = -2, 5.$

> Dividing by 6 and factorising.

This shows we have turning points where $x = -2$ and $x = 5$. Our next step is to find the type of each turning point.

Using $\dfrac{dy}{dx} = 6x^2 - 18x - 60$

we differentiate again to get

$$\frac{d^2y}{dx^2} = 12x - 18.$$

This is a general expression for the second derivative. We need to know whether it is positive or negative at our specific values of x. Substituting for $x = -2$,

$$\frac{d^2y}{dx^2} = 12(-2) - 18 = -42, \quad \text{i.e. negative.}$$

This means we have a maximum turning point where $x = -2$.

Also substituting for $x = 5$,

$$\frac{d^2y}{dx^2} = 12(5) - 18 = 42, \quad \text{i.e. positive.}$$

This means we have a minimum turning point where $x = 5$.

For maximum and minimum turning points there are exceptions to the rules for the second derivative. When this fails to be either positive or negative we need to try our first method, the gradient test.

Example 13.22

Find the type (nature) of turning point of the curve $y = x^4 + 2$.

Using $y = x^4 + 2$

we differentiate once,

i.e. $\dfrac{dy}{dx} = 4x^3.$

At a turning point $\dfrac{dy}{dx} = 0$ so that

$4x^3 = 0$

i.e. $x^3 = 0$

i.e. $x = 0.$

> Dividing by 4.

This shows we have a turning point where $x = 0$.

Using $\quad \dfrac{dy}{dx} = 4x^3$

we differentiate again to get

$$\frac{d^2y}{dx^2} = 12x^2.$$

Substituting for $x=0$ into our general expression we get

$$\frac{d^2y}{dx^2} = 12(0)^2 = 0.$$

This is neither positive nor negative yet Fig. 13.11 shows we have a minimum turning point.

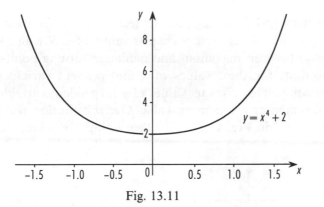

Fig. 13.11

We return to our gradient test using values of x slightly below and slightly above $x=0$.

Using $\quad \dfrac{dy}{dx} = 4x^3$

we substitute for $x=-0.2$ to get

$$\frac{dy}{dx} = 4(-0.2)^3 = -0.032.$$

Also for $x=0.2$ we get

$$\frac{dy}{dx} = 4(0.2)^3 = 0.032.$$

Looking at our results in order we have

x	−0.2	0	0.2
$\dfrac{dy}{dx}$	−0.032	0	0.032
	(−)		(+)

As x increases the gradient changes from negative through zero to positive. This means we have a minimum turning point at $x=0$, confirmed by Fig. 13.11.

■■■■ **EXERCISE 13.11** ■■■■■■■■■■■■■■■■■■■■■■■■■

In each question find the nature (type) of the turning points using the i) gradient test, and ii) second differential.

1 $y = 3x^2 + 2x - 6$

2 $y = 7 - 2x - 4x^2$

3 $y = 4x^3 - 3x^2 - 36x$

4 $y = x^3 + x^2 - 1.75x + 0.5$

5 $y = 2 - 6x + x^3$

Maximum and minimum values

This is the third and final stage in our technique. So far we have found that turning points can exist at particular values of x. We also know how to distinguish between maximum and minimum turning points. Now we simply substitute for these values of x (independent variable) into our original equation. This gives us values of y (dependent variable) that are either a maximum or a minimum value. Our substitution will give the y values at A and B in Fig. 13.12. Together with the x values we will have their coordinates.

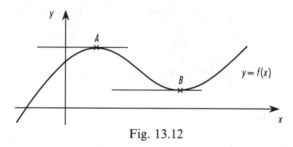

Fig. 13.12

■■■■■■ **Example 13.23** ■■■■■■■■■■■■■■■■■■■■■■■■

Find the local maximum and minimum values of $y = 10 + 30x - 2.25x^2 - x^3$. Also write down the coordinates of these turning points.

Stage 1   *(To find that turning points exist and their values of x.)*

Using $y = 10 + 30x - 2.25x^2 - x^3$

we differentiate once,

i.e. $\dfrac{dy}{dx} = 30 - 4.5x - 3x^2.$

At a turning point $\dfrac{dy}{dx} = 0$ so that

$$30 - 4.5x - 3x^2 = 0.$$

This quadratic equation has solutions
$$x = -4, 2.5,$$
i.e. we have turning points where $x = -4$ and $x = 2.5$.

Stage 2   *(To find the type (nature) of the turning points.)*

Using $\dfrac{dy}{dx} = 30 - 4.5x - 3x^2$

we differentiate again to get

$$\dfrac{d^2y}{dx^2} = -4.5 - 6x.$$

Substituting for $x = -4$,

$$\dfrac{d^2y}{dx^2} = -4.5 - 6(-4) = 19.5, \quad \text{i.e. positive.}$$

This means we have a minimum turning point where $x = -4$.

Also substituting for $x = 2.5$,

$$\dfrac{d^2y}{dx^2} = -4.5 - 6(2.5) = -19.5, \quad \text{i.e. negative.}$$

This means we have a maximum turning point where $x = 2.5$.

Stage 3   *(To find the local maximum and minimum values of y.)*

We need to know the values of y at these turning points.

Using $y = 10 + 30x - 2.25x^2 - x^3$

we substitute for $x = -4$ to get
$$y_{min} = 10 + 30(-4) - 2.25(-4)^2 - (-4)^3 = -82.$$
The local minimum value is -82, occurring at $(-4, -82)$.

Also we substitute for $x = 2.5$ to get

$$y_{max} = 10 + 30(2.5) - 2.25(2.5)^2 - (2.5)^3 = 55.3.$$
The local maximum value is 55.3, occurring at $(2.5, 55.3)$.

■■■■■ EXERCISE 13.12 ■■■■■

Find the local maximum and minimum values of y, where appropriate, in each question. Also write down their coordinates, distinguishing between them.

1   $y = 2x^2 - 3x$

2   $y = 4 - 10x - 5x^2$

3   $y = -24x - 3x^2 + x^3$

4   $y = 2 - 7x + 4x^2 + 4x^3$

5   $y = 3x^3 + 4.5x + 1.5$

Practical applications of maxima and minima

![Example 13.24]

We have an open cistern with a square base of side x and of height, h. The area of material used is 12 m². What dimensions make the volume of this cistern a maximum? Calculate the maximum volume.

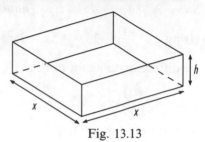

Fig. 13.13

The volume, V, is given by $V = x^2h$. So far in our differentiation we have only used one independent and one dependent variable. We need to eliminate an independent variable, either x or h. We can do this by using our information about the area of material and substituting for h. The cistern has 4 sides, each of area xh and a square base of area x^2. This means we have

$$\text{Area} = 4xh + x^2 = 12$$

i.e.

$$h = \frac{12 - x^2}{4x}.$$

Then

$$V = x^2 \frac{(12 - x^2)}{4x}$$

i.e.

$$V = 3x - 0.25x^3.$$

We differentiate once to get

i.e.

$$\frac{dV}{dx} = 3 - 0.75x^2.$$

At a turning point $\dfrac{dV}{dx} = 0$ so that

$$3 - 0.75x^2 = 0$$

i.e.

$$3 = 0.75x^2$$

i.e.

$$x^2 = 4$$

$$\therefore \quad x = \pm\sqrt{4} = \pm 2.$$

Our practical example involves distances so $x = -2$ has no practical use. We continue our method with just $x = 2$.

Using

$$\frac{dV}{dx} = 3 - 0.75x^2$$

we differentiate again to get

$$\frac{d^2V}{dx^2} = -1.5x.$$

Substituting for $x=2$ we get

$$\frac{d^2V}{dx^2} = -1.5(2) = -3, \quad \text{i.e. negative.}$$

This means we have a local maximum for a value of $x=2$.

Using $\qquad h = \dfrac{12 - x^2}{4x}$

and substituting for $x=2$ we get

$$h = \frac{12 - 2^2}{4(2)} = 1.$$

Dimensions of 2 m for the base and 1 m for the height give us a maximum volume for the cistern.

Also using $V = x^2h$ and substituting for $x=2$ and $h=1$ we get $V_{max} = 2^2 \times 1 = 4\,\text{m}^3$.

███████ **Example 13.25** ███████

We have a site of area 7200 m² to be developed. It is bounded on one side by a shallow river. Find the minimum length of fencing needed to enclose this rectangular area.

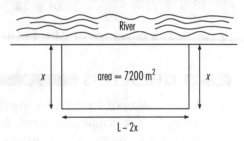

Fig. 13.14

We label one of the sides x and the complete length of fencing L. This means that the other side is $L - 2x$. Linking together our area information for a rectangle we have

$$\text{Area} = (L - 2x)x = 7200$$

i.e. $\qquad\qquad L = \dfrac{7200}{x} + 2x \quad \text{or} \quad 7200x^{-1} + 2x.$

We differentiate once to get

i.e. $\qquad\qquad \dfrac{dL}{dx} = -7200x^{-2} + 2.$

At a turning point $\dfrac{dL}{dx}=0$ so that

$$-7200x^{-2} + 2 = 0$$

i.e. $\qquad\qquad 2 = \dfrac{7200}{x^2}$

i.e. $\qquad\qquad x^2 = 3600$

$\therefore \qquad\qquad x = \pm\sqrt{3600} = \pm 60.$

Our practical example involves distances so $x = -60$ has no practical use. We continue our method with just $x = 60$.

Using $\dfrac{dL}{dx} = -7200x^{-2} + 2$

we differentiate again to get

$$\frac{d^2L}{dx^2} = 14\,400x^{-3}.$$

Substituting for $x = 60$ we get

$$\frac{d^2L}{dx^2} = \frac{14\,400}{(60)^3} = 0.0\bar{6}, \quad \text{i.e. positive.}$$

This means we have a local minimum for a value of $x = 60$.

Using $L = \dfrac{7200}{x} + 2x$

and substituting for $x = 60$ we get

$$L_{\min} = \frac{7200}{60} + 2(60) = 240,$$

i.e. we need a minimum of 240 m of fencing.

■■■■ EXERCISE 13.13 ■■■■■■■■■■■■■■■

1   A rigid beam of length 10 m carries a uniformly distributed load of 10 000 N. The bending moment, M, is related to its distance from one end of the beam, x, according to $M = -10000 + 5000x - 500x^2$. Using differentiation show that the maximum bending moment occurs at the mid-point. What is the value of this maximum bending moment?

2   Sketch the graph of $y = \sin\theta$ in the range $0 \leqslant \theta \leqslant 2\pi$. By differentiating find the coordinates of the local maximum and minimum points. Confirm these with your sketch. Sketch the graph of $y = \cos\theta$ over the same range. Similarly find the turning points for this curve, distinguishing between them. Finally calculate the turning points of the curve $y = \sin\theta - \cos\theta$, again over the same range.

3   A disc is spun from rest. It spins through $y°$ in t seconds according to $y = 100t - 5t^3$. Has this relationship any turning points? If there are any, after what time do they occur and what type are they?

4   One particular machine in an engineering workshop costs £C to lease each week according to the formula $C = 200t - \dfrac{t^3}{30}$. t is the number of hours/week worked by the machine. What is the maximum weekly lease cost?

5   A rectangular sheet of metal has sides of 2000 mm and 3000 mm. From each corner are cut squares of side x mm as shown in the diagram. The metal is now folded as shown.

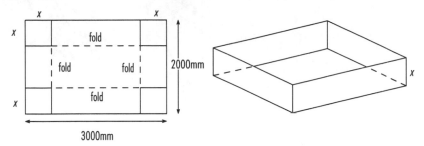

Write down the dimensions of the open rectangular box in terms of x. Similarly write down the volume of this box. Using differentiation find the dimensions that maximise the volume and calculate this maximum volume.

Summary of differentiation rules

y	$\dfrac{dy}{dx}$	
ax^n	anx^{n-1}	where a and n are numbers
$\sin x$	$\cos x$	
$\sin(ax + b)$	$a\cos(ax + b)$	all angles in radians
$\cos x$	$-\sin x$	
$\cos(ax + b)$	$-a\sin(ax + b)$	
e^x	e^x	
e^{ax+b}	ae^{ax+b}	
$\ln x$	$\dfrac{1}{x}$	
$\ln(ax + b)$	$\dfrac{a}{ax + b}$	

▉▉▉▉ MULTI-CHOICE TEST 13 ▉▉▉▉

1 On the curve $y = 2x^2$ the gradient of the chord PQ is:

A) 1
B) 8
C) 10
D) 28

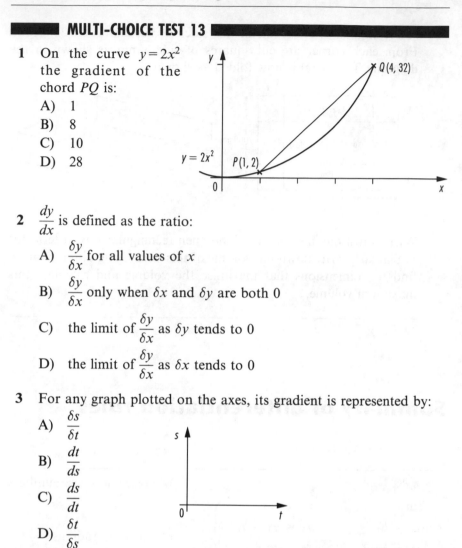

2 $\dfrac{dy}{dx}$ is defined as the ratio:

A) $\dfrac{\delta y}{\delta x}$ for all values of x

B) $\dfrac{\delta y}{\delta x}$ only when δx and δy are both 0

C) the limit of $\dfrac{\delta y}{\delta x}$ as δy tends to 0

D) the limit of $\dfrac{\delta y}{\delta x}$ as δx tends to 0

3 For any graph plotted on the axes, its gradient is represented by:

A) $\dfrac{\delta s}{\delta t}$

B) $\dfrac{dt}{ds}$

C) $\dfrac{ds}{dt}$

D) $\dfrac{\delta t}{\delta s}$

Questions **4** and **5** refer to the accompanying diagram and information.

$y = 3x^2$ is the equation of the curve. P is any point on the curve with the coordinates (x, y). Very close to P, also on the curve, is $R\,(x + \delta x, y + \delta y)$.

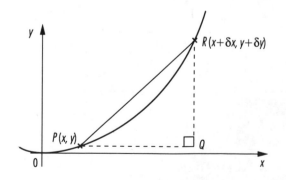

4   The length of RQ is:
   A) δx
   B) $(\delta x)^2$
   C) $6x\delta x + 3(\delta x)^2$
   D) $3(x^2 + 2x\delta x + (\delta x)^2)$

5   Given the length of PQ to be δx, the gradient of PR is:
   A)   1
   B) δx
   C) $6x + 3\delta x$
   D) $3\left(\dfrac{x^2}{\delta x} + 2x + \delta x\right)$

6   The first derivative with respect to x of bx^a is:
   A) ax^{a-1}
   B) abx^{a-1}
   C) abx^{a+1}
   D) $\dfrac{bx^{a+1}}{a}$

7   If $y = 2x^3$ then $\dfrac{dy}{dx} =$:
   A) $3x^2$
   B) $6x^2$
   C) $\dfrac{x^4}{2}$
   D) $\dfrac{2}{3}x^4$

8   Given $y = 3x^2 + 2x$ then $\dfrac{dy}{dx} =$:
   A) $6x + 2$
   B) $x^3 + x^2$
   C) $6x^2 + 2$
   D) $6x^3 + 2x^2$

9   The gradient of the curve $y = x^3 - x^2$ at the point $(-1, -2)$ is:
   A) -2
   B)   0
   C)   5
   D)   8

10   The expression for $\dfrac{dy}{dx}$ on the curve $y = x^{1/2}$ is:

A) $\dfrac{1}{2} x^{1/2}$

B) $\dfrac{1}{2} x^{-1/2}$

C) $x^{-1/2}$

D) $-\dfrac{1}{2} x^{-1/2}$

11   For $y = \dfrac{1}{2\sqrt{x}}$ $\dfrac{dy}{dx} = :$

A) $\dfrac{1}{x^{-1/2}}$

B) $\dfrac{2}{x^{-1/2}}$

C) $-\dfrac{1}{4x^{3/2}}$

D) $\dfrac{1}{x^{1/2}}$

12   The combination of 2 cyclic waveforms is represented by $y = \cos 3t + \sin 2t.\ \dfrac{dy}{dt} = :$

A) $3 \sin 3t - 2 \cos 2t$

B) $3 \sin 2t + 2 \cos t$

C) $3 \cos 2t + 2 \sin t$

D) $-3 \sin 3t + 2 \cos 2t$

13   The curve X is a cosine curve. It is reflected in the horizontal axis to obtain the curve Y. The graph of the gradient of curve Y is curve Z. Decide whether each of the statements is True (T) or False (F).

   i)   Curve Y is a negative cosine curve.

   ii)   Curve Z is a negative sine curve.

Which option best describes the two statements?

A)   i) T   ii)   T

B)   i) T   ii)   F

C)   i) F   ii)   T

D)   i) F   ii)   F

14   If $y = \ln 3x$ then $\dfrac{dy}{dx} = :$

A) $\dfrac{1}{3x}$ B) $\dfrac{1}{x}$

C) $\dfrac{3}{x}$ D) $\ln 3$

15 An exponential decay is given by $y = 2e^{-4t}$. $\dfrac{dy}{dx} =$:

A) $-8e^{-4t-1}$

B) $-8e^{-4t}$

C) $-2e^{-4t}$

D) $-8e^{-5t}$

16 At a maximum turning point $\dfrac{dy}{dx} =$:

A) -1

B) 0

C) 1

D) $\dfrac{d^2y}{dx^2}$

17 The gradient of a curve at a particular point is parallel to the horizontal axis. At this point the gradient is

A) -1

B) 0

C) 1

D) ∞

18 We are considering the graph of a parabola.
Decide whether each of the statements is True (T) or False (F).

i) It has one turning point.

ii) A graph of the gradient is a straight line.

Which option best describes the two statements?

A) i) T   ii)   T

B) i) T   ii)   F

C) i) F   ii)   T

D) i) F   ii)   F

19 The conditions for a local minimum turning point are:

A) $\dfrac{dy}{dx} < 0$ and $\dfrac{d^2y}{dx^2} > 0$

B) $\dfrac{dy}{dx} = 0$ and $\dfrac{d^2y}{dx^2} < 0$

C) $\dfrac{dy}{dx} = 0$ and $\dfrac{d^2y}{dx^2} > 0$

D) $\dfrac{dy}{dx} > 0$ and $\dfrac{d^2y}{dx^2} = 0$

20 The second derivative of $y = 4x^5$ is:

A) $4x^3$

B) $20x^3$

C) $80x^3$

D) $80x^7$

14 Calculus: II – Integrating

Introduction

We introduced Calculus in Chapter 13 through differentiation. In this chapter we look at the opposite process to differentiation. **Integration** is that reverse process. Our early examples each start with the gradient of a graph and aim to find that graph's equation. Later we look to apply the technique to find the area under a curve.

420

■■■ ASSIGNMENT ■■■■■■■■■■■■■■■■■■■■■

Our Assignment for this chapter looks again at the curve $y = 4 - x^2$. We find, using integration, the irregular area under the curve. In this case between the vertical lines $x = 1$ and $x = 2$. We have shaded the area in Fig. 14.1.

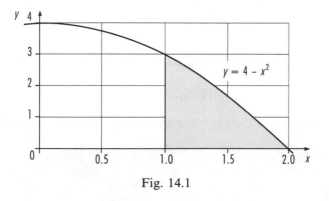

Fig. 14.1

Geometrical interpretation

Suppose we know the gradient of a graph. For example the gradient might be 5. Based on Chapter 2 we know the graph must be a straight line with $m = 5$, i.e. $y = 5x + c$.

Remember c is the intercept on the vertical axis of our graph. We know there are many parallel lines with gradients of 5. In each case they cut the vertical axis at a different point. Fig. 14.2 shows a few examples.

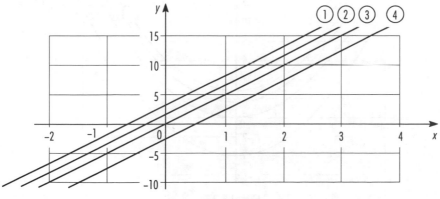

Fig. 14.2

From the many possibilities we can decide upon our particular line if we know a point lying on it.

▬▬▬▬▬ **Example 14.1** ▬▬▬▬▬

Find the equation of the graph with gradient 5 passing through the point $(2, 13)$.

Already we know the graph is a straight line with $y = 5x + c$. Because it passes through the point $(2, 13)$ we can substitute for $x = 2$ and $y = 13$ to get

$$13 = 5(2) + c$$

i.e. $13 - 10 = c$

i.e. $3 = c.$

This means the equation of the straight line is $y = 5x + 3$.

c is going to appear often in integration. It is the **constant of integration**.

From Chapter 13 we know, for example, the curves

$$y = x^2$$
$$y = x^2 - 4$$
$$y = x^2 + 3$$

and $y = x^2 + 1.8$ all have $\dfrac{dy}{dx} = 2x.$

If we start with $\dfrac{dy}{dx} = 2x$

or $\dfrac{dy}{dx} = 2x + 0$

we integrate to get $y = x^2 + c.$

c, the constant of integration, should appear because when we differentiate a constant we get 0.

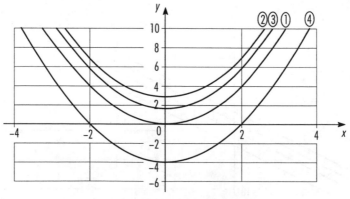

Fig. 14.3

In fact $y = x^2 + c$ represents a family of curves (Fig. 14.3) all with gradient $2x$.

Integration by rule

When we differentiated we saw a pattern of movement of numbers. For integration we have the reverse pattern of movement.

Generally if $\quad \dfrac{dy}{dx} = x^n$

then $\qquad\qquad y = \dfrac{x^{n+1}}{n+1} + c \qquad n \neq -1$

where n is a numerical value.

Notice the important exception to this rule when $n \neq -1$. We cannot allow the possibility of -1 because this would create a divisor of 0. Remember we cannot divide by 0 in Mathematics. We say this integration rule as "increase the power by 1, and divide by the new power".

Remember integration is the reverse process to differentiation. You can check for yourself the differentiation of $y = \dfrac{x^{n+1}}{n+1} + c$ gives the correct answer for $\dfrac{dy}{dx}$ above.

There is an alternative notation,

$$\int x^n \, dx = \frac{x^{n+1}}{n+1} + c \qquad n \neq -1$$

$\int$ is the symbol for integration.

dx means we are going to integrate with respect to x, i.e. x is the variable being processed.

The term(s) between $\int$ and dx is the section to be integrated.

At present we have no values to substitute to find c. Where we include a constant of integration we are performing **indefinite integration**.

▬▬▬▬ Examples 14.2 ▬▬▬▬

With respect to x integrate i) 5, ii) $2x$, iii) $2x + 5$, iv) $5 - 2x$.

In each case we apply our rule "increase the power by 1, and divide by the new power".

i) $\quad \int 5 \, dx = \int 5 \times 1 \, dx$

$\qquad\qquad = \int 5x^0 \, dx$

$\qquad\qquad = \dfrac{5x^{0+1}}{0+1} + c$

$\qquad\qquad = 5x + c.$

$\boxed{x^0 = 1.}$

$\boxed{x^1 = x.}$

ii) $\int 2x\,dx = \int 2x^1\,dx$

$$= \frac{2x^{1+1}}{1+1} + c$$

$$= \frac{2x^2}{2} + c$$

$$= x^2 + c.$$

In this example 2 is a scalar multiplier. We can always move this type of multiplier out of the integral. *We can never do it to a variable.*

Let us have another look at this example.

ii)   Again $\int 2x\,dx = 2\int x\,dx$

$$= 2 \times \frac{x^{1+1}}{1+1} + c$$

$$= 2 \times \frac{x^2}{2} + c$$

$$= x^2 + c \qquad \text{as before.}$$

We can add/subtract our integrals, integrating each term individually. In this case we need only include one constant of integration, c.

iii) $\int 2x + 5\,dx = x^2 + 5x + c.$

iv) $\int 5 - 2x\,dx = 5x - x^2 + c.$

Examples 14.3

With respect to x integrate   i) $3x^7 + 2x^2 - \dfrac{4}{x^3}$

$$\text{and} \quad \text{ii)} \quad \frac{2x^3 - 3x^2}{x}.$$

i)   We need to re-arrange $\dfrac{4}{x^3}$ before we integrate our expression term by term. $\dfrac{4}{x^3}$ needs to be re-written as $4x^{-3}$.

Then $\int 3x^7 + 2x^2 - 4x^{-3}\,dx = \dfrac{3x^{7+1}}{7+1} + \dfrac{2x^{2+1}}{2+1} - \dfrac{4x^{-3+1}}{-3+1} + c$

$$= \frac{3x^8}{8} + \frac{2x^3}{3} - \frac{4x^{-2}}{-2} + c$$

$$= \frac{3x^8}{8} + \frac{2x^3}{3} + 2x^{-2} + c.$$

ii)   We need to simplify our expression before attempting to integrate, again term by term, i.e.

$$\frac{2x^3 - 3x^2}{x} = \frac{2x^3}{x} - \frac{3x^2}{x}$$

$$= 2x^2 - 3x.$$

Then $\int \dfrac{2x^3 - 3x^2}{x}\, dx = \int 2x^2 - 3x\, dx$

$$= \frac{2x^{2+1}}{2+1} - \frac{3x^{1+1}}{1+1} + c$$

$$= \frac{2x^3}{3} - \frac{3x^2}{2} + c.$$

EXERCISE 14.1

Integrate the following expressions.

1 $\int 7\, dx$

2 $\int 5x\, dx$

3 $\int (7 - 5x)\, dx$

4 $\int 3x^2\, dx$

5 $\int (3t^2 + 5t - 7)\, dt$

6 $\int (7 - 5x - 3x^2)\, dx$

7 $\int (x^4 + 2x)\, dx$

8 $\int (4x^6 - 2x^3)\, dx$

9 $\int (5t^{-2} + 3t^{-4})\, dt$

10 $\int \dfrac{3}{t^3}\, dt$

11 $\int \left(\dfrac{5}{x^2} - \dfrac{3}{x^4}\right) dx$

12 $\int \left(2t^3 - \dfrac{3}{t^2}\right) dt$

13 $\int \left(\dfrac{3}{2x^5} - 3x^5\right) dx$

14 $\int \left(\dfrac{3x + 4x^3}{x}\right) dx$

15 $\int \left(\dfrac{2x^2 - x}{x^4}\right) dx$

So far we have concentrated on whole number (integer) powers. We follow exactly the same rule with fractional and decimal powers, i.e. "increase the power by 1, and divide by the new power".

Examples 14.4

i) $\int \sqrt{x}\, dx = \int x^{1/2}\, dx$

$$= \frac{x^{1/2+1}}{\frac{1}{2}+1} + c$$

$$= \frac{x^{3/2}}{\frac{3}{2}} + c$$

$$= \frac{2}{3}x^{3/2} + c.$$

> Dividing by a fraction: invert and multiply.

ii) $\int \dfrac{2}{5\sqrt{x}}\,dx = \dfrac{2}{5} \times \int x^{-1/2}\,dx$

$\boxed{\dfrac{1}{\sqrt{x}} = \dfrac{1}{x^{1/2}} = x^{-1/2}.}$

$= \dfrac{2}{5} \times \dfrac{x^{-1/2+1}}{-\frac{1}{2}+1} + c$

$= \dfrac{2}{5} \times \dfrac{x^{1/2}}{\frac{1}{2}} + c$

$\boxed{\dfrac{2}{5} \times \dfrac{1}{\frac{1}{2}} = \dfrac{2}{5} \times \dfrac{2}{1} = \dfrac{4}{5}.}$

$= \dfrac{4}{5}\, x^{1/2} + c.$

■■■ EXERCISE 14.2 ■■■

Integrate the following expressions.

1 $\int 6\sqrt{x}\,dx$

2 $\int 5x^{0.5}\,dx$

3 $\int 8t^{1/3}\,dt$

4 $\int \dfrac{1}{5}x^{1/4}\,dx$

5 $\int 4t^{1/2}\,dx$

6 $\int (\sqrt{x^3} - 6)\,dx$

7 $\int (\sqrt[3]{x^2} - 4)\,dx$

8 $\int (4t^{0.2} + 2t^{0.4})\,dx$

9 $\int (9x^{0.5} + 4x^{-0.5})\,dx$

10 $\int \dfrac{3}{2x^{0.5}}\,dx$

11 $\int \left(\dfrac{2}{5}x^{-0.5} + \dfrac{1}{5}x^{-1.5}\right)\,dx$

12 $\int \left(\dfrac{5}{2\sqrt[3]{x}} + \dfrac{2}{\sqrt{x}}\right)\,dx$

13 $\int \left(3\sqrt{t} - \dfrac{4}{\sqrt{t}}\right)\,dt$

14 $\int \left(8\sqrt{x} + \dfrac{1}{4\sqrt{x}}\right)\,dx$

15 $\int \left(\dfrac{\sqrt{x} + 4}{2x^2}\right)\,dx$

Definite integration

Earlier we saw the indefinite integral $\int 2x\,dx = x^2 + c$, not knowing the value of c.

In contrast, for **definite integration** we get a value for the integral rather than a general expression. This value comes from the numerical substitution of **limits**.

The general rule is

$$\int_{a}^{b} x^n\,dx = \left[\dfrac{x^{n+1}}{n+1}\right]_{a}^{b}.$$

a is the **lower limit** of x and b is the **upper limit** of x. We expect $a < b$.

We integrate to get some function of x, say $F(x)$. Next we substitute for $x = a$ and $x = b$, and subtract, according to

$$\left[F(x)\right]_a^b = F(b) - F(a).$$

In the following examples we show how to substitute for the values in the a and b positions.

████ **Example 14.5** ██

Find the value of $\int_1^3 (5 - 2x)\, dx$.

We know from Examples 14.2 that

$$\int (5 - 2x)\, dx = 5x - x^2 + c.$$

Thus $\int_1^3 (5 - 2x)\, dx = \left[5x - x^2 + c\right]_1^3$

We substitute for $x = 3$ and for $x = 1$ and subtract to get

$$(5(3) - 3^2 + c) - (5(1) - 1^2 + c)$$
$$= (15 - 9 + c) - (5 - 1 + c)$$
$$= 6 + c - 4 - c$$
$$= 2.$$

Notice how the cs cancel out. Because this always happens we do *not* need to include them.

████ **Example 14.6** ██

Find the value of $\int_{0.5}^3 \left(3\sqrt{t} - \dfrac{4}{\sqrt{t}}\right) dt$.

We know from Exercise 14.2, Question 13, that

$$\int \left(3\sqrt{t} - \frac{4}{\sqrt{t}}\right) dt = 2t^{3/2} - 8t^{1/2} + c.$$

Thus $\int_{0.5}^3 \left(3\sqrt{t} - \dfrac{4}{\sqrt{t}}\right) dt = \left[2t^{3/2} - 8t^{1/2}\right]_{0.5}^3.$

We substitute for $t = 2$ and $t = 0.5$ and subtract to get

$$= \left[2(3^{3/2}) - 8(3^{1/2})\right] - \left[2(0.5^{3/2}) - 8(0.5^{1/2})\right]$$
$$= 2 \times 5.196 - 8 \times 1.732$$
$$\quad\quad -2 \times 0.353\ldots + 8 \times 0.707\ldots$$
$$= 10.392 - 13.856 - 0.707 + 5.657$$
$$= 1.49.$$

▬▬▬ EXERCISE 14.3 ▬▬▬

Find the values of the following integrals.

1 $\int_1^2 2x^3 \, dx$

2 $\int_2^3 3x^2 \, dx$

3 $\int_0^3 (3x - x^4) \, dx$

4 $\int_{0.5}^1 (2t - 7) \, dt$

5 $\int_{-1}^1 (6 - 3x - x^2) \, dx$

6 $\int_6^{12} \frac{3}{t^2} \, dt$

7 $\int_{0.5}^{1.5} (2t^{-2} + 3t^{-4}) \, dt$

8 $\int_1^3 \left(\frac{4}{5x^2} + 2x \right) dx$

9 $\int_{0.5}^{1.5} (4x^3 - \sqrt{x}) \, dx$

10 $\int_2^4 \left(6\sqrt{x} + \frac{2}{3\sqrt{x}} \right) dx$

Integration of sine and cosine

Remembering that integration is the reverse process to differentiation we have

$$\int \cos \theta \, d\theta = \sin \theta + c$$

and $\qquad \int \sin \theta \, d\theta = -\cos \theta + c$

provided θ is in radians.

You need to be careful with the $+/-$ signs. Do *not* confuse differentiation and integration.

More generally we quote the rules:

$$\int \cos (a\theta + b) \, d\theta = \frac{1}{a} \sin (a\theta + b) + c$$

and $\quad \int \sin (a\theta + b) \, d\theta = -\frac{1}{a} \cos (a\theta + b) + c.$

▬▬▬ Examples 14.7 ▬▬▬

i) $\displaystyle\int \cos 2x \, dx = \frac{1}{2} \sin 2x + c.$ $\boxed{a = 2.}$

ii) $\displaystyle\int (3 \cos 2\theta + \sin 4\theta) \, d\theta = \frac{3}{2} \sin 2\theta - \frac{1}{4} \cos 4\theta + c.$

Examples 14.8

i) $\int \cos\frac{1}{2}x\,dx = \frac{1}{\frac{1}{2}}\sin\frac{1}{2}x + c$

$a = \frac{1}{2}$ and $\frac{1}{\frac{1}{2}} = 2.$

$= 2\sin\frac{1}{2}x + c.$

ii) $\int \sin\frac{t}{3}\,dt = -\frac{1}{\frac{1}{3}}\cos\frac{t}{3} + c$

$a = \frac{1}{3}$ and $\frac{t}{3}$ is $\frac{1}{3}t.$

$= -3\cos\frac{t}{3} + c.$

iii) $\int \frac{\sin 4x}{3}\,dx = \int \frac{1}{3}\sin 4x\,dx$

$a = 4$ and the whole function is divided by 3.

$= -\frac{1}{3} \times \frac{1}{4}\cos 4x + c$

$= -\frac{1}{12}\cos 4x + c.$

Now we combine the integration of trigonometric functions with definite integration. Remember the limits are in radians because the integrals are defined in radians.

Example 14.9

Find the value of $\displaystyle\int_{3\pi/4}^{7\pi/8}\left(4 + \sin\frac{2\theta}{3}\right)d\theta.$

Using the standard rules we integrate and then substitute for our limits so that

$$\int_{3\pi/4}^{7\pi/8}\left(4 + \sin\frac{2\theta}{3}\right)d\theta = \left[4\theta - \frac{1}{\frac{2}{3}}\cos\frac{2\theta}{3}\right]_{3\pi/4}^{7\pi/8}$$

$a = \frac{2}{3}.$

$$= \left[4\theta - \frac{3}{2}\cos\frac{2\theta}{3}\right]_{3\pi/4}^{7\pi/8}$$

$\frac{1}{\frac{2}{3}} = \frac{3}{2}.$

$$= \left[4\left(\frac{7\pi}{8}\right) - \frac{3}{2}\cos\frac{2}{3}\left(\frac{7\pi}{8}\right)\right]$$

$$- \left[4\left(\frac{3\pi}{4}\right) - \frac{3}{2}\cos\frac{2}{3}\left(\frac{3\pi}{4}\right)\right]$$

$$= \left[\frac{7\pi}{2} - \frac{3}{2}\cos\frac{7\pi}{12}\right] - \left[3\pi - \frac{3}{2}\cos\frac{\pi}{2}\right].$$

At this stage we can work out each bracket. Alternatively we can gather together similar terms, perhaps removing any common factors to get

$$\frac{7\pi}{2} - 3\pi - \frac{3}{2}\cos\frac{7\pi}{12} - -\frac{3}{2}\cos\frac{\pi}{2}$$

$$= \left(\frac{7}{2} - 3\right)\pi - \frac{3}{2}\left(\cos\frac{7\pi}{12} - \cos\frac{\pi}{2}\right)$$

$$= 1.571 - 1.5(-0.259 - 0)$$

$$= 1.96.$$

EXERCISE 14.4

Integrate the following sines and cosines by rule.

1 $\int 3\sin\theta\, d\theta$ **4** $\int 4\cos 3t\, dt$

2 $\int 5\cos\theta\, d\theta$ **5** $\int 4\cos(3t+7)\, dt$

3 $\int \sin 4\theta\, d\theta$

Find the values of the following integrals.

6 $\displaystyle\int_0^\pi 5\cos\theta\, d\theta$ **8** $\displaystyle\int_\pi^{2\pi}(2\cos 2x - 3\sin 6x)\, dx$

7 $\displaystyle\int_{\pi/4}^{\pi/2}(6\sin 3t + 4\cos 2t)\, dt$ **9** $\displaystyle\int_0^{3\pi/2}\left(12\cos\frac{4x}{3} + 2\sin\frac{x}{4}\right)\, dx$

10 $\displaystyle\int_{\pi/3}^{2\pi/3}\left(6\sin\left(\frac{3\theta}{2}+\pi\right) - 21\cos\left(\frac{3\theta}{2}+\frac{\pi}{2}\right)\right)\, d\theta$

Integration of exponential functions

The integration is as simple as the differentiation, i.e.

$$\int e^{ax}\, dx = \frac{1}{a}e^{ax} + c.$$

Examples 14.10

i) $\displaystyle\int e^{3x}\, dx = \frac{1}{3}e^{3x} + c.$ $a = 3.$

ii) $\displaystyle\int 4e^{3x}\, dx = \frac{4}{3}e^{3x} + c.$

iii) $\displaystyle\int 14e^{x/2}\, dx = \frac{14}{\frac{1}{2}}e^{x/2} + c = 28e^{x/2} + c.$ $a = \frac{1}{2}.$

iv) $\int 6e^{1.5x}\, dx = \dfrac{6}{1.5}e^{1.5x} + c = 4e^{1.5x} + c.$

$\boxed{a = 1.5.}$

Example 14.11

Find the indefinite integral $\int e^{3x+2}\, dx$.

The power of this exponential contains the addition of $3x$ and 2. Using a law of indices we may rewrite this as $e^{3x} \times e^2$. Now e^2 is a number, a constant, with a calculator value of 7.389... We know we can remove a multiplying constant from the integral so that

$$\int e^{3x+2}\, dx = \int e^{3x} \times e^2\, dx$$

$$= e^2 \times \int e^{3x}\, dx$$

$$= e^2 \times \frac{1}{3}e^{3x} + c$$

$\boxed{a = 3.}$

$$= \frac{1}{3}e^{3x} \times e^2 + c$$

$$= \frac{1}{3}e^{3x+2} + c.$$

We see that $+2$ in the power, because it is a pure number, has no effect on the overall integration.

Generally we can write $\displaystyle\int e^{ax+b}\, dx = \dfrac{1}{a}e^{ax+b} + c$ where a and b are constants.

Example 14.12

Find the value of $\displaystyle\int_0^{1.5} \left(2e^{3x} - \dfrac{6}{5e^x}\right) dx$.

We rearrange the second exponential to get

$$\int_0^{1.5} \left(2e^{3x} - \frac{6}{5}e^{-x}\right) dx = \left[\frac{2}{3}e^{3x} - \frac{6}{5(-1)}e^{-x}\right]_0^{1.5}$$

$$= \left[\frac{2}{3}e^{3x} + \frac{6}{5}e^{-x}\right]_0^{1.5}$$

$$= \left[\frac{2}{3}e^{3\times1.5} + \frac{6}{5}e^{-1.5}\right] - \left[\frac{2}{3}e^{3\times0} + \frac{6}{5}e^{-0}\right]$$

$$= \frac{2}{3}e^{4.5} + \frac{6}{5}e^{-1.5} - \frac{2}{3}e^0 - \frac{6}{5}e^0$$

$$= 60.01 + 0.27 - 0.\overline{66} - 1.20$$

$$= 58.4.$$

■ EXERCISE 14.5 ■

Write down and simplify where necessary the following indefinite integrals.

1 $\int (e^{2x} + e^{-2x}) \, dx$

2 $\int (2e^x - 3e^{-x}) \, dx$

3 $\int (e^{x/4} + 2e^{x/2}) \, dx$

4 $\int \left(e^{3t} + \dfrac{4}{e^{-3t}} \right) dt$

5 $\int (2e^{2x+1} + 5e^{1-2x}) \, dx$

Find the values of the following definite integrals.

6 $\displaystyle\int_{0}^{1} \left(\dfrac{e^{4x}}{8} - 4e^{2x} \right) dx$

7 $\displaystyle\int_{-1}^{1} (4e^{0.5x} - 2e^{-1.5x}) \, dx$

8 $\displaystyle\int_{-1.5}^{0.5} \left(\dfrac{e^{2x}}{5} - 5e^{-2x} \right) dx$

9 $\displaystyle\int_{-0.5}^{1.5} \left(2e^t + \dfrac{3}{e^{2t}} \right) dt$

10 $\displaystyle\int_{1.4}^{2.6} 3e^{2x-4} \, dx$

Integration of $\dfrac{a}{x}$

You will remember our general rule for algebraic functions had the exception of $n \neq -1$. Now we will deal with that case where $n = -1$. Remembering the connection between differentiation and integration we have

$$\int \frac{a}{x} \, dx = a \int \frac{1}{x} \, dx = a \ln x + c.$$

You should check back to the differentiation of natural logarithmic functions. This will confirm that integration is the reverse process of differentiation. Notice that we deal only with natural logarithms. Any common logarithms must be converted, changing their base from 10 to e.

■ Examples 14.13 ■

i) $\displaystyle\int \frac{4}{3x} \, dx = \frac{4}{3} \ln x + c.$

ii) $\displaystyle\int \left(1 + \frac{5}{2x} + \frac{1}{x^2} \right) dx = \int \left(1 + \frac{5}{2x} + x^{-2} \right) dx$

$$= x + \frac{5}{2} \ln x + \frac{x^{-1}}{-1} + c$$

$$\text{or} \quad x + 2.5 \ln x - x^{-1} + c.$$

Remember also we can write $x^{-1} = \dfrac{1}{x}$.

■ **Example 14.14** ■

Find the value of $\int_{1.8}^{3.2} \dfrac{3}{4x}\, dx$.

Remember that a logarithmic function is only defined for positive values of x. Both our limits in this example are positive. At this level of mathematics we respect our definition.

Now $\int_{1.8}^{3.2} \dfrac{3}{4x}\, dx = \left[\dfrac{3}{4}\ln x\right]_{1.8}^{3.2}$

> $\dfrac{3}{4}$ or 0.75 is a common factor.

$$= 0.75[\ln 3.2 - \ln 1.8]$$

$$= 0.75[1.163 - 0.588]$$

$$= 0.43.$$

Generally we can write $\int \dfrac{a}{ax+b}\, dx = \ln(ax+b) + c$ where a and b are constants.

■ **EXERCISE 14.6** ■

Find the values of the following definite integrals.

1 $\int_{2.6}^{3.9} \dfrac{6}{x}\, dx$

2 $\int_{0.50}^{1.75} \dfrac{1}{5x}\, dx$

3 $\int_{2.5}^{4.5} \dfrac{6}{5x}\, dx$

4 $\int_{0.25}^{0.6} \left(\dfrac{10}{x} - \dfrac{1}{2x}\right) dx$

5 $\int_{1}^{2} \left(4 - \dfrac{5}{2x} + \dfrac{3}{x^2}\right) dx$

Area under a curve

Integration has many applications. Finding the area under a curve is one simple case.

Fig. 14.4 shows a general curve. We wish to find the area under the curve bounded by the x-axis and the lines $x=a$ and $x=b$. To do this we split the area into a series of thin parallel strips each of width δx.

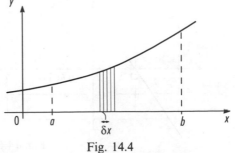

Fig. 14.4

Let us magnify one of those strips. Suppose A is the point (x, y). Remember a small change in x, δx, causes a small change in y, δy. Because B is a small distance away we know it is the point $(x + \delta x, y + \delta y)$. We approximate this strip to a rectangle. As the strip width gets narrower and narrower (i.e. as $\delta x \rightarrow 0$) the accuracy of the approximation improves. This means the extra area above the rectangle becomes very small. Then

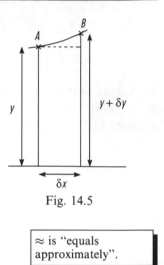

Fig. 14.5

Area of thin strip $\approx$ Area of rectangle

$$\approx y\delta x.$$

$\approx$ is "equals approximately".

The area between $x = a$ and $x = b$ is the sum of all such strips, i.e

$$\text{Area} = \lim_{\delta x \rightarrow 0} \sum_{x=a}^{x=b} y\delta x.$$

In practice this sum to find the area would be difficult to calculate. We can reach the same result more easily using the integration formula

$$\text{Area} = \int_a^b y \, dx.$$

We easily apply this formula in the next examples.

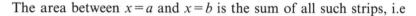

Example 14.15

Find the area under the curve $y = (1 + x)(4 - x)$ from $x = 0.6$ to $x = 3$.

We assume the area under the curve is bounded by the horizontal axis. The shading in Fig. 14.6 shows our area of interest.

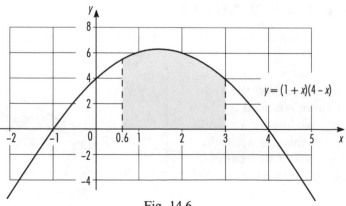

Fig. 14.6

Before attempting the integration we need to multiply out the brackets, i.e.

$$(1+x)(4-x) = 4 + 3x - x^2$$

Now using our formula

$$\text{Area} = \int_a^b y\,dx$$

we substitute for $a = 0.6$, $b = 3$ and y to get

$$\text{Area} = \int_{0.6}^{3} (4 + 3x - x^2)\,dx$$

$$= \left[4x + \frac{3x^2}{2} - \frac{x^3}{3}\right]_{0.6}^{3}$$

$$= \left(4(3) + \frac{3(3)^2}{2} - \frac{3^3}{3}\right) - \left(4(0.6) + \frac{3(0.6)^2}{2} - \frac{0.6^3}{3}\right)$$

$$= 16.500 - 2.868$$

$$= 13.63 \text{ unit}^2.$$

Because we have been given no units (e.g. mm or m) we use unit^2 (e.g. mm^2 or m^2) for the area.

Examples 14.16

Find the area bounded by the curve $y = \cos x$ and the x-axis between

i) $x = 0$ and $x = \dfrac{\pi}{2}$,

ii) $x = \dfrac{\pi}{2}$ and $x = \pi$

and iii) $x = 0$ and $x = \pi$.

Fig. 14.7 is a reminder of the cosine curve.

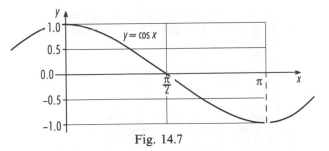

Fig. 14.7

Notice that some of the area is above and some below the x-axis. This will be noted by our area formula. Areas below the x-axis are given as negative and we need to interpret that sign. We apply our area formula to these questions.

i)   Area $= \displaystyle\int_0^{\frac{\pi}{2}} \cos x \, dx$

$= \Big[\sin x \Big]_0^{\frac{\pi}{2}}$

$= \sin\dfrac{\pi}{2} - \sin 0$

$= 1 \text{ unit}^2.$

ii)   Area $= \displaystyle\int_{\frac{\pi}{2}}^{\pi} \cos x \, dx$

$= \Big[\sin x \Big]_{\frac{\pi}{2}}^{\pi}$

$= \sin \pi - \sin\dfrac{\pi}{2}$

$= -1$

i.e. the area is 1 unit^2 *below* the x-axis.

iii)   Because the graph crosses the horizontal axis it splits the areas. We need to deal with each section separately. If we fail to do this then the positive and negative areas may cancel out each other.

We consider the area from $x=0$ to $x=\pi$ as

the area from $x=0$ to $x=\dfrac{\pi}{2}$

and the area from $x=\dfrac{\pi}{2}$ to $x=\pi$.

Remember we know the meaning of the negative sign and do *not* have to worry about it.

Total area $= 1+1 = 2 \text{ unit}^2.$

■■■■■■■ ASSIGNMENT ■■■■■■■

We have needed quite an amount of work before being able to look at our assignment. Fig. 14.8 shows the shaded area we are going to calculate.

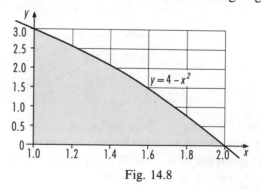

Fig. 14.8

Using our formula

$$\text{Area} = \int_a^b y\,dx$$

we substitute for $a=1$, $b=2$ and $y=4-x^2$ to get

$$\text{Area} = \int_1^2 (4-x^2)\,dx$$

$$= \left[4x - \frac{x^3}{3}\right]_1^2$$

$$= \left(4(2) - \frac{2^3}{3}\right) - \left(4(1) - \frac{1^3}{3}\right)$$

$$= 5.\bar{3} - 3.\bar{6}$$

$$= 1.\bar{6} \text{ unit}^2.$$

EXERCISE 14.7

1   Find the area bounded by the curve $y=x^3+2$ and the x-axis from $x=0$ to $x=1.5$.

2   $y = \dfrac{x^4+2}{x^2}$. Find the area lying between this curve and the x-axis from $x=1$ to $x=2$.

3   Sketch the curve $y=e^x$. Find the area bounded by both the axes, the curve and the line $x=2.75$.

4   From $x=5$ to $x=10$ find the area between the x-axis and the curve $y=\dfrac{1}{x}$.

5   From $\theta=\pi$ to $\theta=\dfrac{3\pi}{2}$ find the area between the curve $y=\cos\theta$ and the horizontal axis. With a sketch of the graph confirm that your answer should be negative.

6   What is the area between the curve $y=\dfrac{4}{x}$ and the x-axis from $x=\frac{1}{2}$ to $x=1$?

7   Over one cycle find the area bounded by the curve $y=4\sin\theta$ and the horizontal axis.

8   Find the area bounded by the curve $y=x^3$ and the x-axis from $x=-2$ to $x=2$.

9　a)　Construct a table of values from $x = -3$ to $x = 3$ at intervals of 0.5 for $y = (1 - x)(2 - x)$. Now plot y vertically and x horizontally.
　　　　Where does this graph cross the horizontal axis?
　　　　Find the area enclosed by the curve and the horizontal axis.
　　b)　Repeat the tasks for $y = (x - 1)(x - 2)$ and notice any similarities/differences between your answers.

10　For values of $x = 0$ to $x = 5$ at intervals of 0.5 construct a table of values for $y = x^2 - 3x$. Hence plot a fully labelled graph of y vertically against x horizontally.
　　On the same axes plot the straight line $y = x$.
　　Show that these two graphs intersect at the points $(0, 0)$ and $(4, 4)$.
　　From $x = 0$ to $x = 4$ find the area between
　　　i)　$y = x^2 - 3x$ and the x-axis,
　　　ii)　$y = x$ and the x-axis,
　　　iii)　the two graphs.

Now we can apply our area formula to a further practical problem before you attempt a final short exercise.

▌▌▌ Example 14.17 ▌▌▌

The area under a velocity-time graph represents displacement. Fig. 14.9 shows the curve $v = \dfrac{1}{10}(4 - t)(2 + t)$ where v (ms^{-1}) is the velocity at time, t (s). For the first 6 seconds find the
　　　i)　distance travelled
and　ii)　displacement from the starting point.

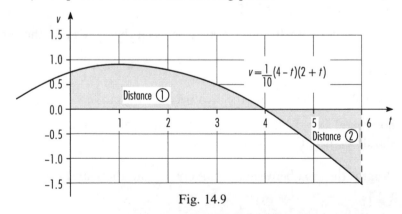

Fig. 14.9

Because our area is split above and below the horizontal axis we need to be careful. When this graph crosses the horizontal axis the velocity is 0. There the velocity changes from being positive to negative. For the area below the horizontal axis the distance is negative, being in the opposite direction, i.e. re-tracing part of the original path.

Before attempting to find the distance using our area formula we need to multiply out the brackets, i.e.

$$\frac{1}{10}(4-t)(2+t) = \frac{1}{10}(8+2t-t^2).$$

Applying our formula of

$$\text{Area} = \int_a^b y \, dx$$

we get

i)   Distance ① = Area above the horizontal axis

$$= \frac{1}{10}\int_0^4 (8+2t-t^2)\,dt$$

$$= 0.1\left[8t + t^2 - \frac{t^3}{3}\right]_0^4$$

$$= 0.1\left(\left(8(4) + 4^2 - \frac{4^3}{3}\right) - (0)\right)$$

$$= 0.1\,(26.\bar{6})$$

$$= 2.\bar{6} \text{ m}.$$

Distance ② = Area below the horizontal axis

$$= \frac{1}{10}\int_4^6 (8+2t-t^2)\,dt$$

$$= 0.1\left[8t + t^2 - \frac{t^3}{3}\right]_4^6$$

$$= 0.1\left(8(6) + 6^2 - \frac{6^3}{3}\right) - \left(8(4) + 4^2 - \frac{4^3}{3}\right)$$

$$= 0.1\,(12 - 26.\bar{6})$$

$$= -1.4\bar{6} \text{ m}.$$

The negative sign confirms the distance is in the reverse direction.

Total distance $= 2.\bar{6} + 1.4\bar{6} = 4.1\bar{3}$ m.

ii) Displacement is a vector quantity (Chapter 7). It accepts that a distance in one direction can cancel a distance in an opposite direction. This means we must include our negative sign in $-1.4\bar{6}$ m. Then we get

Displacement $= 2.\bar{6} - 1.4\bar{6} = 1.20$ m.

EXERCISE 14.8

1   $v = 30 - 4t$ relates speed, v (ms^{-1}), to time, t (s). The area under a speed-time graph represents the distance travelled. By integration find the distance travelled in the first 5 seconds.
Plot this straight line graph of v against t for the first 5 seconds. Using the formula for the area of a trapezium check your first answer for the distance travelled.

2   Hooke's law relates the tension in an elastic spring, T (N), according to $T = \dfrac{\lambda x}{l}$. λ (N) is the coefficient of elasticity, l (m) is the natural length and x (m) is the extension from that natural length. If $\lambda = 30$ N and $l = 0.75$ m show that $T = 40x$. For T plotted vertically and x plotted horizontally the area under the graph represents the work done (J) in stretching the spring. This means

$$\text{Work done} = \int_a^b T \, dx.$$

Find the work done in stretching the spring from its natural length to 1.05 m.

3   In electrical theory we may plot voltage, V, vertically against current, I (A), horizontally. The area under such a curve is associated with power, P (W). This is

$$\text{Power} = 2\int_a^b V \, dI.$$

For a resistor of 470 Ω we have $V = 470I$. Find the power associated with an increase in current from 1×10^{-3} A to 1.08×10^{-3} A.

4   A variable force, F (N), is related to distance, s (m), by $F = 5s^2 + 2s$. The work done (J) by this force is given by

$$\text{Work done} = \int_a^b F \, ds.$$

Calculate the work done by the force as s increases from 3.50 m to 4.75 m.

5   The pressure, p (Nm^{-2}), and volume, V (m^3) of a gas are related by $pV^{1.5} = k$. If $p = 150 \times 10^3$ Nm^{-2} and $V = 0.5$ m^3 evaluate k.
Re-arrange your general formula to make p the subject.
The work done (J) expanding the gas from this volume to 0.75 m^3 is given by the integral formula

$$\text{Work done} = \int_{0.50}^{0.75} p \, dV.$$

Calculate this work done.

Summary of integration rules

y	$\int y\,dx$
ax^n	$\dfrac{ax^{n+1}}{n+1}$

where a and n are numbers and $n \neq -1$.

$\sin x$	$-\cos x + c$
$\sin(ax+b)$	$-\dfrac{1}{a}\cos(ax+b) + c$
$\cos x$	$\sin x + c$
$\cos(ax+b)$	$\dfrac{1}{a}\sin(ax+b) + c$

all angles in radians.

e^x	$e^x + c$
e^{ax+b}	$\dfrac{1}{a}e^{ax+b} + c$
$\dfrac{1}{x}$	$\ln x + c$
$\dfrac{a}{ax+b}$	$\ln(ax+b) + c$

▰▰▰ MULTI-CHOICE TEST 14 ▰▰▰

1 The diagram shows the graph of the gradient of a particular function. That function is represented by a:
   A) horizontal line
   B) vertical line
   C) parabola
   D) logarithmic graph

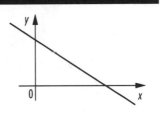

2 Every trigonometric function has a gradient. Our diagram shows the gradient of that function. That function's graph is a:
   A) cosine wave
   B) straight line
   C) negative cosine wave
   D) sine wave

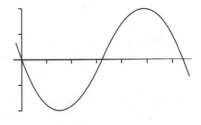

3 $\int (4 + 6x)\,dx$ is:
   A) $6 + c$
   B) $4x + 3x^2 + c$
   C) $3x^2 + c$
   D) $\dfrac{10x^2}{2} + c$

4 $\int \dfrac{5}{x^3}\, dx$ is:

A) $\dfrac{5}{2}\ln x^2 + c$

B) $-\dfrac{5}{4}x^{-4} + c$

C) $\dfrac{5}{2}\ln x^3 + c$

D) $-\dfrac{5}{2}x^{-2} + c$

5 $\int\left(\dfrac{2x - 4x^3}{3}\right) dx$ is:

A) $2 - 12x^2 + c$

B) $\dfrac{x^2 - x^4}{3} + c$

C) $\dfrac{x^2 - x^4}{3x} + c$

D) $\dfrac{2 - 12x^2}{3} + c$

6 $\int \sqrt[4]{x^3}\, dx$ is:

A) $\dfrac{4}{7}x^{7/4} + c$

B) $\dfrac{7}{4}x^{7/4} + c$

C) $\dfrac{3}{7}x^{7/3} + c$

D) $\dfrac{7}{3}x^{7/3} + c$

7 $\int \dfrac{2}{3\sqrt{x}}\, dx$ is:

A) $-4x^{-3/2} + c$

B) $-\dfrac{4}{9}x^{-3/2} + c$

C) $\dfrac{1}{3}x^{1/2} + c$

D) $\dfrac{4}{3}x^{1/2} + c$

8 $\int 4 \sin(5 - 3\theta) \, d\theta$ is:

A) $\dfrac{4}{3} \cos(5 - 3\theta) + c$

B) $-\dfrac{4}{3} \cos(5 - 3\theta) + c$

C) $\dfrac{4}{5} \cos(5 - 3\theta) + c$

D) $-\dfrac{4}{5} \cos(5 - 3\theta) + c$

9 The value of $\displaystyle\int_0^{\pi/2} 2 \sin\dfrac{t}{2} \, dt$ is given by:

A) $\left[4 \cos\dfrac{t}{2} \right]_0^{\pi/2}$

B) $4 \cos 0 + 4 \cos\dfrac{\pi}{4}$

C) $-4 \cos 0 + 4 \cos\dfrac{\pi}{4}$

D) $-4 \cos\dfrac{\pi}{4} + 4 \cos 0$

10 The gradient of a particular curve is the negative exponential decay. The original function is $y =$:
A) $-e^x + c$
B) $e^x + c$
C) $e^{-x} + c$
D) $\ln x + c$

11 $\displaystyle\int \dfrac{1}{2e^{x/2}} \, dx$ is:

A) $-e^{-x/2} + c$
B) $-4e^{-x/2} + c$
C) $e^{-x/2} + c$
D) $\dfrac{1}{4} e^{-x/2} + c$

12 $\displaystyle\int \left(\dfrac{2}{x} + \dfrac{1}{x^2} \right) dx$ is:

A) $2 \ln x + \ln x^2 + c$

B) $\ln 2x - \dfrac{1}{x} + c$

C) $2 \ln x - \dfrac{1}{3x^3} + c$

D) $2 \ln x - \dfrac{1}{x} + c$

13 The general integral $\int bx^a \, dx$ is:

A) $abx^{a+1} + c$

B) $\dfrac{ax^{a-1}}{b} + c$

C) $\dfrac{bx^{a+1}}{a} + c$

D) $\dfrac{bx^{a+1}}{a+1} + c$

14 The general exponential integral $\int e^{a-bt} \, dt$ is:

A) $\dfrac{1}{a} e^{a-bt} + c$

B) $-be^{a-bt} + c$

C) $-\dfrac{1}{b} e^{a-bt} + c$

D) $ae^{a-bt} + c$

15 $\displaystyle\int \dfrac{m}{n+px} \, dx$ is:

A) $mp \ln(n+px) + c$

B) $mn \ln(n+px) + c$

C) $\dfrac{m}{n} \ln(n+px) + c$

D) $\dfrac{m}{p} \ln(n+px) + c$

16 The shaded area is given by:

A) $\displaystyle\int_a^b t \, dt$

B) $\displaystyle\int_a^b t \, dy$

C) $\displaystyle\int_a^b y \, dy$

D) $\displaystyle\int_a^b y \, dt$

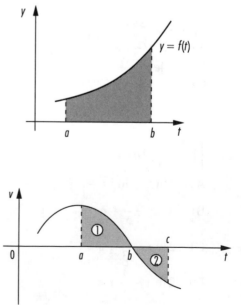

Questions **17** and **18** refer to the following information.
The area under a velocity-time graph represents displacement. Area ① is positive. Area ② is negative.

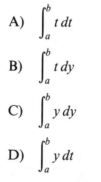

17 From $t=a$ to $t=c$ the displacement is given by:

A) Area ① $-$ Area ②

B) |Area ①| $+$ |Area ②|

C) Area ① $\times$ Area ②

D) Area ① $+$ Area ②

18 From $t=a$ to $t=c$ the distance travelled is given by:

A) |Area ①| $+$ |Area ②|

B) |Area ①| $-$ |Area ②|

C) $\dfrac{\text{Area ①}}{\text{Area ②}}$

D) Area ① $+$ Area ②

19 The formula to find the shaded area is:

A) $\displaystyle\int_c^d x\,dy$

B) $(b-a)(d-c) - \displaystyle\int_a^b y\,dx$

C) $(b-a)d - \displaystyle\int_a^b y\,dx$

D) $\dfrac{1}{2}(d-c)(b-a)$

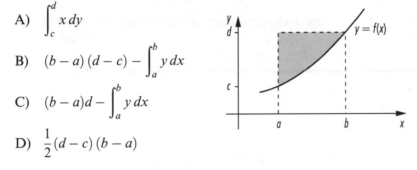

20 The shaded area is found using:

A) $\displaystyle\int_a^0 y_1\,dx + \int_0^b y_2\,dx$

B) $-\displaystyle\int_a^0 y_1\,dx + \int_0^b y_2\,dx$

C) $\displaystyle\int_a^b (-y_1 + y_2)\,dx$

D) $\displaystyle\int_a^b (y_1 + y_2)\,dx$

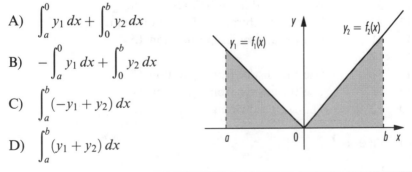

Answers to Exercises

Chapter 1

Exercise 1.1

1	90	
2	−90	Care with the position of the minus sign within the multiplication.
3	71	
4	−71	Care with the position of the minus sign within the division.
5	71	(−) ÷ (−) gives +.
6	90	Work out the brackets first.
7	90	
8	22	
9	18	
10	426	
11	5	
12	9	Use the answer from Question **11**.
13	−4	
14	5	Work out addition before division.
15	24	Work out subtraction before division.
16	31	Use the answer from Question **15**.
17	17	
18	51	Use the answer from Question **17**.
19	32	Use the answer from Question **11**.
20	32	Think after dividing by a fraction.

Exercise 1.2

1	81	
2	243	
3	243	Only 9 is squared.
4	243	Apply the minus sign to 9 before squaring.
5	729	Square each number before multiplying.
6	729	Multiply and then square. Link with Question **5**.
7	36	Add, then divide and finally square.
8	43	Use the answer from Question **7**.
9	193	Work out the bracket, square it and then add 49.

10	16	Only the bracket is squared.
11	±3	
12	±9	
13	±9	Square and square root 'cancel' out each other.
14	±27	
15	±1	Subtract, square root and then divide.
16	±2	Subtract, divide and then square root.
17	12, 16	Use the answers from Question **16** to give $14 + 2$ and $14 - 2$.
18	13, 15	
19	−23, 31	Use the answers from Question **14**.
20	7, 9	Use the answers from Question **15** to give $4 + 4 + 1$ and $4 + 4 - 1$.
21	1296	
22	1296	Apply the minus sign before raising to the power 4.
23	−1296	Apply the power 4 before the minus sign.
24	1298	Use the answer from Question **21**.
25	−1294	
26	54	Apply the power 3 before multiplying by 2.
27	384	Apply the power 5 to 4, then divide and multiply.
28	−33	Use the answer from Question **26**.
29	13845	Do *not* confuse with Question **28**.
30	15	

Exercise 1.3

1	0.99	**2**	0.3885	**3**	17.39	**4**	5.7
5	21.40	**6**	4	**7**	10		

8 220 000   Trailing zeroes are *not* significant.
9 0.00376   Leading zeroes are *not* significant.

10	3200	**11**	62.0	**12**	6.2	**13**	20

14 620.4   Included zeroes are significant.

15	620.5	**16**	100	**17**	5490	**18**	1 000 000
19	4200	**20**	0.071				

Take care when adjusting the decimal point and the powers of 10.

21	9.92×10^{-1}	**22**	3.8847×10^{-1}
23	1.738847×10	**24**	5.69
25	2.1399×10	**26**	4.32
27	1.432×10	**28**	2.17×10^5
29	3.7602×10^{-3}	**30**	3.165×10^3
31	6.204×10	**32**	6.204
33	2.03×10	**34**	6.20401×10^2
35	6.20461×10^2	**36**	9.98×10
37	5.48897×10^3	**38**	1.4×10^6
39	4.23906×10^3	**40**	7.0629×10^{-2}

Exercise 1.4

i) **Estimated answers**:

Remember these are estimates. Be bold with your approximate cancellations. Do *not* aim to get exactly the textbook answers.

1	24	**2**	6.5	**3**	14 or 15	**4**	0.5
5	30	**6**	42	**7**	67	**8**	43
9	68	**10**	−25	**11**	25	**12**	$\frac{1}{4}$
13	4000	**14**	4000	**15**	1200	**16**	1.5
17	75	**18**	$\frac{1}{2}$	**19**	$2\frac{1}{2}$	**20**	24

ii) **Calculator answers**:

We do not know the accuracy of the original values. These answers are cautiously rounded.

1	24.6	(1 dp)		**2**	7	(1 sf)
3	14.5	(1 dp)		**4**	0.41	(2 dp)
5	30	(1 sf)		**6**	48	(2 sf)
7	67	(2 sf)		**8**	46.7	(1 dp)
9	68	(2 sf)		**10**	−50	(1 sf)
11	25.7	(1 dp)		**12**	0.3	(1 sf)
13	4436	(4 sf)		**14**	4436	(4 sf)
15	1157	(4 sf)		**16**	1.5	(2 sf)
17	75.9	(3 sf)		**18**	0.4	(1 dp)
19	3	(1 sf)		**20**	27	(2 sf)

Exercise 1.5

1	37.8	(3 sf)	
2	112	(3 sf)	Only 2.96 is cubed.
3	3.8	(2 sf)	Square each number and then add.
4	18.4	(3 sf)	Use the raw answer from Question **3**, then multiply and round.
5	36.8	(3 sf)	Add within the bracket. Only the bracket is squared.
6	±2.9	(2 sf)	± because the root is an even root.
7	±2.9	(2 sf)	Compare with Question **6** – alternative notation.
8	3.22	(3 sf)	Odd roots have the same sign as the original.
9	−3.22	(3 sf)	The original is negative, so the odd root must be negative too.
10	−0.28	(2 sf)	Use earlier, raw, unrounded answers.
	and −6.2	(2 sf)	
11	6.2	(2 sf)	
	and 0.28	(2 sf)	
12	2	(1 sf)	Use the raw answer from Question **3**.
13	0.159	(3 dp)	
14	0.1	(1 dp)	Addition before 'one over', i.e. before the reciprocal.

15 0.9   (1 dp)   2 multiplied by π. Now subtract 4.9, then reciprocal, finally square root.

16 2   (1 sf)   Cube root of 119.6, then divide.

17 0.5   (1 dp)   Reciprocal of Question **16**.

18 1.3   (1 dp)   Work out the numerator and denominator separately. Finally divide and then round the answer.

19 1.8   (1 dp)   Use the raw answer from Question **17**.

20 3.7   (1 dp)

Exercise 1.6

1 $0.2\,\text{m}^2$ (1 dp)

2 $1500\,\text{mm}^2$ (2 sf)   Only r is squared.

3   12 N, 51 N   (2 sf)

4   71.3 m   (1 dp)   Only t is squared.

5   0.356 m   (3 dp)   Do *not* approximate the decimal places too early.

6   1070   (3 sf)

7   4.7 W   (1 dp)

8   140 mm   (2 sf)   Use the calculator memory. Do *not* approximate until the end.

9 $2\,\Omega$ (1 sf)

10   0.2 m   (1 sf)

11   131 m   (3 sf)

12 $1.1\,\text{ms}^{-2}$ (2 sf)   Square values before subtracting.

13 $126\,000\,\text{mm}^2$ (3 sf)

14   2.5 s   (2 sf)

15 $22\,\text{ms}^{-1}$ (2 sf)

16 $67.5\,\text{mm},\ 780\,\text{mm}^2$ Work out s. Work out the brackets, multiply, finally square.

17   7.1, 300.3°K

18   7.9, $300\,\Omega$

19

v	0	2	4	6	8	10	12
Q	0	0.01	0.02	0.03	0.04	0.05	0.06

20

t	0	10	20	30	40	50	60
v	1.00	25.5	50.0	74.5	99.0	123.5	148.0

Multi-choice Test 1

1	C	**2**	C	**3**	C	**4**	B
5	D	**6**	A	**7**	D	**8**	A
9	C	**10**	A	**11**	A	**12**	D
13	D	**14**	B	**15**	D	**16**	D
17	B	**18**	D	**19**	D	**20**	D

Chapter 2

Exercise 2.1

1	31	**2**	−16	**3**	31	**4**	7
5	−6	**6**	2	**7**	−3	**8**	12
9	$\gamma - \alpha$	**10**	$\alpha - \beta$				

Exercise 2.2

1   0.8   **2**   1.3

3   −1.8   The minus sign is attached to the 2 during division: no change to it.

4	−10.7	**5**	28	**6**	0.33	**7**	4.26
8	$2.2\overline{6}$	**9**	−2.5	**10**	−0.34		

Exercise 2.3

1   1.5   Move the 2 and then the 6.

2   5.3

3   20   Multiply throughout by 5 to remove the fraction.

4   −20

5   2.8   Gather $2y$ and $3y$ on the left. Gather 17 and 3 on the right. Now you should be able to divide by 5.

6	9	**7**	18	**8**	13.125	**9**	−11
10	2.4						

Exercise 2.4

1   −0.625   Multiply out the bracket.

2 $1.8\overline{3}$ Care with the minus signs when multiplying the brackets.

3 $-1.\overline{3}$ Multiply throughout by 20 to remove the fractions.

4 $7.\overline{3}$

5   20

6   −20   Multiply throughout by 6 to remove the fractions, then multiply out the bracket.

7   6.6

8   15   Multiply throughout by 12.

9 $-0.0\overline{9}$

10 $-1.1\overline{36}$ Remove the fractions, then multiply out each bracket.

Exercise 2.5

1–10

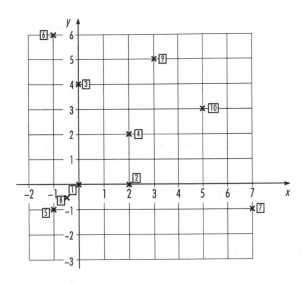

11	Lie on a line at 45° to the horizontal.
12	Lie on a line at 45° to the horizontal.
13	Lie on a line through $y = 1$.
14	Lie on a vertical line through $x = 3$.
15	No pattern for them all.

Exercise 2.6

	Gradient	Intercept	Equation
1	3	15	$y = 3x + 15$
2	-1.25	12.5	$y = -1.25x + 12.5$
3	-2	-10	$y = -2x - 10$
4	5	0	$y = 5x$
5	-25	550	$y = -25x + 550$

Care with the intercept – there is no origin.

6	-1	4	$y = -x + 4$

Generally work out the gradient and substitute for m in $y = mx + c$. Substitute a pair of the coordinates to find the intercept, c.

7	$\dfrac{2}{3}$	$-\dfrac{1}{3}$	$y = \dfrac{2}{3}x - \dfrac{1}{3}$ or $3y = 2x - 1$
8	-1	8	$y = -x + 8$
9	1	3	$y = x + 3$
10	$\dfrac{3}{13}$	$-\dfrac{8}{13}$	$y = \dfrac{3}{13}x - \dfrac{8}{13}$ or $13y = 3x - 8$

Exercise 2.7

2   2, −1; 15, 2.5
3   3, 50; 440 N   Not all the points lie on a straight line. Don't expect your answers to match these exactly.
4   125 s, 3.19
5   *a, b*; 1.8, 32; 20
6   400, −1200; −500 N, 3 m
7   13.5, gradient
8    $v = mt + c$; 0.24, 1.7; 16.1 ms^{-1}, 58.0 kmh^{-1}, 36.2 mph
     Not all the points lie on a straight line. Don't expect your answers to match these exactly.
9   39 Ω
10   0.2, 1.7 × 10^{-5}; 586°C   Care with the decimal places on the vertical scale.

Multi-choice Test 2

1	C	2	C	3	D	4	C
5	D	6	A	7	C	8	C
9	B	10	A	11	D	12	B
13	B	14	B	15	A	16	D
17	D	18	D	19	C	20	A

Chapter 3

Exercise 3.1

Replace the proportional symbol with '$= k$'
1   0.4, 0.6
2   192
3   10, care, question asks for *x* rather than *y*, 1.17
4   7.$\overline{27}$
5   3.266, *k*
6   ±6.36
7   15, cannot find the square root of a negative number.
8   Read the question carefully. 50
9   150, 17.36
10   $d = kt$, 672 m
11   33.$\overline{3}$ N
12   $V = kr^3$, 0.46 m^3
13   1.96 s
14   $d = kv^2$, 17.15 ms^{-1}
15   $T = \dfrac{k}{\omega}$, 0.57 s

Exercise 3.2

1 $x = -\alpha + \beta$

2 $x = -\alpha + \beta + \gamma$

3 $x = -\alpha + \beta + \gamma$

4 $x = \dfrac{\beta + \gamma}{\alpha}$ Move β then α.

5 $x = \alpha(\delta - \beta)$ Move β then α.

6 $x = \left(\dfrac{\delta}{\gamma} - \beta\right)\alpha$ Move γ, β and then α.

7 $x = \dfrac{\gamma}{\beta - \alpha}$ First gather the x terms on the left.

8 $x = \dfrac{ac}{1 - ab}$ Multiply through by a. Gather the x terms.

9 $x = \dfrac{ab(c - d)}{b - a}$ Multiply through by ab first.

10 $x = \dfrac{1}{b - a}$

11 $x = \dfrac{a}{b + c}$

12 $x = \dfrac{bc}{1 - ab}$

13 $x = \dfrac{a - b}{d}$ Multiply throughout by x.

14 $P = \dfrac{100I}{RT}$

15 $d = \dfrac{C}{\pi}$

16 $b = \dfrac{2A}{h}$

17 $F = \dfrac{9C}{5} + 32$ Dividing by 9 becomes multiplying. Multiplying by 5 becomes dividing.

18 $h = \dfrac{2A}{a + b}$ Multiplying by $\dfrac{1}{2}$ becomes dividing by $\dfrac{1}{2}$, i.e. multiplying by 2.

19 $x = \dfrac{S + 2}{W}$

20 $t = \dfrac{2s}{u + v}$

21 $I_1 = -(I_2 + I_3 + I_4)$

22 $r = \dfrac{C}{2\pi}$

23 $R = \dfrac{100I}{PT}$

24 $h = \dfrac{2A}{b}$

25 $c_1 = \dfrac{Cc_2}{c_2 - C}$ Multiply throughout by $c_1 c_2 C$.

26 $v = \dfrac{fu}{u - f}$

27 $T = \dfrac{pV}{mR}$

28 $p_2 = \dfrac{p_1 V_1}{V_2}$ The first move is $p_1 V_1 = p_2 V_2$

29 $t = \left(\dfrac{R}{R_0} - 1\right)\dfrac{1}{\alpha}$ Move R_0, 1 and then α.

30 $R = N(1 - S)$ Multiply throughout by N.

31 $a = \dfrac{2A}{h} - b$

32 $T = \dfrac{100I}{PR}$

33 $u = \dfrac{2s}{t} - v$

34 $N = \dfrac{4V}{\pi D^2 f}$

35 $b = \dfrac{2A}{h} - a$

36 $v = \dfrac{Ft}{m} + u$

37 $N = \dfrac{R}{1 - S}$

38 $\alpha = \dfrac{\Omega - \omega}{t}$

39 $t = \left(\dfrac{l}{l_0} - 1\right)\dfrac{1}{\alpha}$

40 $\alpha = \dfrac{R_1 - R_2}{R_2 t_1 - R_1 t_2}$ Multiply throughout by $R_2(1 + \alpha t_2)$.

Exercise 3.3

1 $r = \pm\sqrt{\dfrac{A}{\pi}}$ Find r^2, then square root.

2 $v = \pm\sqrt{\dfrac{32Pr}{W}}$

3 $b = \pm\sqrt{a^2 - c^2}$

4 $r = \pm\sqrt{\dfrac{V}{\pi h}}$

5 $\quad x = \pm\sqrt{\dfrac{2aW}{\lambda}}$

6 $\quad \omega = \pm\dfrac{v}{r}$ 　　　Perfect square: square roots 'cancel' with the squares.

7 $\quad x = \pm\sqrt{2(1 - C)}$ 　　　Multiply throughout by 2, make x^2 the subject, then square root.

8 $\quad x = \pm\sqrt{\dfrac{y - b}{a}}$

9 $\quad h = \dfrac{v^2}{2g}$

10 $\quad x = \left(\dfrac{y - b}{a}\right)^2$ 　　　Move b then a. Now square.

11 $\quad x = \pm\sqrt{2(Ch - 1)}$

12 $\quad r = 100\left(1 \pm \sqrt{\dfrac{A}{P}}\right)$ 　　　Move P, now square root. Move 1, then 100.

13 $\quad b = \pm\sqrt{a^2 - c^2}$

14 $\quad d = \pm\sqrt{(A - 320)\dfrac{4}{\pi}}$ 　　　Move 320, then π and 4. Now square root.

15 $\quad V_R = \pm\sqrt{V^2 - (V_L b - V_C)^2}$

　　　Square throughout, make V_R^2 the subject, then square root.

16 $\quad r - 200\left(\pm\sqrt{\dfrac{A}{P}} - 1\right)$

17 $\quad u = \pm\sqrt{v^2 - \dfrac{2Ck}{m}}$

18 $\quad t = \dfrac{v^2 - gr\upsilon}{gr + v^2\upsilon}$ 　　　Square throughout. Multiply throughout by $(1 - \upsilon t)$. Gather together the t terms.

19 $\quad s = c \pm \dfrac{1}{\sqrt{Lf}}$ 　　　Multiply throughout by $(s - c)^2$, square root, make s the subject.

20 $\quad t = \pm\sqrt{\dfrac{1 - C}{1 + C}}$ 　　　Multiply throughout by $(1 + t^2)$, gather the t^2 terms, make t^2 the subject, then square root.

Exercise 3.4

1 $\quad x = \sqrt[3]{\dfrac{y - b}{a}}$ 　　　Move b, then a. Now cube root.

2 $\quad x = \sqrt[5]{\dfrac{y - c}{m}}$

3 $x = \pm \sqrt[4]{\dfrac{-y-c}{m}}$ Move y and c, then m. Now fourth root.

4 $x = \left(\dfrac{y-b}{a}\right)^3$ Move b, then a. Now cube throughout to remove the cube root.

5 $x = \left(\dfrac{y-c}{m}\right)^4$

6 $x = \left(\dfrac{-y-c}{m}\right)^5$

7 $c = \pm z^{1/4} - ts$ Fourth root, move ts.

8 $R = 100\left(\left(\dfrac{A}{P}\right)^{1/10} - 1\right)$ Move P, tenth root, move 1, move 100.

9 $T = \left(\dfrac{c}{S}\right)^7$

10 $D = \left(\dfrac{32z}{\pi}\right)^{1/3}$

11 $V = \left(\dfrac{c}{p}\right)^{1/\gamma}$

12 $d = \left(\dfrac{12I}{b}\right)^{1/3}$

13 $T = \left(\dfrac{c}{S}\right)^{10}$

14 $s = \dfrac{c - z^{1/3}}{t}$ Cube root, move c, divide by $-s$, adjust signs.

15 $R = 100\left(1 - \left(\dfrac{A}{P}\right)^{1/7}\right)$

16 $D = \left(\dfrac{T}{kf^{0.75}}\right)^{5/9}$ 1.8 is $\dfrac{9}{5}$. Moving the power gives $\dfrac{1}{\frac{9}{5}} = \dfrac{5}{9}$.

17 $D = \pm\left(\dfrac{32I}{\pi}\right)^{1/4}$

18 $f = \left(\dfrac{T}{kD^{1.8}}\right)^{4/3}$

19 $D = \left(\dfrac{12I + bd^3}{B}\right)^{1/3}$

20 $d = \pm\left(D^4 - \dfrac{32Dz}{\pi}\right)^{1/4}$ Multiply by $32D$ and divide by π. Move D^4 and adjust the minus signs. Fourth root.

Multi-choice Test 3

1	C	**2**	D	**3**	B	**4**	B
5	C	**6**	A	**7**	B	**8**	D
9	B	**10**	A	**11**	C	**12**	B
13	A	**14**	A	**15**	D	**16**	D
17	D	**18**	A	**19**	B	**20**	D

Chapter 4

Exercise 4.1

Questions **1–20**: the unit is the radian.
For each angle in degrees multiply by π and divide by 180.

1	$\dfrac{\pi}{18}$	**2**	$\dfrac{\pi}{4}$	**3**	$\dfrac{\pi}{3}$	**4**	$\dfrac{5\pi}{6}$
5	$\dfrac{\pi}{3}$	**6**	$\dfrac{7\pi}{6}$	**7**	$\dfrac{7\pi}{4}$	**8**	2π
9	3π	**10**	6π	**11**	0.44	**12**	1.28
13	2.92	**14**	3.93	**15**	4.45	**16**	5.17
17	5.37	**18**	6.02	**19**	8.73	**20**	10.65

Questions **21–30**: the unit is the degree.
For each angle in radians multiply by 180 and divide by π.

21	401.07	**22**	260.70	**23**	297.94	**24**	120.00
25	71.62	**26**	779.22	**27**	225.00	**28**	108.86
29	300.00	**30**	352.37				

Exercise 4.2

1	0.5446	**2**	0.7986	**3**	−0.8572	**4**	0.7986
5	−7.1154	**6**	2.1445				
7	−0.8572	The $+/-$ should be used with the angle before applying the trig function.					
8	−0.9925	**9**	−2.1445	**10**	−0.6691	**11**	0
12	0.4027	**13**	1.6977	**14**	−1.6003	**15**	0.9455
16	−0.9063	**17**	0.4877	**18**	−0.8746	**19**	−0.4067
20	1.0355	**21**	−4.3315	**22**	−0.7632	**23**	−0.6494
24	−0.7986	**25**	−0.3256	**26**	1.4281	**27**	0.9397
28	0.3256	**29**	0.5446	**30**	−0.8572		

Exercise 4.3

Use the radian mode on your calculator, then there is *no* need to convert any angles.

1 0.9093   **2** 0.7248   **3** 0.9425   **4** 0.2203
5 −1   3 multiplied by π and then divided by 4 before applying tan.
6 −0.9995   **7** −0.4461   **8** −0.8660
9 −0.8660   5 multiplied by π and divided by 6. Now use the +/−| key, before applying cos.
10 −1.2602

Exercise 4.4

Use the degrees mode on your calculator. Your calculator will give one answer. In the chapter we saw how to find the second answer (where it does exist).

1 60° using 0.866| inv| sin|, 120°   **2** 45°, 135°
3 60°, 300°   **4** 76.35°, 283.65°
5 45°, 315°
6 210°, 330°   Apply the +/−| key to the value before inv| sin|.
7 240°, 300°   **8** 36.41°, 143.59°
9 0°, 360°   **10** 153.25°, 206.75°
11 82.38°, 277.62°
12 90°   Check the graph to confirm there is only one answer.
13 0°, 180°, 360°
14 No solution because cosine must lie between −1 and 1.
15 120°, 240°   **16** 270°
17 No solution   **18** 90°, 270°
19 180°   **20** 225°, 315°

Exercise 4.5

Use the degrees mode on your calculator. Your calculator will give one answer. In the chapter we saw how to find the second answer.

1 26.57° using 0.5| inv| tan|, 206.57°
2 56.31°, 236.31°
3 112.20°, 292.20°   Apply the +/−| key to the value before inv| tan|.
4 80.95°, 260.95°   **5** 45°, 225°
6 120°, 300°   **7** 123.69°, 303.69°
8 72.90°, 252.90°   **9** 107.10°, 287.10°
10 96.83°, 276.83°

Exercise 4.6

1   24.62° using sine
3   33.69° using tangent
5   32.23° using sine
7   65.98°
9   36.66°

2   26.11° using cosine
4   33.06° using sine
6   35.54°, 54.46°
8   20.19°
10   65.38°

Exercise 4.7

1   80 mm using sine
3   37 mm using tangent
5   1.2 m using tangent
7   0.5 m
9   3.2 m

2   94 mm using cosine
4   2.4 m using sine
6   4.2 m
8   63 mm
10   3.4 m

Exercise 4.8

1   3.0 m using sine
3   336 m using tangent
5   617 mm using sine
7   767 mm
9   404 mm

2   5.2 m using cosine
4   6.7 m using cosine
6   4.2 m
8   6.9 m
10   772 mm

Exercise 4.9

1   1.9 m using cosine
3   23.9° using tangent
5   3.4 m using sine
6   5.739°, 11.478°, 19.9%; care to use many decimal places
7   8.4 N, 14.6 N
8   $1.35 - 1.33 = 0.02$ m, 10.7°
9   1.24 m, 1.24 m, $(1.24 \times 2) + 1.47 = 3.9$ m
10   10 m, 8 m, 11.0 m using sine, 9.8 m using cosine.

2   68.2°, 21.8° using tangent
4   2.45 m using tangent

Exercise 4.10

1   2.366 using $\dfrac{1}{\cos 65°}$

2   1 using $\dfrac{1}{\tan 45°}$

3   -2 using $\dfrac{1}{\sin(-150)°}$

4   0.577

5   1.155

6   -1.604

Remember to use the radian mode on your calculator.

7   2.670
9   -1.013

8   -12.64
10   -1.082

Exercise 4.11

Most answers should simply be checking both sides of identities.

5   Use $\dfrac{1}{\tan}=\cot.$

9 $\dfrac{1}{0}$ is not defined; 180° and 360° for cosecant and cotangent, 270° for secant.

10   Yes

Multi-choice Test 4

1	D	**2**	B	**3**	A	**4**	A
5	C	**6**	B	**7**	C	**8**	D
9	D	**10**	C	**11**	D	**12**	D
13	C	**14**	D	**15**	A	**16**	B
17	D	**18**	C	**19**	B	**20**	A

Chapter 5

Exercise 5.1

In Questions **1–5** first use the sum of the angles of a triangle is 180°. In Questions **6–10** this is used later in each solution.

1   Isosceles triangle means you only need to apply the sine rule once.

The style is $b=\dfrac{0.25\times\sin 70°}{\sin 40°}$

$b=0.37\,\text{m}$, $c=0.37\,\text{m}$, $\angle C=70°$

2 $a=0.124\,\text{m}$, $b=0.119\,\text{m}$, $\angle B=52°$

3 $b=1.74\,\text{m}$, $c=1.46\,\text{m}$, $\angle A=118°$

4 $a=8.90\,\text{m}$, $b=5.08\,\text{m}$, $\angle C=80°$

5 $b=0.54\,\text{m}$, $c=0.56\,\text{m}$, $\angle A=86°$

6   From $\sin B=0.9510$ there are two possible angles between 0° and 180° that will form triangles.

The style is $\sin B=\dfrac{0.36\times\sin 50°}{0.29}$

This is followed by $c=\dfrac{0.29\times\sin 58°}{\sin 50°}$

$c=0.32\,\text{m}$, $\angle B=72°$, $\angle C=58°$;

$c=0.14\,\text{m}$, $\angle B=108°$, $\angle C=22°$

7 $a=0.72\,\text{m}$, $\angle A=132.8°$, $\angle C=21.7°$

8 $b=1.2\,\text{m}$, $\angle A=22.5°$, $\angle B=37.5°$

9 $b=0.36\,\text{m}$, $\angle B=81°$, $\angle C=58°$;
   $b=0.11\,\text{m}$, $\angle B=17°$, $\angle C=122°$
10 $c=1.05\,\text{m}$, $\angle A=27°$, $\angle C=105°$

Exercise 5.2

In Questions **1–5** find the third side, then use the other version of the cosine rule to find a second angle. Finally use the sum of the angles of a triangle is 180°.

1 $c=0.83\,\text{m}$, $\angle A=53°$, $\angle B=65°$
2 $a=0.64\,\text{m}$, $\angle B=116°$, $\angle C=21°$
3 $b=0.53\,\text{m}$, $\angle A=14°$, $\angle C=95°$
4 $c=0.63\,\text{m}$, $\angle A=106°$, $\angle B=49°$
5 $b=0.70\,\text{m}$, $\angle A=25.5°$, $\angle C=45.5°$

In Questions **6–10** find any angle and then a second angle. Finally use the sum of the angles of a triangle is 180°.

6 $\angle A=15°$, $\angle B=112°$, $\angle C=53°$
7 $\angle A=20°$, $\angle B=97°$, $\angle C=63°$
8 $\angle A=33°$, $\angle B=38°$, $\angle C=109°$
9 No triangle is possible because $a+b<c$
10 $\angle A=126°$, $\angle B=22°$, $\angle C=32°$

Exercise 5.3

Use Area $=\frac{1}{2}\times$ base $\times$ altitude.

1 $0.096\,\text{m}^2$ **2** $1.96\,\text{m}^2$ **3** $0.15\,\text{m}^2$ **4** $0.032\,\text{m}^2$
5 $0.032\,\text{m}^2$

Use Area $=\frac{1}{2}\times a\times b\times \sin C$ or similar.

6 $0.28\,\text{m}^2$ **7** $0.10\,\text{m}^2$ **8** $0.04\,\text{m}^2$ **0** $0.34\,\text{m}^2$
10 $0.08\,\text{m}^2$

Use Area $=\sqrt{s\,(s-a)\,(s-b)\,(s-c)}$ where $s=\frac{1}{2}(a+b+c)$.

11 $5.6\,\text{m}^2$ **12** $7.6\,\text{m}^2$ **13** $0.31\,\text{m}^2$
14 No triangle is possible because $a+b<c$
15 $6.46\,\text{m}^2$

Exercise 5.4

1 $3\,000\,\text{m}^2$ using the area of a triangle, $16\,000\,\text{m}^2$ using the area of a rectangle minus the area of the triangle.
2 Find the length of the sloping base using Pythagoras' theorem. The sides are trapezia. Separately find the areas of the sides, ends and base. $121.4\,\text{m}^2$, $17.5\,\text{m}^2$, 14.4%
3 From the area of the sector subtract the area of the triangle. $2680\,\text{mm}^2$

4   Use trig to find the bases and altitudes. From the area of the larger triangle subtract the area of the smaller triangle. $0.16\,m^2$

5   Split the trapezium into a triangle and a rectangle. Use trig to find the adjacent side. Subtract the area of the circle from the area of the trapezium. $0.52\,m^2$

6   The area of a parallelogram is twice the area of its constituent triangles. $0.56\,m^2$

7   Think about your solution to Question **5**. $0.50\,m^2$

8   Accurately draw the triangle on graph paper. Create a large right-angled triangle and subtract two smaller right-angled triangles and a square. 13.5 unit^2

9   Use the sine rule to find the length *BD*. $4172\,m^2$

10   Think about your solution to Question **8**. $2.43\,m^2$

Exercise 5.5

1   $0.97\,m$, $14°$
2   $0.46°$
3   $9.8°$, $8.2°$
4   $10\,m$.   Create a right-angled triangle, labelling the opposite side $t + h$.
5   $24.33\,m$, $6.13\,m$.   Create two right-angled triangles with opposite sides of $t + v$.

Exercise 5.6

1   i) 69.3, ii) $75°$ Find *PR*. Use Pythagoras' theorem and cosine.
2   $7.3°$, $30.5°$ Use Pythagoras' theorem, sine and tangent.
3   $0.91\,m$, $62.1°$, $69.4°$ Split the base into right-angled isosceles triangles. Find the length of a hypotenuse. Use tangents.
4   $0.42\,m^2$, $0.39\,m^2$, $28°$ Use the cosine rule to find the angle.
5   $55.2°$, $1.62\,m$ Use cosine and Pythagoras' theorem.
6   $0.82\,m^2$
7   $144\,m^2$, $32\,m$ Use cosine. $AB = 14 - 4 - 4 = 6\,m$.
8   $8.19°$, $8.21°$ Use Pythagoras' theorem and tangent.
9   i) $53.1°$, ii) $57°$; $2.65\,m^2$.   Use the hint for triangles.
10   i) $26.6°$, ii) $26.6°$, iii) $26.6°$, iv) $30°$, v) $30°$.   Think about tangent.

Multi-choice Test 5

1	B	2	C	32	A	4	C
5	A	6	D	7	B	8	C
9	B	10	A	11	C	12	B
13	B	14	A	15	B	16	A
17	C	18	B	19	C	20	C

Chapter 6

Exercise 6.1

1 i) 1,   ii) 3,   iii) 120°, 2.09 rad   iv) 0.48 Hz

2 i) 5,   ii) 2,   iii) 180°, 3.14 rad   iv) 0.32 Hz

3 i) 1,   ii) 3,   iii) 120°, 2.09 rad,   iv) 0.48 Hz

4 i) 2,   ii) 3,   iii) 120°, 2.09 rad,   iv) 0.48 Hz

5 i) 1,   ii) 1,   iii) 360°, 6.28 rad,   iv) 0.16 Hz

6 1, 2, 3

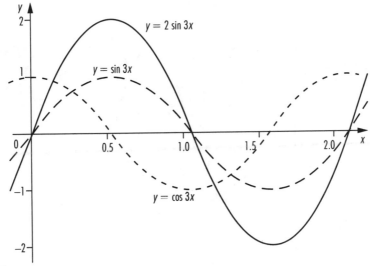

 4, 5

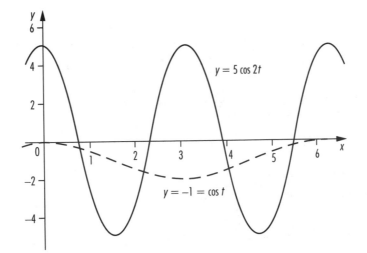

7 $y = 4 \sin 15.7t$

8 $y = 2 \sin 6t$

9

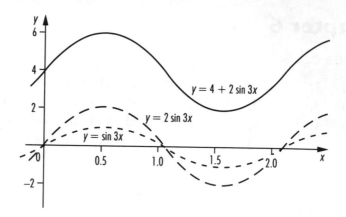

$y = 4 + 2 \sin 3x$

$y = 2 \sin 3x$

$y = \sin 3x$

11

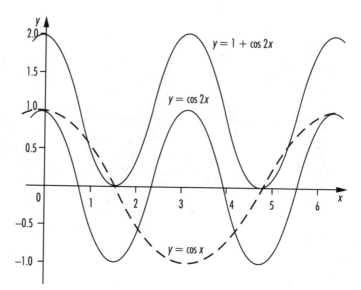

$y = 1 + \cos 2x$

$y = \cos 2x$

$y = \cos x$

14 π rad

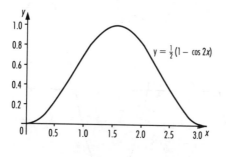

$y = \frac{1}{2}(1 - \cos 2x)$

15

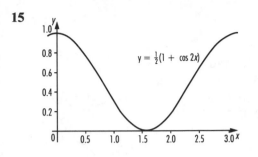

$y = \frac{1}{2}(1 + \cos 2x)$

Exercise 6.3

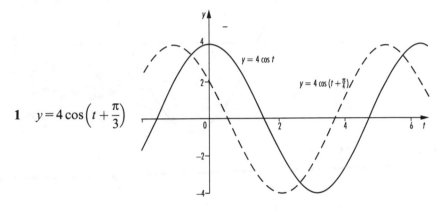

$y = 4\cos t$

$y = 4\cos\left(t + \frac{\pi}{4}\right)$

1 $y = 4\cos\left(t + \dfrac{\pi}{3}\right)$

2 $y = 3\sin\left(2t - \dfrac{\pi}{6}\right)$

3 $y = \sin(x + 90°)$

4 i_i leads i_2 by $\dfrac{\pi}{6}$

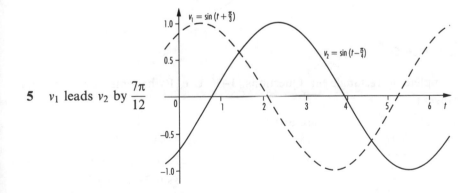

$v_1 = \sin\left(t + \frac{\pi}{3}\right)$

$v_2 = \sin\left(t - \frac{\pi}{4}\right)$

5 v_1 leads v_2 by $\dfrac{7\pi}{12}$

Exercise 6.4

1　$y = 13 \sin(x + 67.4°)$, lead 67.4°
2　$y = \sqrt{5} \sin(2t + 1.107)$, lead 1.107 rad
3　$y = 16.97 \sin(x + 45°)$, lead 45°
4　$y = 6.71 \sin(2t + 1.107)$, lead 1.107 rad
5　$y = 10.82 \sin(3x + 33.7°)$, lead 33.7°

1

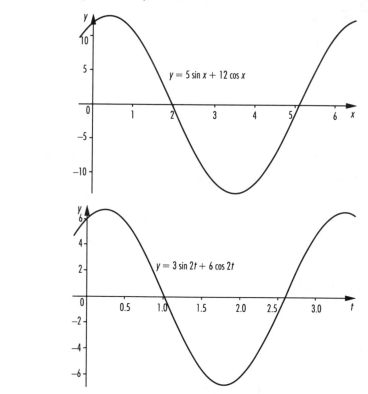

4

Exercise 6.5

Complete a rectangle in Questions **1–3**. Use Pythagoras' theorem and tangent.

	amplitude	phase angle	
1	16.97	45°	$y = 16.97 \sin(x + 45°)$
2	7.21	33.7°	$y = 7.21 \sin(2t + 33.7°)$
3	7.83	26.6°	$y = 7.83 \sin(3x + 26.6°)$

Complete a parallelogram in Questions **4** and **5**. Use the cosine and sine rules.

4	1.93	$\dfrac{\pi}{12}$	$i = 1.93 \sin\left(2t + \dfrac{\pi}{12}\right)$
5	1.22	$\dfrac{\pi}{24}$	$y = 1.22 \sin\left(t + \dfrac{\pi}{24}\right)$

Exercise 6.6

1

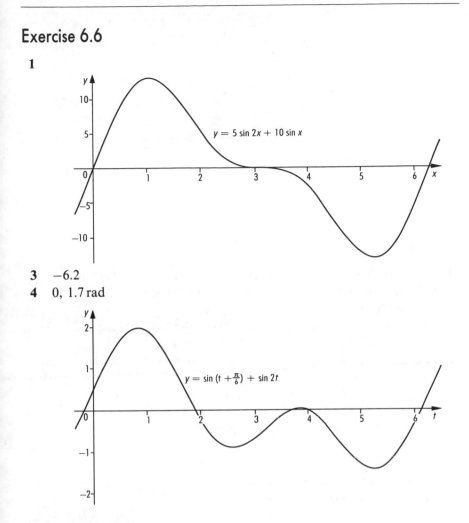

$y = 5 \sin 2x + 10 \sin x$

3 -6.2

4 0, 1.7 rad

$y = \sin\left(t + \frac{\pi}{6}\right) + \sin 2t$

Multi-choice Test 6

1	A	**2**	B	**3**	D	**4**	C
5	D	**6**	C	**7**	C	**8**	D
9	A	**10**	C	**11**	D	**12**	A
13	C	**14**	B	**15**	B	**16**	B
17	A	**18**	B	**19**	C	**20**	C

Chapter 7

Exercise 7.1

1 Think about direction. i) scalar, ii) vector, iii) vector, iv) vector, v) scalar

2 2 m due North

3 i) *CB*, ii) *CB*, iii) *AC*, iv) *AD* + *CD* + *CF*, v) *BE* + *FE*

4 i)   ii)

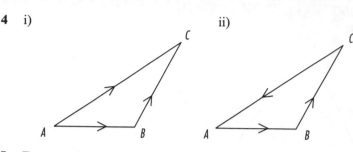

5 Draw and label a square first. i) true, ii) true, iii) false, iv) false,
v) true

Exercise 7.2

1 i) 5 N, 8.66 N;   ii) 4.95 ms^{-1}, 4.95 ms^{-1};
iii) 1.20 m, 3.29 m;   iv) 5.64 ms^{-1}, 2.05 ms^{-1};
v) 12 N, 0 N

2 Write in the angles between the forces and the inclined planes. i) 4.33 N,
2.50 N; ii) 2.72 N, 1.27 N; iii) 2.57 N, 3.06 N

3 Complete the rectangle to show the direction of the resultant vector.
i)   5 N, 36.9° above horizontal;
ii)   9.43 N, 32.0° above horizontal;
iii) 6.05 ms^{-1}, 34.2° below horizontal;
iv)   5.83 N, 59.0° below horizontal;
v)   7.27 m, 55.6° below horizontal;
vi)   7.43 ms^{-1}, 49.6° above horizontal;
vii) 8.49 N vertically upwards
viii) 4.88 m, 65.0° above horizontal

Exercise 7.3

Resolve each vector into horizontal and vertical components. We give the
sum of those, without units, as an aid.
i) 8.893, 2.668; 9.28 N, 16.7° above horizontal;
ii) 6.457, 2.407; 6.89 ms^{-1}, 20.4° above negative horizontal;
iii) 8.642, 16.289; 18.44 m, 62.1° below negative horizontal;
iv) 2.287, 7.793; 8.12 N, 73.6° below horizontal

Exercise 7.4

Answer by accurate scale drawing using a ruler and protractor. Remember a
negative sign reverses the direction of a vector.

Exercise 7.5

Apply any scalar multiplication before simplifying to reach the final answer. Take care with minus signs applying to all components of a vector.

1   $i - j - 4k$ 2 $4i + j - 6k$
3   $-i + 7j - k$ 4 $5(i - 2k)$
5   $-3i + 27j - 8k$ 6 $-5(i - 2k)$
7   $i + 7.5j - 4k$ 8 $20.5j - 10k$
9   $3.6i + 4.4j - 6.9k$ 10 $2i + 40j - 43k$

Exercise 7.6

1   Use Pythagoras' theorem. 4.12, 10.44, 7.21
2   $MN = -i + 3j - 9k$, 9.54
3   $3i + 3j + 5k$, 6.56
4   Equate components $-8, -2$
5   7.81, 10.82, 37.17
6   Draw the forces and complete the rectangle. 8.06 N, 60.3° to vertical
7   2.69 ms^{-1}, 42° to vertical
8   12.81 ms^{-1}, 051.3°; Use ratios, 160 m
9   191.3°, 31.50 ms^{-1}; find the sum of the horizontal and vertical components, 4.243 N, 31.213 N; find an acute angle and interpret as a bearing, 187.7°
10   Separate diagrams will aid your solution. $-100i$, $-35.36(i + j)$, $-120j$, $-135.36i - 155.36j$, 206 m, 9.64°

Multi-choice Test 7

1	A	2	C	3	B	4	C
5	C	6	D	7	C	8	D
9	D	10	D	11	A	12	B
13	D	14	D	15	B	16	A
17	D	18	D	19	B	20	A

Chapter 8

Exercise 8.1

1   $x^2 + 7x + 10$ 2 $10x^2 + 7x + 1$
3   $x^2 + 7x + 10$ Notice this answer agrees with Question 1.
4   $10x^2 + 26x + 12$ 5 $3x^2 + 17x + 10$
6   $12x^2 + 26x + 10$ 7 $x^2 - 7x + 10$
8   $12 - 7x + x^2$ 9 $x^2 - 14x + 40$
10   $12x^2 - 17x + 6$ 11 $12x^2 - 29x + 14$

12 $6x^2 - 5x + 1$ **13** $x^2 - 3x - 10$

14 $x^2 - 5x - 36$ **15** $4x^2 - 2x - 12$

16 $2x^2 + 3x - 2$ **17** $49x^2 - 14x - 3$

18 $66x^2 + x - 45$ **19** $x^2 + 3x - 10$

20 $9x^2 + x - 10$ **21** $6x^2 + 20x - 26$

22 $22x^2 + 86x - 8$ **23** $2x^2 + x - 10$

24 $6x^2 + 5x - 6$

25 $x^2 + 2x + 1$ Write as $(x + 1)(x + 1)$ and then multiply out the brackets.

26 $12x^2 - 7x - 10$ **27** $16 - 8x + x^2$

28 $x^2 - 2x + 1$ **29** $x^2 - 8x + 16$

30 $x^2 + 10x + 25$ **31** $4x^2 + 28x + 49$

32 $49 - 28x + 4x^2$

33 $x^2 - 4$ Notice the original question and that the answer has no term in x.

34 $4x^2 - 169$ **35** $6x^2 - 6$

36 $7x + 5x^2$ **37** $28x - 4x^2$

38 $-14x^2 - 63x$ **39** $10x^2 + 23x + 12$

40 $8x^2 - 10x - 33$ **41** $3 - 16x - 12x^2$

42 $9 - 82x + 9x^2$ **43** $2x + x^2$

44 $9 - 61x - 14x^2$ **45** $6x^2 - 10.5x - 3$

46 $15 - t - 2t^2$ **47** $6t^2 - 7t - 49$

48 $8x - 12x^2$ **49** $6t^2 - 7t - 49$

50 $14t^2 + 61t - 9$

Exercise 8.2

1

x	−3.5	−3.0	−2.5	−2.0	−1.5	−1.0	−0.5	0	0.5	1.0
$2x^2$	24.5	18.0	12.5	8.0	4.5	2.0	0.5	0	0.5	2.0
$4x$	−14.0	−12.0	−10.0	−8.0	−6.0	−4.0	−2.0	0	2.0	4.0
-5	−5.0	−5.0	−5.0	−5.0	−5.0	−5.0	−5.0	−5.0	−5.0	−5.0
y	5.5	1.0	−2.5	−5.0	−6.5	−7.0	−6.5	−5.0	−2.5	1.0

2 Only some tables and graphs are given.

i)

x	−4	−3	−2	−1	0	1	2	3
y	16	9	4	1	0	1	4	9

iii)

x	−2.0	−1.5	−1.0	−0.5	0	0.5	1.0	1.5	2.0
$3x^2$	12.00	6.75	3.00	0.75	0	0.75	3.00	6.75	12.00
$5x$	−10.00	−7.50	−5.00	−2.50	0	2.50	5.00	7.50	10.00
y	2.00	−0.75	−2.00	−1.75	0	3.25	8.00	14.25	22.00

i) Graph for Question **8.2i)**

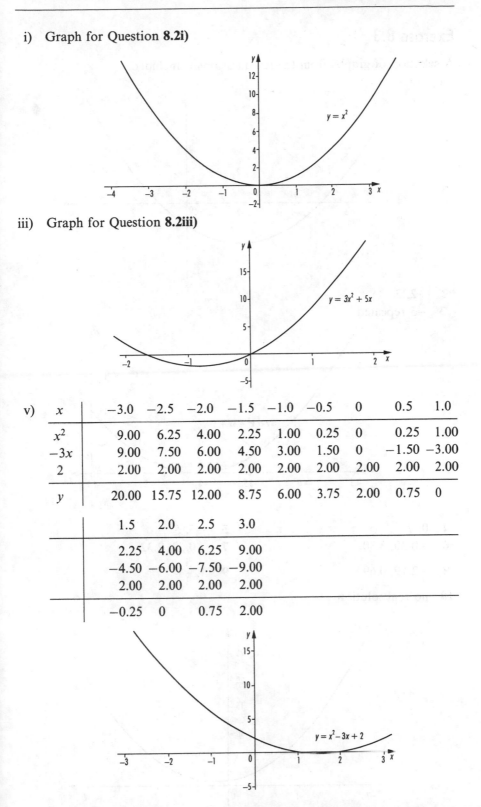

iii) Graph for Question **8.2iii)**

v)

x	-3.0	-2.5	-2.0	-1.5	-1.0	-0.5	0	0.5	1.0
x^2	9.00	6.25	4.00	2.25	1.00	0.25	0	0.25	1.00
$-3x$	9.00	7.50	6.00	4.50	3.00	1.50	0	-1.50	-3.00
2	2.00	2.00	2.00	2.00	2.00	2.00	2.00	2.00	2.00
y	20.00	15.75	12.00	8.75	6.00	3.75	2.00	0.75	0

	1.5	2.0	2.5	3.0
	2.25	4.00	6.25	9.00
	-4.50	-6.00	-7.50	-9.00
	2.00	2.00	2.00	2.00
	-0.25	0	0.75	2.00

Exercise 8.3

A selection of graphs from the ten questions is included.

1 −5, 1

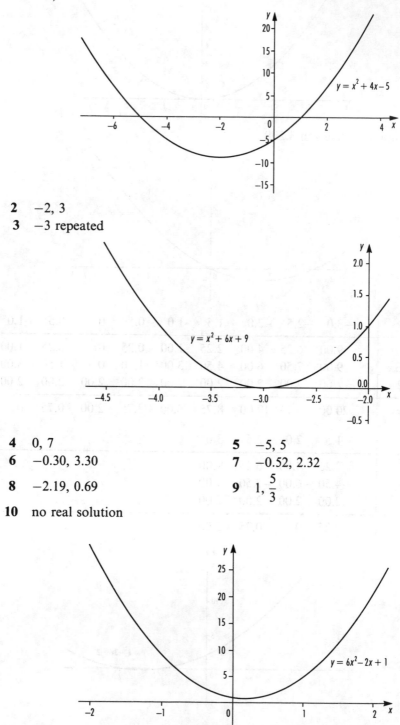

$y = x^2 + 4x - 5$

2 −2, 3

3 −3 repeated

$y = x^2 + 6x + 9$

4 0, 7

5 −5, 5

6 −0.30, 3.30

7 −0.52, 2.32

8 −2.19, 0.69

9 $1, \dfrac{5}{3}$

10 no real solution

$y = 6x^2 - 2x + 1$

Exercise 8.4

Remember the order of signs from the original quadratic equation through to those in the brackets.

Signs in the brackets are both $+$.

1	$-5, -1$	**2**	$-3, -2$
3	$-4, -3$	**4**	$-2, -1$
5	-7 repeated	This is because of $(x + 7)^2 = 0$	

Signs in the brackets are both $-$. Here the constant term is positive and the x term is negative.

6	2, 3	**7**	3, 9
8	5, 6	**9**	5 repeated
10	2, 6		

Different signs in the brackets. Here the constant term is negative.

11	$-5, 1$	**12**	$-6, 2$
13	$-4, 2$	**14**	$-9, 2$
15	$-16, 3$	**16**	$-3, 5$
17	$-2, 5$	**18**	$-6, 8$
19	$-7, 8$	**20**	$-7, 9$
21	$-10, 4$	**22**	$-10, 7$

From here there are no further clues.

23	-10 repeated	**24**	3 repeated
25	$-8, 9$	**26**	$-3, 4$
27	$-12, 6$	**28**	$-9, 4$
29	$-2, 20$	**30**	$-4, 15$

Exercise 8.5

Here the coefficient of x^2 is not 1. These questions need care together with trial and error.

1	$-1, 2/3$	**2**	1/2 repeated
3	$-1/2, 2/3$	**4**	$-2/3, 1$
5	$-2, -3/2$	**6**	$-5, -3/2$
7	$-3, 1/3$	**8**	$3/2, 2$
9	$-5, -1/2$	**10**	$-5/2, 4/3$
11	$-1/2, 7/3$	**12**	$2/3, 3/2$
13	$-1/3, 2/5$	**14**	$-4, -1/3$
15	3/2 repeated	**16**	3/5 repeated
17	$-3/2, 5/4$	**18**	$-3/2, -2/3$
19	$-2/3$ repeated	**20**	$-3/8, 6$

Exercise 8.6

1   −4, 0   Factorise with a simple x
2   0, 8/3
3   0, 4
4   −10/3, 10/3   Find x^2 and then square root.
5   0, 1
6   −10/3, 0
7   0, 3   Gather both terms on the left.
8   No real solution because of the square root of a negative number.
In Questions 9 and 10 find the square root first.
9   3
10   −3, 2

Exercise 8.7

These are standard applications of the formula. Work in stages rather than inputting all the values at once into your calculator.

1   −21.576, 1.576   2   −0.618, 1.618
3   −1.772, 6.772   4   −0.629, 28.629
5   −3.562, 0.562   6   −0.464, 6.464
7   6   8   −4.519, −0.148
9   No real solution because of the square root of a negative number.
10   0.147, 0.453

Exercise 8.8

The formula is used in most questions. Think about your answers in each case. Some of them may not be possible practically.

1   0.291 A, 1.376 A   2   7.42 s
3   2; 6   4   18.3%
5   11.22 V   6   −6.32 ms^{-1}, 6.32 ms^{-1}
7   6.60   8   4 m, 6 m; 2.76 m, 7.24 m
9   0.283 m; 1.595 m   10   5.38 Ω, 15.62 Ω

Multi-choice Test 8

1	D	2	B	3	C	4	A
5	D	6	A	7	C	8	B
9	A	10	B	11	B	12	B
13	A	14	D	15	B	16	C
17	D	18	B	19	C	20	D

Chapter 9

Exercise 9.1

1   $x = 6, y = 5$ Re-arrange the equations to get $y = 11 - x$ and $y = x - 1$. Try plots based on $x = 0, 5, 10$.

2   $x = 5, y = 1$ Re-arrange the equations to get $y = (8 - x)/3$ and $y = (x - 3)/2$. Try plots based on $x = 2, 6, 10$.

3   $x = 5, y = 2$ Re-arrange the equations to get $y = (x + 1)/3$ and $y = (20 - 2x)/5$. Try plots based on $x = 2, 6, 10$

4   $x = 1/2, y = 1$ Re-arrange the equations to get $y = 2x$ and $y = (4x + 3)/5$. Try plots based on $x = -2, 0, 2$.

5   $x = -1/5, y = 1$ Re-arrange the equation to get $y = (5x + 4)/3$. Try plots based on $x = -1.5, 0, 1.5$.

6   $x = 3.8, y = 0.6$ Re-arrange the equations to get $y = (5 - x)/2$ and $y = 2x - 7$. Try plots based on $x = 0, 3, 6$.

7   $x = 0, y = 1$ Re-arrange the equation to get $y = (3 - 2x)/3$. Try plots based on $x = -1, 2, 5$.

8   no solution   Parallel lines

9   $x = -10/7, y = 29/7$ Re-arrange the equations to get $y = (4 - 3x)/2$ and $y = (11 - x)/3$. Try plots based on $x = -4, -1, 2$.

10   no solution   Only one equation.

Exercise 9.2

1   $x = 10, y = -9$ Add to eliminate y, then substitute for y in either equation.

2   $x = -2, y = 13$ Subtract to eliminate y, then substitute for y in either equation.

3   $x = 3, y = 3$ Add to eliminate y. From the second equation $x = y$.

4   $x = -2, y = 1$ Add to eliminate y or subtract to eliminate x.

5   $x = 4, y = 10$ Multiply equation ① by 2, add to eliminate y.

6   $x = -7, y = -2$ Multiply equation ① by 5, multiply equation ② by 8, add to eliminate y.

7   $x = -1, y = 2.5$ Multiply equation ② by 3, subtract to eliminate x.

8   $x = 24, y = 4$ Multiply equation ① by 2, add to eliminate y; or multiply equation ② by 3, subtract to eliminate x.

9   $x = 2, y = -1$ Multiply equation ② by 5, subtract to eliminate y.

10   $x = 3, y = -0.5$ Multiply equation ② by 2, add to eliminate y.

Exercise 9.3

Plot these graphs with care. Smooth curves improve accuracy.
1   $x=-1$, $y=-2$ and $x=4$, $y=18$
2   $x=-1.77$, $y=8.66$ and $x=0.94$, $y=4.59$
3   $x=-2$, $y=-9$ repeated   The straight line is a tangent to the curve.
4   no real solution   No intersection.
5   $x=0.56$, $y=2.72$ and $x=2.69$, $y=1.65$

Exercise 9.4

1   $I_1=0.4\,\text{A}$, $I_2=0.2\,\text{A}$
    Multiply out the brackets, gather together like terms. Multiply equation
    ① by 9, multiply equation ② by 5, add to eliminate I_2.
2   $M_{max}=200\,\text{Nm}$, $x=10\,\text{m}$; $M_{max}=225\,\text{Nm}$, $x=7.5\,\text{m}$; $x=0\,\text{m}$, $10\,\text{m}$;
    $M=0\,\text{Nm}$, $200\,\text{Nm}$
    Plot smooth curves. the maximum occurs at the turning point/peak
    position. Look for intersections and read off the common values.
3   $99°\text{C}$; i) $15.5°\text{C}$, ii) $36355\,\text{k}\Omega$
    Plot a straight line (3 points only are required) and a smooth curve, look
    for the intersection.
4   $13.5\,\text{s}$, $173.\bar{3}\,\text{m}$, $9.75\,\text{s}$
5   Supply is $y=100+10x$, Demand is $y=350-8x$; £13.89; £13.20
    For supply and demand think about price rises and falls in terms of
    gradient.

Multi-choice Test 9

1	C	2	C	3	A	4	C
5	B	6	A	7	B	8	D
9	B	10	B	11	B	12	A
13	A	14	D	15	C	16	D
17	C	18	D	19	C	20	B

Chapter 10

Exercise 10.1

Use common logarithms, i.e. log| on the calculator.

1	0.760	2	1.127	3	3	4	1.428
5	2.301	6	−0.301	7	6	8	−0.602

9   Logarithms are *not* defined for negative numbers.
10   0.246

Use inv| log| on the calculator, i.e. 10^x.

11	17	**12**	0.282	**13**	4	**14**	1.5
15	288	**16**	5	**17**	0.0562	**18**	45.5
19	4900	**20**	88.9				

Exercise 10.2

Use natural logarithms, i.e. ln| on the calculator.

1	2.757	**2**	1.224	**3**	6.908	**4**	1.917
5	5.521	**6**	−1.897	**7**	6	**8**	−1

9 Logarithms are *not* defined for negative numbers.
10 5.165

Use inv| ln| on the calculator, i.e. e^x.

11	3.42	**12**	0.577	**13**	1.83	**14**	1.19
15	11.7	**16**	2.01	**17**	0.287	**18**	5.25
19	40	**20**	7.02				

Exercise 10.3

Use either common or natural logarithms.
1 3
2 1.64 $x \log 5 = \log 14$
3 1.21   Divide by 2 then $x \log 5 = \log 7$
4 4.42
5 4.40
6 3.42 $2^x \times 3^x = (2 \times 3)^x = 6^x$
7 7.13
8 −9.81 $x \log 2 + (x + 2) \log 5 = (x - 1) \log 6.$
9 1.16   Let $q = 4^x$ so that $4^{2x} = (4^x)^2 = 4^x \times 4^x$.
               Then $q^2 - 5q = 0$ factorises to $q(q - 5) = 0$. Later inverse log to find x from q.
10 1.00, 1.77   Let $q = 3^x$ so that $3^{2x} = (3^x)^2$.
               Then $q^2 - 10q + 21 = 0$ factorises. Later inverse log to find x from q.

Exercise 10.4

Use $\dfrac{\log N}{\log e}$ and substitute for $\log e = \log 2.718 \ldots = 0.434 \ldots$

1	1.09	**2**	2.25	**3**	−0.15	**4**	2.00
5	0.49						

Use $\dfrac{\ln N}{\ln 10}$ and substitute for $\ln 10 = 2.3025 \ldots$

6	5.76	**7**	−1.50	**8**	11.51	**9**	16.99
10	6.91						

Exercise 10.5

1   17.78 dB, 30 W. Use $20 = 10\log\left(\dfrac{P_0}{300/10^3}\right)$

2   $n = \dfrac{\log c - \log S}{\log T}$

3   0.13

4   $h = \dfrac{1}{a}\ln\dfrac{c}{p}$ from $\ln p + ah\ln e = \ln c$, 6320 m

5   1.36 rad, 0.42. Change 60° to radians.

6   8.50 by taking logarithms of $2 = 1.085^n$, 6.94 using $2 = 1.105^n$, 14.87% involving fifth roots.

7   0.34, 16.2 mm

8   25%

9   Involve $\ln 1 = 0$.

10   0.025, $t = -T\ln\left(1 - \dfrac{Ri}{E}\right)$, 0.047 s

Multi-choice Test 10

1	C	2	D	3	B	4	D
5	A	6	A	7	B	8	C
9	A	10	B	11	B	12	B
13	A	14	C	15	B	16	B
17	D	18	C	19	A	20	C

Chapter 11

Exercise 11.1

1   5.474

2   1636

3   0.3679

4   1   Any number to the power zero is always 1.

5   3272

6   597.7 ⎱   Apply the laws of indices to simplify, then use your
7   53732 ⎰   calculator.

8   54.68

9   1.001 ⎱
10   1.001 ⎰   Compare the questions and answers.

11   54.68

12   3.158

13   34.81

14   109.9 ⎱   Compare the questions and answers.
15   109.9 ⎰

16 2749 }
17 549.7 }   Notice how these involve earlier results and factors of 5.
18 3.158
19 0.04344
20 1.439

Exercise 11.2

A selection of graphs for Questions **3, 4, 5, 6, 7, 8, 11, 12, 13** shows the changes to the original exponential growth and decay.

3

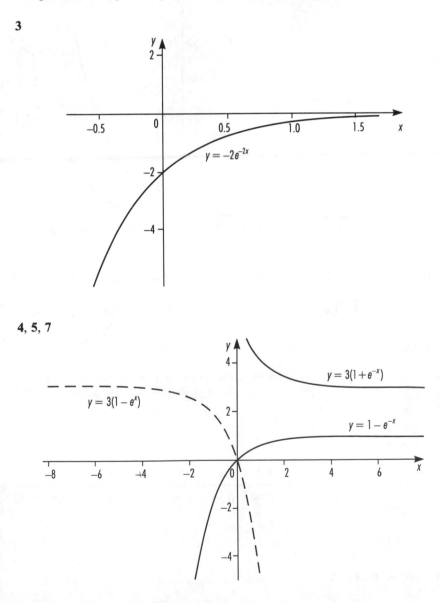

4, 5, 7

11, 12

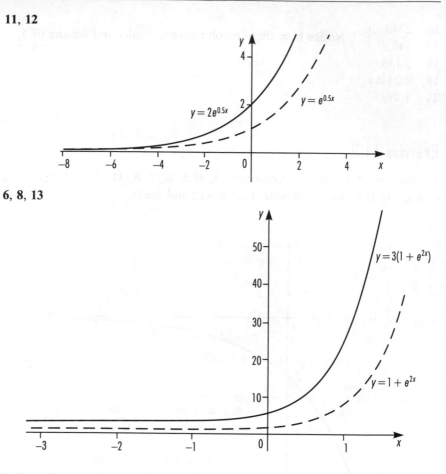

6, 8, 13

In Questions **17–20** accurate drawing will give answers close to the quoted ones.

17 1.06   **18** 2.12   **19** 1328   **20** 1328

Exercise 11.3

Accurate drawing will give answers close to the quoted ones.

1 4.29×10^{-3}

2 0.02, -88.8, 0.014 s

3 2.61×10^3, 1 hr 52 min

4 0.3648

t	0	0.5	1.0	1.5	2.0	2.5	3.0	3.5	4.0
N	1.00	0.83	0.69	0.58	0.48	0.40	0.33	0.28	0.23
	4.5	5.0							
	0.19	0.16							

5 0.05 A, 0.8829 As^{-1}

Exercise 11.4

1 $\ln p = -kH + \ln p_0$. Plot $\ln p$ vertically and H horizontally. Gradient is negative. Intercept is $\ln p_0$. $p_0 = 75.8$, $k = -\dfrac{1}{15\,000}$

2 $E = 5.61$, $T = \dfrac{1}{63}$, $0.011\,\text{s}$

3 $\ln T = \gamma \ln p + \ln \alpha$. Plot $\ln T$ vertically and $\ln p$ horizontally. Gradient is γ. Intercept is $\ln \alpha$. $\alpha = 515$, $\gamma = 0.254$

4 -1.95

5 $\ln V = R \ln a$. Plot $\ln V$ vertically and R horizontally. Gradient is $\ln a$. Intercept is the origin. 1.06

6 $B_0 = 1.60 \times 10^{-3}$, $c = 0.75$

7 1.525

8 0.3648

9 $0.0417\,\text{A}$

10

y	1	2	3	4	5	6
P	150	180	216	259	311	373

, yes, $P_1 = 125$, $r = 1.2$

Multi-choice Test 11

1	C	2	D	3	C	4	C
5	B	6	B	7	A	8	A
9	A	10	B	11	B	12	A
13	B	14	D	15	C	16	C
17	A	18	B	19	D	20	C

Chapter 12

Exercise 12.1

1, 2, 5

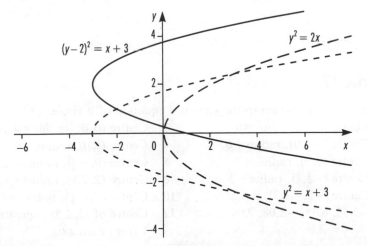

3

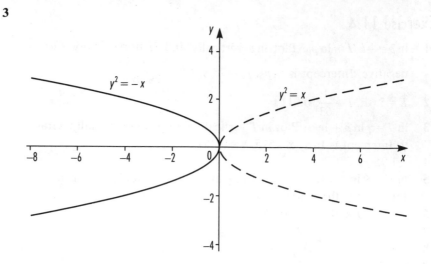

4

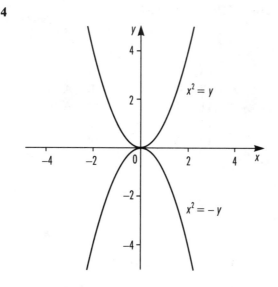

Exercise 12.2

Compare each equation to the standard equation of a circle.

1   Centre $(0, 0)$, radius $= 6$
2   Centre $(0, 0)$, radius $= 4.24$
3   Centre $(-\frac{1}{2}, 0)$, radius $= 6$
4   Centre $(\frac{1}{2}, 0)$, radius $= 5$
5   Centre $(0, 3)$, radius $= 7$
6   Centre $(0, -\frac{1}{4})$, radius $= 2.5$
7   Centre $(-\frac{1}{4}, \frac{1}{3})$, radius $= 3$
8   Centre $(2, 2.5)$, radius $= 1$
9   Centre $(-2, -1.7)$, radius $= 2$
10   Centre $(\frac{2}{5}, -\frac{3}{4})$, radius $= 1$
11   $-2.33, 6.33; -2.08, 7.08$
12   Centre of $(2, 2.5)$ is greater than 1 from each axis.

Exercise 12.4

Use your calculator. Check your answers with a sketch.

1	(1.75, 0.40)	**2**	(−2.04, 4.01)
3	(0.60, −2.43)	**4**	(−8.06, −8.06)
5	(0, −2.7)	**6**	(5.39, 21.8°)
7	(9.22, 310.6°)	**8**	(3.16, 108.4°)
9	(5, 233.1°)	**10**	(21.21, 225°)

Exercise 12.5

We include only graphs for Questions **1, 5, 7, 10, 11, 15** as a check.

1

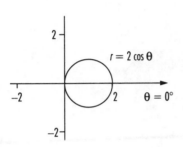

5

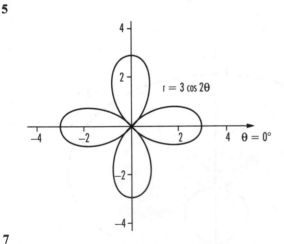

7

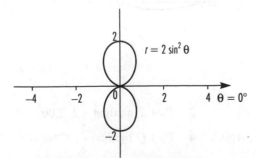

10

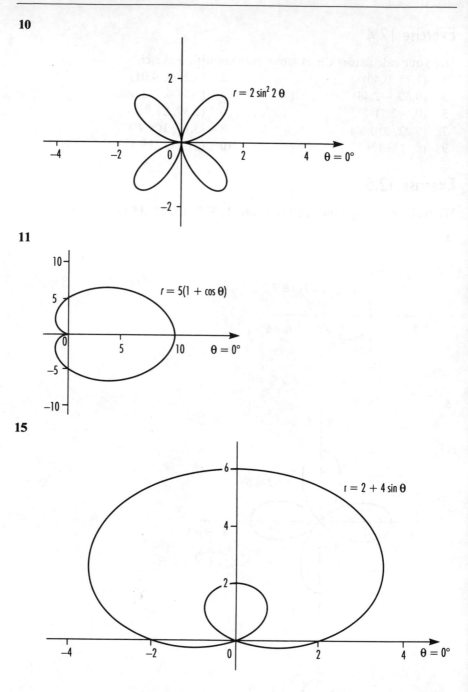

$r = 2 \sin^2 2\theta$

11

$r = 5(1 + \cos\theta)$

15

$r = 2 + 4\sin\theta$

Exercise 12.6

1 Plot s against t^2, 1.5

2 Plot T against $\sqrt{l}$, 2.00

3 Plot B against $\dfrac{1}{\sqrt{t}}$, 2300, -160

4 Plot Q against $\dfrac{1}{x}$, 1260, 0

5 Plot C against t^3, 0.05, 500

6 Plot P against I^2. Expect $R = 3$.

7 2000

8 0.04, 0.16

9 0.48, -2.25

10 Plot I against D^4, 0.049, -0.00125

Multi-choice Test 12

1	B	**2**	C	**3**	A	**4**	C
5	A	**6**	C	**7**	A	**8**	D
9	B	**10**	C	**11**	C	**12**	B
13	C	**14**	C	**15**	C	**16**	C
17	A	**18**	D	**19**	D	**20**	D

Chapter 13

Exercise 13.1

1 2, 4.5

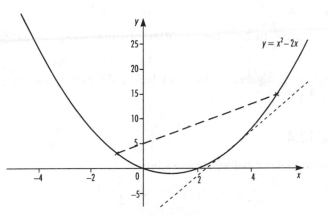

2 2.1, -2.1; 2.4, -1.6

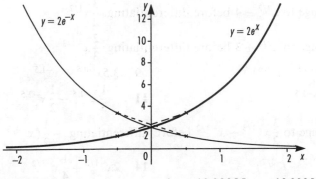

3 0, 0.64, -0.64, 0

4 $-10\,000\,\text{Nm}$, $-10\,000\,\text{Nm}$, 0

5 $-0.22°\text{Cs}^{-1}$

Exercise 13.2

We give expressions for δy (or δs) before each answer.

1 δx; 1

2 $3\delta x$; 3

3 $2x(\delta x) + (\delta x)^2$; $2x$

4 $4t(\delta t) + 2(\delta t)^2$; $4t$

5 $2t(\delta t) + (\delta t)^2 + \delta t$; $2t + 1$

6 $4x(\delta x) + 2(\delta x)^2 - 3\delta x$; $4x - 3$

7 $3x^2(\delta x) + 3x(\delta x)^2 + (\delta x)^3$; $3x^2$

8 $-\dfrac{2\delta x}{(x + \delta x)x}$; $-\dfrac{2}{x^2}$

9 $\dfrac{-8t(\delta t) - 4(\delta t)^2}{(t + \delta t)^2 t^2}$; $-\dfrac{8}{t^3}$

10 $-\dfrac{2\delta x}{(x + \delta x)x} - 3\delta x$; $-\dfrac{2}{x^2} - 3$

Exercise 13.3

1 0

2 3

3 3

4 $8x$

5 $8t - 3$

6 $-3 - 8x$

7 $2x + 1$

8 $4x - 3$

9 $-6t^{-2} - 4t^{-3}$

10 $-8t^{-3}$

11 Re-arrange $\dfrac{2}{x}$ to $2x^{-1}$ before differentiating. $-12x^{-3} - 2x^{-2}$

12 $3 + 2x^{-2}$

13 $-12x^{-7} - 6x$

14 Simplify the algebra to $3x^2 + 2x$ before differentiating. $6x + 2$

15 $-3x^{-4} + 2x^{-3}$

Exercise 13.4

1 Change to $4x^{1/2}$ before differentiating. $2x^{-1/2}$

2 $1.75x^{-0.5}$

3 $\dfrac{8}{3}x^{-2/3}$

4 $\dfrac{1}{24}x^{-7/8}$

5 $2t^{-1/2}$

6 Change to $x^{3/2} - 4$ before differentiating. $\dfrac{3}{2}x^{1/2}$

7 Change to $x^{2/3} + 3$ before differentiating. $\dfrac{2}{3}x^{-1/3}$

8 $0.8(t^{-0.8} + t^{-0.6})$

9 $3.5x^{-0.5} + x^{-1.5}$

10 $-\dfrac{1}{3}x^{-1.5}$

11 $-\dfrac{1}{3}x^{-1.5} - \dfrac{1}{2}x^{-2.5}$

12 Change to $\dfrac{3}{2}x^{-1/3} + x^{-1/2}$ before differentiating. $-\dfrac{1}{2}(x^{-4/3} + x^{-3/2})$

13 $\dfrac{1}{2}(t^{-1/2} + t^{-3/2})$

14 $2x^{-1/2} + \dfrac{1}{4}x^{-3/2}$

15 Simplify the algebra before differentiating. $-\dfrac{3}{2}x^{-5/2} - 2x^{-3}$

Exercise 13.5

The derivatives are a mixture; with respect to x or t or θ. Remember the angle remains the same during differentiation.

1 $2\cos\theta$ | 2 $-3\sin\theta$
3 $2\cos 2\theta$ | 4 $-6\sin 2t$
5 $12\cos(3t-2)$ | 6 $-4\sin(\theta+2)$
7 $12\cos(3t-2)-6\sin 2t$ | 8 $-6(\sin 3x+\cos(2x+1))$

9 $-4.5\sin\dfrac{3x}{4}+\cos\dfrac{x}{2}$

10 $6\cos\left(\dfrac{2\theta}{3}-1\right)+21\sin\left(\dfrac{3\theta}{2}+1\right)$

Exercise 13.6

The derivatives are a mixture; with respect to x or t or θ. Remember the exponential power remains the same during differentiation.

1 $12e^{3x}$ | 2 $12e^{3x-2}$
3 $-22.5e^{-3t}$
4 Re-arrange to $5e^{-3x}$ and then differentiate. $-15e^{-3x}$
5 $-15e^{-3t-6}$ | 6 $0.25\times-4=-1.\ -e^{7-4x}$
7 $2.5-0.5e^{3+t}$ | 8 $8(e^{2\theta}-\sin 4\theta)$
9 $3\cos\theta+2e^{4+2\theta}$ | 10 $6(\cos(2\theta-3)+e^{3\theta-2})$

Exercise 13.7

1 $\dfrac{2.5}{x}$ | 2 $\dfrac{5}{x}$

3 Use the laws of logs to simplify, $y=-\ln x.\ \dfrac{-1}{x}$

4 $-\dfrac{1}{x}$ | 5 $\dfrac{8}{4x+3}$

6 $\dfrac{8}{4x-3}$ | 7 $-\dfrac{20}{3-4x}$

8 $6\cos 2\theta+\dfrac{1}{\theta-2}$ | 9 $8e^{4t-3}+\dfrac{3}{3t+4}$

10 $\dfrac{12}{1.5x+5}+\dfrac{2}{1-2x}$

Exercise 13.8

1 15.4 | 2 $2x^{-3.5}$, -0.18
3 2.866 | 4 $\dfrac{10}{\pi}x^{-2}-1$, -1.3
5 -2.53 | 6 $15.6\,\text{ms}^{-1}$, $7.36\,\text{s}$
7 $2000\,\text{Nm}^{-1}$, $-2000\,\text{Nm}^{-1}$ | 8 $0.06\,\text{radN}^{-1}$
9 $406.25°\text{s}^{-1}$ | 10 $\dfrac{3t^2}{20}$, 31.6 hour

Exercise 13.9

1   $6x^2 + 10x$, $12x + 10$
2   Simplify the algebra first, $y = x^{-3} + 2x^{-2}$; $-3x^{-4} - 4x^{-3}$, $12(x^{-5} + x^{-4})$
3   $y = 2t^{1/2} + 4t^{-1/2}$; $t^{-1/2} - 2t^{-3/2}$, $-0.5t^{-3/2} + 3t^{-5/2}$
4   $-6\sin 2x$, $-12\cos 2x$
5   $6\cos 3\theta + 8\sin 2\theta$, $-18\sin 3\theta + 16\cos 2\theta$
6   $6e^{2x} - 4e^{-x}$, $12e^{2x} + 4e^{-x}$
7   $3 - 4e^{-2x}$, $8e^{-2x}$
8   $2x^{-1}$, $-2x^{-2}$
9   $-0.5e^{-0.5x} + 2x^{-1}$, $0.25e^{-0.5x} - 2x^{-2}$
10 $5e^x + 4\cos 2x$, $5e^x - 8\sin 2x$
11 Simplify the algebra first, $y = 2x^{-1} - 6x^{-2}$, $2.\bar{6}$, -5.93
12 -3.64
13 $-e^{-x} + 2x^{-1}$, $e^{-x} - 2x^{-2}$, i) 1.63, -1.63; ii) 0.72, -0.24
14 i) $-5 + 4t - 12t^3$, ii) $4 - 36t^2$
15 $10t^{-1}$, $-10t^{-2}$
16 $2t$, 2
17 $5\cos 2t$, $-10\sin 2t$, 4
18 i) $70 - 44 = 26\,\text{m}$, ii) $24\,\text{ms}^{-1}$, $28\,\text{ms}^{-1}$, $4\,\text{ms}^{-1}$
19 $\dfrac{2\pi}{3}\cos\dfrac{\pi t}{6} - \dfrac{\pi}{2}\sin\dfrac{\pi t}{6}$; i) $2.09\,\text{ms}^{-1}$, ii) $1.62\,\text{ms}^{-1}$;

$-\pi^2\left(\dfrac{1}{9}\sin\dfrac{\pi t}{6} + \dfrac{1}{12}\cos\dfrac{\pi t}{6}\right)$; $-1.18\,\text{ms}^{-2}$
20 $49(1 - e^{-2t})$

Exercise 13.10

We give the first derivatives as well as the numerical answers.
1   $2x - 5$; 2.5
2   $3 - 2x$; 1.5
3   $3(2x^2 - 3x + 1)$ to factorise; $\dfrac{1}{2}$, 1
4   $12x^2 - 6x - 11$ needs the formula; -0.74, 1.24
5   $3x^2 - 8.8x + 5.6$ needs the formula; $0.9\bar{3}$, 2

Exercise 13.11

We give $\dfrac{d^2y}{dx^2}$ as well.

1   6; $-\dfrac{1}{3}$, min   2 -8; $-\dfrac{1}{4}$, max

3   $24x - 6$; -1.5, max; 2, min   4 $6x + 2$; $-\dfrac{7}{6}$, max; $\dfrac{1}{2}$, min

5   $6x$; $-\sqrt{2}$, max; $\sqrt{2}$, min

Exercise 13.12

We give expressions for both $\dfrac{dy}{dx}$ and $\dfrac{d^2y}{dx^2}$ as well.

1. $4x - 3$, 4; $y_{min} = -1.125$, $(0.75, -1.125)$
2. $-10 - 10x$, -10; $y_{max} = 9$, $(-1, 9)$
3. $-24 - 6x + 3x^2$, $-6 + 6x$; $y_{max} = 28$, $(-2, 28)$; $y_{min} = -80$, $(4, -80)$
4. $-7 + 8x + 12x^2$, $8 + 24x$; $y_{max} = 9.26$, $(-1.1\overline{6}, 9.26)$; $y_{min} = 0$, $(0.5, 0)$
5. No max/min because we cannot find the square root of a negative number.

Exercise 13.13

1. 2500 Nm
2. Max at $\left(\dfrac{\pi}{2}, 1\right)$, min at $\left(\dfrac{3\pi}{2}, -1\right)$; max at $(0, 1)$, min at $(\pi, -1)$, max at $(2\pi, 1)$; max at $\left(\dfrac{3\pi}{4}, \sqrt{2}\right)$, min at $\left(\dfrac{7\pi}{4}, -\sqrt{2}\right)$
3. Yes, 2.58 s, max at $(2.58, 172°)$
4. £5963
5. x, $2000 - 2x$, $3000 - 2x$; $x(2000 - 2x)(3000 - 2x)$; 392 mm, 1215 mm, 2215 mm, 1.056×10^9 mm^3

Multi-choice Test 13

1	C	2	D	3	C	4	C
5	C	6	B	7	B	8	A
9	C	10	B	11	C	12	D
13	B	14	B	15	B	16	B
17	B	18	A	19	C	20	C

Chapter 14

Exercise 14.1

1. $7x + c$
2. $\dfrac{5}{2}x^2 + c$
3. $7x - \dfrac{5}{2}x^2 + c$
4. $x^3 + c$
5. $t^3 + \dfrac{5}{2}t^2 - 7t + c$
6. $7x - \dfrac{5}{2}x^2 - x^3 + c$
7. $\dfrac{x^5}{5} + x^2 + c$
8. $\dfrac{4}{7}x^7 - \dfrac{1}{2}x^4 + c$

9 $-5t^{-1} - t^{-3} + c$

10 Before integrating re-arrange to $3t^{-3}$; $-\dfrac{3}{2}t^{-2} + c$

11 $-5x^{-1} + x^{-3} + c$

12 $\dfrac{1}{2}t^4 + 3t^{-1} + c$

13 $-\dfrac{3}{8}x^{-4} - \dfrac{1}{2}x^6 + c$

14 Before integrating simplify the algebra to $(3 + 4x^2)$; $3x + \dfrac{4}{3}x^3 + c$

15 $-2x^{-1} + \dfrac{1}{2}x^{-2} + c$

Exercise 14.2

Re-write all surds as fractional powers before integrating.

1   Re-write as $6x^{1/2}$ before integrating. $4x^{3/2} + c$

2 $3.\bar{3}x^{1.5} + c$ **3** $6t^{4/3} + c$

4 $\dfrac{4}{25}x^{5/4} + c$ **5** $2.\bar{6}t^{3/2} + c$

6 $\dfrac{2}{5}x^{5/2} - 6x + c$

7   Re-write as $(x^{2/3} - 4)$ before integrating. $\dfrac{3}{5}x^{5/3} - 4x + c$

8 $3.\bar{3}t^{1.2} + 1.43t^{1.4} + c$ **9** $6x^{1.5} + 8x^{0.5} + c$

10 Re-write as $\dfrac{3}{2}x^{-0.5}$ before integrating. $3x^{0.5} + c$

11 $0.8x^{0.5} - 0.4x^{-0.5} + c$ **12** $\dfrac{15}{4}x^{2/3} + 4x^{1/2} + c$

13 $2t^{3/2} - 8t^{1/2} + c$ **14** $\dfrac{16}{3}x^{3/2} + \dfrac{1}{2}x^{1/2} + c$

15 Simplify the algebra before integrating. $-x^{-1/2} - 2x^{-1} + c$

Exercise 14.3

1 $[0.5x^4]$, 7.5 **2** $[x^3]$, 19

3 $[1.5x^2 - 0.2x^5]$, -35.1 **4** $[t^2 - 7t]$, -2.75

5 $\left[6x - \dfrac{3x^2}{2} - \dfrac{x^3}{3}\right]$, $11.\bar{3}$ **6** $\left[-\dfrac{3}{t}\right]$, 0.25

7 $\left[-\dfrac{2}{t} - \dfrac{1}{t^3}\right]$, 10.37 **8** $\left[-\dfrac{4}{5x} + x^2\right]$, $8.5\bar{3}$

9 $\left[x^4 - \dfrac{2x^{3/2}}{3}\right]$, 4.01 **10** $\left[4x^{3/2} + \dfrac{4}{3}x^{1/2}\right]$, 21.47

Exercise 14.4

Remember the angle remains the same during integration.

1 $-3\cos\theta + c$ **2** $5\sin\theta + c$

3 $-0.25\cos 4\theta + c$ **4** $\dfrac{4}{3}\sin 3t + c$

5 $\dfrac{4}{3}\sin(3t + 7) + c$ **6** 0

7 $[-2\cos 3t + 2\sin 2t], -3.414$ **8** $[\sin 2x + 0.5\cos 6x], 0$

9 $\left[9\sin\dfrac{4x}{3} - 8\cos\dfrac{x}{4}\right], 4.94$

10 $\left[-4\cos\left(\dfrac{3\theta}{2} + \pi\right) - 14\sin\left(\dfrac{3\theta}{2} + \dfrac{\pi}{2}\right)\right], 10$

Exercise 14.5

Remember the exponential power remains the same during integration.

1 $\dfrac{1}{2}(e^{2x} - e^{-2x}) + c$ **2** $2e^x + 3e^{-x} + c$

3 $4(e^{x/4} + e^{x/2}) + c$ **4** $\dfrac{5}{3}e^{3t} + c$

5 $e^{2x+1} - 2.5e^{1-2x} + c$ **6** -11.10

7 2.66 **8** -49.03

9 11.75 **10** 4.53

Exercise 14.6

1 2.43
2 0.25
3 0.71
4 Think about the algebra before attempting to integrate, 8.32
5 This is a mixture of techniques, 3.77

Exercise 14.7

1 $[0.25x^4 + 2x], 4.27$ unit^2 **2** Simplify the algebra and then integrate, $3.\overline{3}$ unit^2

3 $[e^x], 14.64$ unit^2 **4** $[\ln x], 0.693$ unit^2

5 $[\sin\theta], -1$ unit^2 **6** $[4\ln x], 2.77$ unit^2

7 $[-4\cos\theta], 16$ unit^2 **8** 8 unit^2

9 a) $x = 1, 2; -0.1\overline{6}$ unit^2 b) $x = 1, 2; 0.1\overline{6}$ unit^2

10 i) $6.\overline{3}$ unit^2, ii) 8 unit^2, iii) $10.\overline{6}$ unit^2

Exercise 14.8

1	100 m	**2**	1.8 J
3	7.82×10^{-5} W	**4**	117.5 J
5	5.303×10^4, 2.75×10^4 J		

Multi-choice Test 14

1	C	**2**	A	**3**	B	**4**	D
5	B	**6**	A	**7**	D	**8**	A
9	D	**10**	C	**11**	A	**12**	D
13	D	**14**	C	**15**	D	**16**	D
17	D	**18**	A	**19**	C	**20**	A

Index